Advanced Hydrologic Modeling in Watershed Scales

Advanced Hydrologic Modeling in Watershed Scales

Editors

Dengfeng Liu
Hui Liu
Xianmeng Meng

MDPI • Basel • Beijing • Wuhan • Barcelona • Belgrade • Manchester • Tokyo • Cluj • Tianjin

Editors

Dengfeng Liu
School of Water Resources and Hydropower
Xi'an University of Technology
Xi'an
China

Hui Liu
China Institute of Water Resources and Hydropower Research
Beijing
China

Xianmeng Meng
School of Environmental Studies
China University of Geosciences
Wuhan
China

Editorial Office
MDPI
St. Alban-Anlage 66
4052 Basel, Switzerland

This is a reprint of articles from the Special Issue published online in the open access journal *Water* (ISSN 2073-4441) (available at: www.mdpi.com/journal/water/special_issues/hydrologic_modeling_watershed_scales).

For citation purposes, cite each article independently as indicated on the article page online and as indicated below:

LastName, A.A.; LastName, B.B.; LastName, C.C. Article Title. *Journal Name* **Year**, *Volume Number*, Page Range.

ISBN 978-3-0365-7115-7 (Hbk)
ISBN 978-3-0365-7114-0 (PDF)

Contents

About the Editors vii

Preface to "Advanced Hydrologic Modeling in Watershed Scales" ix

Dengfeng Liu, Hui Liu and Xianmeng Meng
Advanced Hydrologic Modeling in Watershed Scale
Reprinted from: *Water* **2023**, *15*, 691, doi:10.3390/w15040691 1

Xin Jin, Yanxiang Jin, Xufeng Mao, Jingya Zhai and Di Fu
Modelling the Impact of Vegetation Change on Hydrological Processes in Bayin River Basin, Northwest China
Reprinted from: *Water* **2021**, *13*, 2787, doi:10.3390/w13192787 5

Selim Şengül and Muhammet Nuri İspirli
Predicting Snowmelt Runoff at the Source of the Mountainous Euphrates River Basin in Turkey for Water Supply and Flood Control Issues Using HEC-HMS Modeling
Reprinted from: *Water* **2022**, *14*, 284, doi:10.3390/w14030284 25

Zheng Mu, Guanpeng Liu, Shuai Lin, Jingjing Fan, Tianling Qin and Yunyun Li et al.
Base Flow Variation and Attribution Analysis Based on the Budyko Theory in the Weihe River Basin
Reprinted from: *Water* **2022**, *14*, 334, doi:10.3390/w14030334 47

Junli Liu, Yun Zhang, Lei Yang and Yuying Li
Hydrological Modeling in the Chaohu Lake Basin of China—Driven by Open-Access Gridded Meteorological and Remote Sensing Precipitation Products
Reprinted from: *Water* **2022**, *14*, 1406, doi:10.3390/w14091406 63

Ahsen Maqsoom, Bilal Aslam, Nauman Khalid, Fahim Ullah, Hubert Anysz and Abdulrazak H. Almaliki et al.
Delineating Groundwater Recharge Potential through Remote Sensing and Geographical Information Systems
Reprinted from: *Water* **2022**, *14*, 1824, doi:10.3390/w14111824 85

Wenhua Wan, Hang Zheng, Yueyi Liu, Jianshi Zhao, Yingqi Fan and Hongbo Fan
Ecological Compensation Mechanism in a Trans-Provincial River Basin: A Hydrological/Water-Quality Modeling-Based Analysis
Reprinted from: *Water* **2022**, *14*, 2542, doi:10.3390/w14162542 115

Xue Yang, Chengxi Yu, Xiaoli Li, Jungang Luo, Jiancang Xie and Bin Zhou
Comparison of the Calibrated Objective Functions for Low Flow Simulation in a Semi-Arid Catchment
Reprinted from: *Water* **2022**, *14*, 2591, doi:10.3390/w14172591 135

Lifeng Yuan and Kenneth J. Forshay
Evaluating Monthly Flow Prediction Based on SWAT and Support Vector Regression Coupled with Discrete Wavelet Transform
Reprinted from: *Water* **2022**, *14*, 2649, doi:10.3390/w14172649 153

Jiashuai Yang, Chaowei Xu, Xinran Ni and Xuantong Zhang
Study on Urban Rainfall–Runoff Model under the Background of Inter-Basin Water Transfer
Reprinted from: *Water* **2022**, *14*, 2660, doi:10.3390/w14172660 173

Haizhe Wu, Dengfeng Liu, Ming Hao, Ruisha Li, Qian Yang and Guanghui Ming et al.
Identification of Time-Varying Parameters of Distributed Hydrological Model in Wei River Basin on Loess Plateau in the Changing Environment
Reprinted from: *Water* **2022**, *14*, 4021, doi:10.3390/w14244021 . . . . **195**

About the Editors

Dengfeng Liu

Dr. Dengfeng Liu is working as a Professor at the School of Water Resources and Hydropower, Xi'an University of Technology, in China. His research primarily focuses on watershed hydrological modeling, hydrological models, ecohydrological modeling, socio-hydrological modeling, and ecohydrology. He proposed the ecohydrological evolution model on riparian vegetation (ERV model) in hyperarid regions by coupling groundwater movement and vegetation dynamics to critically examine the state-of-the-art assumptions and dynamic equations used in the evolution study of an ecohydrological system. He developed a simplified conceptual socio-hydrological evolution model based on logistic growth curves for the Tarim River basin in western China, and this was used to illustrate the explanatory power of such a co-evolutionary model and improve our understanding of the co-evolution and self-organization of socio-hydrological systems driven by interactions and feedback operating at different scales. Currently, he is working on socio-hydrological modeling at the watershed scale to investigate the long-term impact of environmental changes.

Hui Liu

Dr. Hui Liu is working as a Professor at the China Institute of Water Resources and Hydropower Research in China. Her research primarily focuses on hydrological modeling, ecohydrology, remote sensing, drought, and water resources in Lancang–Mekong River. Currently, she is working on the impact of global changes on water resources in trans-boundary rivers, especially in Lancang–Mekong River.

Xianmeng Meng

Dr. Xianmeng Meng is working as an Associate Professor at the School of Environmental Studies, China University of Geosciences (Wuhan) in China. His research primarily focuses on groundwater simulation, hydrological processes, river networks, and hydrological modeling. Currently, he is working on the lithologic control of parameters of a conceptual rainfall–runoff model.

Preface to "Advanced Hydrologic Modeling in Watershed Scales"

Hydrologic modeling at the watershed scale is a key topic in the field of hydrology. The hydrological model is an important tool to understand the impact of climate change and human activities on rainfall–runoff processes, and especially on water resources for human beings in a changing environment. In the last two decades, with the development of satellite remote sensing and artificial intelligence, many new datasets and methods have been introduced into hydrological modeling.

Hydrologic modeling at the watershed scale is an important and fundamental research field in hydrology. Therefore, we proposed a Special Issue entitled "Advanced Hydrologic Modeling in Watershed Scales"in *Water* to publish findings regarding the recent progress in hydrological modeling at the watershed scale against global changes. Before the deadline for the submission of manuscripts to this Special Issue, we received many manuscripts regarding hydrological modeling at the watershed scale. In total, ten articles have been published in this Special Issue.

In the simulation of hydrological processes, the SWAT model was applied in five case studies, which were in the Bayin River basin of China, the Fengle watershed in the middle–lower Yangtze Plain of China, the Tangbai River Basin crossing Henan province and Hubei province in China, twelve hydrological sites in the Illinois River watershed in the U.S., and the Wei River Basin on the Loess Plateau in China. At the same time, runoffs were simulated using other models or methods, such as the Hydrological Engineering Center–Hydrological Modeling System (HEC-HMS), a two-stage annual precipitation partitioning method; the Xin An Jiang model; and the GR3 model. In addition, an integrated approach based on remote sensing and GIS using the influence factor (IF) technique was utilized to delineate potential groundwater recharge zones in Islamabad, Pakistan.

Based on hydrological modeling, these studies promoted have our understanding of the impact of vegetation changes on hydrological processes, the performance of meteorological datasets and precipitation datasets, hydrologic and nutrient cycling, a new hybrid SWAT-WSVR model, the relationship between the SWAT model parameters and the factors, the simulation of snowmelt runoff, the impact of climate change and human activities on the annual total stream flow and base flow, the difference between the calibrated objective functions, the simulation of the hydrological process with the background of inter-basin water transfer, and potential groundwater recharge zones.

Researchers interested in hydrological modeling and the impacts of environment change on water resources may be interested in this reprint. This Special Issue, entitled "Advanced Hydrologic Modeling in Watershed Scales" in *Water*, was handled by the Guest Editors, Dr. Dengfeng Liu, Dr. Hui Liu, and Dr. Xianmeng Meng. We acknowledge the editors of the journal, especially Ms. Sanja Vučić, for her help in processing the articles.

Dengfeng Liu, Hui Liu, and Xianmeng Meng
Editors

water

Editorial

Advanced Hydrologic Modeling in Watershed Scale

Dengfeng Liu [1,*], Hui Liu [2,*] and Xianmeng Meng [3,*]

1 School of Water Resources and Hydropower, Xi'an University of Technology, Xi'an 710048, China
2 China Institute of Water Resources and Hydropower Research, Beijing 100038, China
3 School of Environmental Studies, China University of Geosciences, Wuhan 430074, China
* Correspondence: liudf@xaut.edu.cn (D.L.); liuhui@iwhr.com (H.L.); mengxianmeng2000@163.com (X.M.)

Citation: Liu, D.; Liu, H.; Meng, X. Advanced Hydrologic Modeling in Watershed Scale. *Water* **2023**, *15*, 691. https://doi.org/10.3390/w15040691

Received: 13 January 2023
Revised: 6 February 2023
Accepted: 8 February 2023
Published: 9 February 2023

Hydrologic modeling in the watershed scale is a key topic in the field of hydrology. The hydrological model is an important tool to understand the impact of climate change and human activities on rainfall–runoff processes, and especially on water resources for humans in a changing environment. In traditional hydrological modeling, the precipitation data of in situ rainfall gauges are adopted to force hydrological modeling, and the simulated discharge is used to validate the hydrological model by comparing it with the observed discharge at the hydrological station. In the last two decades, with the development of satellite remote sensing and artificial intelligence, many new datasets and methods have been introduced into hydrological modeling. Multi-source fusion precipitation products (such as GPM (Global Precipitation Mission), MSWEP (Multi-Source Weighted-Ensemble Precipitation), CMFD (China Meteorological Forcing Dataset), and atmospheric assimilation datasets (such as CMADS (China Meteorological Assimilation Driving Datasets)) better display spatial distribution than ground rainfall data and have the potential for a better performance in hydrological modeling on middle and large spatial scales. Additionally, data on evaporation, soil moisture, and water level at the channel from remote sensing may be applied to validate the simulated evaporation, soil moisture, and discharge. Even water storage change can be evaluated by GRACE (Gravity Recovery and Climate Experiment) data. Deep learning models and agent-based models may be used in the process representation and parameter estimation. The interaction of hydrological processes to ecological processes and social processes has also attracted attention in recent years.

When the Special Issue opened, we planned to invite original research articles that contribute to new progress in the hydrological modeling in the watershed scale under global changes. Among the topics of interest for this Special Issue are:

- Application of new datasets and methods in hydrological modeling;
- New process representation in hydrological modeling;
- Progress of parameter estimation;
- Interaction of hydrological processes to ecological processes and social processes and their co-evolution processes;
- Coupled modeling of surface water and groundwater;
- Flood and drought based on hydrological modeling;
- Flux observation in the validation of hydrological modeling;
- Isotopic tracing in the validation of hydrological modeling;
- Role of macropore flow or preferential flow in the hydrological process;
- Sediment and other mass transport in the hydrological process.

Before the deadline for manuscript submissions for this Special Issue, we received many manuscripts on the hydrological modeling in the watershed scale. Finally, ten articles are published in the Special Issue.

In the simulation of hydrological processes, SWAT model was applied in 5 case studies, which are in the Bayin River basin of China [1], the Fengle watershed in the middle–lower Yangtze Plain of China [2], the Tangbai River Basin crossing Henan province and Hubei province of China [3], 12 hydrological sites in the Illinois River watershed of U.S [4], and the Wei River Basin on the Loess Plateau in China [5].

In the Bayin River basin of China, the study aimed to accurately simulate the impact of vegetation change on hydrological processes in an arid endorheic river watershed undergoing revegetation, and LU-SWAT-MODFLOW model was developed by integrating dynamic hydrological response units with a coupled SWAT-MODFLOW model [1].

In the Fengle watershed in the middle–lower Yangtze Plain of China, the study assessed the performance of two well-known gridded meteorological datasets, CFSR (Climate Forecast System Reanalysis) and CMADS (China Meteorological Assimilation Driving Datasets), and three satellite-based precipitation datasets, TRMM (Tropical Rainfall Measuring Mission), CMORPH (Climate Prediction Center morphing technique), and CHIRPS (Climate Hazards Group InfraRed Precipitation with Station data), in driving the SWAT model for streamflow simulation [2].

In the Tangbai River Basin of China, the river basin has been exposed to high doses of fertilizers for a long time and the study simulates hydrologic and nutrient cycling using the SWAT model with limited data available [3].

At 12 hydrological sites in the Illinois River watershed of the U.S, it developed a new hybrid SWAT-WSVR model that integrated the SWAT model with a Support Vector Regression (SVR) calibration method coupled with discrete wavelet transforms (DWT) to better support modeling watersheds with limited data availability [4].

In the Wei River Basin on the Loess Plateau in China, based on the measured data at the ground stations, the temporal and spatial evolution of the ecohydrological and meteorological factors were analyzed, and the SWAT model was used to identify the relationship between the model parameters and the factors, such as precipitation, potential evapotranspiration, NDVI, and the other environmental characterization factors of the river basin [5].

At the same time, the runoff are simulated by other models or methods, such as the Hydrological Engineering Center–Hydrological Modeling System (HEC-HMS) [6], a two-stage annual precipitation partitioning method [7], Xin An Jiang model [8], and the GR3 model [9].

The HEC-HMS was used to simulate snowmelt runoff in the Kırkgöze–Çipak Basin that has a complex topography where altitude differences range from 1823 m to 3140 m above the sea level in eastern Turkey [6].

Using a two-stage annual precipitation partitioning method, the study quantified the impact of climate change and human activities on the annual total stream flow, surface runoff, and base flow in the Weihe River Basin (WRB), wherein the surface runoff and base flow are separated from the measured total flow by using a one-parameter digital filter method for which the common filter parameter value is 0.925 [7].

With the calibrated Xin An Jiang model, the study increased the insight into the difference between the calibrated objective functions by evaluating eight objectives in three different classes (single objectives: KGE(log(Q)) and KGE(1/Q); multi objectives: KGE(Q)+KGE(log(Q)), KGE(Q)+KGE(1/Q), KGE(Qsort)+KGE(log(Qsort)) and KGE(Qsort)+ KGE(1/Qsort); split objectives: split KGE(Q) and split (KGE(Q)+KGE(1/Q))) in Bahe, a semi-arid basin in China [8].

The GR3 model, a rainfall–runoff model, was combined with the background of inter-basin water transfer to simulate the hydrological process of Huangtaiqiao basin in Jinan city, Shandong Province, China for 18 consecutive years with a 1 h time step [9].

In the groundwater recharge case, the study utilized an integrated approach based on remote sensing (RS) and GIS using the influence factor (IF) technique to delineate potential groundwater recharge zones in Islamabad, Pakistan [10].

Based on the hydrological modeling, these studies promoted the understanding of the impact of vegetation change on hydrological processes, the performance of meteorological datasets and precipitation datasets, hydrologic and nutrient cycling, a new hybrid SWAT-WSVR model, the relationship between the SWAT model parameters and the factors, simulation of snowmelt runoff, the impact of climate change and human activities on the annual total stream flow and base flow, the difference between the calibrated objective functions, simulation of the hydrological process with the background of inter-basin water transfer, and potential groundwater recharge zones.

Author Contributions: Conceptualization, D.L., H.L. and X.M.; writing—original draft preparation, D.L.; writing—review and editing, D.L., H.L. and X.M.; project administration, D.L.; funding acquisition, D.L. All authors have read and agreed to the published version of the manuscript.

Funding: This research was supported by the National Natural Science Foundation of China (42071335 and 52279025).

Data Availability Statement: No new data were created or analyzed in this study. Data sharing is not applicable to this article.

Conflicts of Interest: The authors declare no conflict of interest.

References

1. Jin, X.; Jin, Y.; Mao, X.; Zhai, J.; Fu, D. Modelling the Impact of Vegetation Change on Hydrological Processes in Bayin River Basin, Northwest China. *Water* **2021**, *13*, 2787. [CrossRef]
2. Liu, J.; Zhang, Y.; Yang, L.; Li, Y. Hydrological Modeling in the Chaohu Lake Basin of China—Driven by Open-Access Gridded Meteorological and Remote Sensing Precipitation Products. *Water* **2022**, *14*, 1406. [CrossRef]
3. Wan, W.; Zheng, H.; Liu, Y.; Zhao, J.; Fan, Y.; Fan, H. Ecological Compensation Mechanism in a Trans-Provincial River Basin: A Hydrological/Water-Quality Modeling-Based Analysis. *Water* **2022**, *14*, 2542. [CrossRef]
4. Yuan, L.; Forshay, K.J. Evaluating Monthly Flow Prediction Based on SWAT and Support Vector Regression Coupled with Discrete Wavelet Transform. *Water* **2022**, *14*, 2649. [CrossRef]
5. Wu, H.; Liu, D.; Hao, M.; Li, R.; Yang, Q.; Ming, G.; Liu, H. Identification of Time-Varying Parameters of Distributed Hydrological Model in Wei River Basin on Loess Plateau in the Changing Environment. *Water* **2022**, *14*, 4021. [CrossRef]
6. Şengül, S.; İspirli, M.N. Predicting Snowmelt Runoff at the Source of the Mountainous Euphrates River Basin in Turkey for Water Supply and Flood Control Issues Using HEC-HMS Modeling. *Water* **2022**, *14*, 284. [CrossRef]
7. Mu, Z.; Liu, G.; Lin, S.; Fan, J.; Qin, T.; Li, Y.; Cheng, Y.; Zhou, B. Base Flow Variation and Attribution Analysis Based on the Budyko Theory in the Weihe River Basin. *Water* **2022**, *14*, 334. [CrossRef]
8. Yang, X.; Yu, C.; Li, X.; Luo, J.; Xie, J.; Zhou, B. Comparison of the Calibrated Objective Functions for Low Flow Simulation in a Semi-Arid Catchment. *Water* **2022**, *14*, 2591. [CrossRef]
9. Yang, J.; Xu, C.; Ni, X.; Zhang, X. Study on Urban Rainfall–Runoff Model under the Background of Inter-Basin Water Transfer. *Water* **2022**, *14*, 2660. [CrossRef]
10. Maqsoom, A.; Aslam, B.; Khalid, N.; Ullah, F.; Anysz, H.; Almaliki, A.H.; Almaliki, A.A.; Hussein, E.E. Delineating Groundwater Recharge Potential through Remote Sensing and Geographical Information Systems. *Water* **2022**, *14*, 1824. [CrossRef]

 water MDPI

Article

Modelling the Impact of Vegetation Change on Hydrological Processes in Bayin River Basin, Northwest China

Xin Jin [1,2,3], Yanxiang Jin [1,2,3,*], Xufeng Mao [1,2,3], Jingya Zhai [1,2] and Di Fu [1,2]

1 MOE Key Laboratory of Tibetan Plateau Land Surface Processes and Ecological Conservation, Qinghai Normal University, Xining 810016, China; jinx13@lzu.edu.cn (X.J.); xfmao1001@163.com (X.M.); jingyasea@163.com (J.Z.); jinxin201016@gmail.com (D.F.)
2 School of the Geographical Science, Qinghai Normal University, Xining 810016, China
3 Academy of Plateau Science and Sustainability, Qinghai Normal University, Xining 810016, China
* Correspondence: jinyx13@lzu.edu.cn

Abstract: Vegetation change in arid areas may lead to the redistribution of regional water resources, which can intensify the competition between ecosystems and humans for water resources. This study aimed to accurately model the impact of vegetation change on hydrological processes in an arid endorheic river watershed undergoing revegetation, namely, the middle and lower reaches of the Bayin River basin, China. A LU-SWAT-MODFLOW model was developed by integrating dynamic hydrological response units with a coupled SWAT-MODFLOW model, which can reflect actual land cover changes in the basin. The LU-SWAT-MODFLOW model outperformed the original SWAT-MODFLOW model in simulating the impact of human activity as well as the leaf area index, evapotranspiration, and groundwater table depth. After regional revegetation, evapotranspiration and groundwater recharge in different sub-basins increased significantly. In addition, the direction and amount of surface-water–groundwater exchange changed considerably in areas where revegetation involved converting low-coverage grassland and bare land to forestland.

Keywords: revegetation; irrigation; leaf area index; evapotranspiration; groundwater

Citation: Jin, X.; Jin, Y.; Mao, X.; Zhai, J.; Fu, D. Modelling the Impact of Vegetation Change on Hydrological Processes in Bayin River Basin, Northwest China. *Water* **2021**, *13*, 2787. https://doi.org/10.3390/w13192787

Academic Editors: Dengfeng Liu, Hui Liu and Xianmeng Meng

Received: 14 August 2021
Accepted: 29 September 2021
Published: 8 October 2021

Publisher's Note: MDPI stays neutral with regard to jurisdictional claims in published maps and institutional affiliations.

1. Introduction

Vegetation is essential for regional carbon sequestration, soil and water conservation, and climate regulation [1,2]. Arid areas, which account for 40% of the world's land area, are characterised by water shortages and uneven spatiotemporal distributions of water resources [3]. Changes in vegetation and related management practices (e.g., irrigation) in arid areas may lead to the redistribution of regional water resources, which can intensify the competition between ecosystems and humans for water resources [1,4]. In this context, the water demand and water consumption characteristics of vegetation change in arid areas are of particular concern [5,6].

Nowadays, physically based distributed (or semi-distributed) hydrological models can clearly reflect the spatial variability of hydrological processes in a basin, and these models are playing an important role in simulations and predictions of the hydrological cycle in basins [7–9]. Notably, SWAT (Soil and Water Assessment Tools) is a typical distributed hydrological model with a strong physical foundation [10]. It is suitable for simulating surface hydrological processes in a complex basin with a variety of soil types, land use types, slopes, and management practices, and it can be used in data-poor regions [11–13]. Currently, SWAT is a key component of the USDA-Conservation Effect Assessment Project and the USEPA-Hydrologic and Water Quality System [14]. Nevertheless, SWAT has a weak ability to simulate groundwater processes, thereby limiting its application in arid areas with strong surface-water–groundwater exchange [15–17].

The ability of SWAT to simulate groundwater processes can be improved by replacing the groundwater module of SWAT with a well-established groundwater model [15,16].

A relatively well-established practice for this approach is to couple SWAT with a MODFLOW model by using the same temporal and spatial scales for both models, thereby allowing SWAT to calculate and input hydrological response unit (HRU)-based groundwater recharge data to the MODFLOW model and then allowing the MODFLOW model to calculate and return the groundwater flow between the aquifer and river to SWAT [15,16]. The SWAT-MODFLOW code developed by Bailey et al. couples the most recent SWAT code with the MODFLOW-NWT code, which improves the solution of unconfined groundwater flow problems [16,18]. This version of the SWAT-MODFLOW model has recently been developed and is the most widely used. Semiromi and Koch [19] modelled complex interaction of surface–groundwater interactions by MODFLOW in the Gharehsoo River basin, located in Northwest Iran. Mosase et al. [20] used SWAT-MODFLOW to assess the spatial distribution of annual and seasonal groundwater recharge and interactions with surface water in the Limpopo River basin, an arid basin in Africa. Jafari et al. [21] developed a calibration tool for SWAT-MODFLOW and used the model to simulate the runoff and groundwater in Shiraz catchment, located in southwestern Iran. The authors used SWAT-MODFLOW to model the natural water cycle of 'atmosphere–slope–underground–river' components. In this process, the impact of human activities, such as land use/land cover change, is generalised [17]. However, in view of increasingly intense human activities, full consideration of both the impact of human activities and natural factors on the water cycle process in a basin is paramount to ensure that distributed hydrological models can accurately describe the water cycle process [22,23]. Intensive vegetation change is one of the final results of human activities [4]. Vegetation growth in SWAT is a key process to consider in the quantitative modelling of eco-hydrological processes, as it directly affects evapotranspiration (ET), water interception, and soil erosion [23]. Therefore, accurate determination of vegetation change in different HRUs is a key to modelling hydrological processes [24]. SWAT can reflect vegetation changes in a basin by using a land-use update module [23]. However, HRUs, the basic computational units of SWAT, are virtual units, each of which is treated as a lumped unit to achieve the same soil type, land use/cover type, and slope at different spatial sites. This makes it infeasible for SWAT to effectively reflect partial land cover type conversions or land cover types converted to multiple other landcovers within the same HRU. To the best of our knowledge, only a few studies have overcome this limitation of HRUs in SWAT-MODFLOW.

Given the above context, in this study, we developed a LU-SWAT-MODFLOW model by integrating a coupled SWAT-MODFLOW model with dynamic HRUs, which can overcome the limitation of considering the vegetation change compared to the original HRUs for the middle and lower reaches of the Bayin River basin, a typical arid endorheic river, where there are frequent surface-water–groundwater interactions and evident vegetation changes. With the advancement of remote sensing technology, data products with high spatiotemporal resolution such as leaf area index (LAI) and ET, combined with observed hydrological data, were used to calibrate the model [25–27]. The performance of SWAT-MODFLOW and LU-SWAT-MODFLOW were compared first. Later, the hydrological effects of revegetation were analysed based on the simulation results of LU-SWAT-MODFLOW. This study can provide assistance for ensuring revegetation sustainability and rationally allocating water resources in arid areas.

2. Materials and Methods

2.1. Study Area

The Bayin River, which is the fourth largest river in the Qaidam Basin, is situated in the north-western region of China (Figure 1). The basin is in an arid area with annual precipitation of approximately 200 mm. The main land cover types of the area are grassland, shrubland, barren land, and farmland. The Bayin River flows out of the mountains into the middle and lower reaches, where the human population and industrial and agricultural activities are concentrated, and with frequent surface-water–groundwater exchange. The vegetation in the Bayin River basin has been restored substantially with the implementation of a series of ecological restoration measures, such as the Grain-for-Green program, over the past 20 years. However, irrigation has become essential for such artificial revegetation projects because of the arid climate and the heterogeneity in the spatial distribution of water resources.

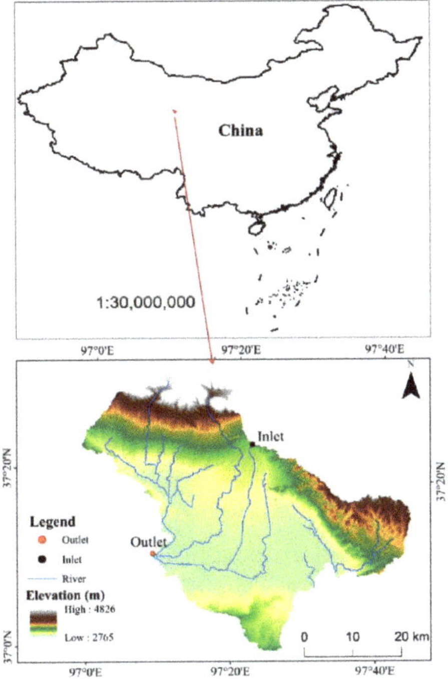

Figure 1. Study area.

2.2. SWAT-MODFLOW Model

In this study, the coupling of SWAT with the MODFLOW model was achieved by using the method of Bailey et al. [16]. Given the inconsistency between the two models in terms of computational spatial units, it was necessary to use a GIS platform to unify the spatial resolution before model coupling (Figure 2). First, specific spatial locations were allocated to the computational units (i.e., HRUs) of SWAT. Second, a mapping relationship was established between the HRUs of SWAT and the computational grid cells of MODFLOW on the same projected coordinate system using a GIS platform [14,15]. Third, the SWAT model was run to simulate the groundwater recharge, evaporation, and extraction with a temporal step of 1 d. Finally, the simulation results were taken as boundary conditions on the corresponding computational grid cells of MODFLOW for groundwater flow modelling [16]. The MODFLOW model was run to simulate the groundwater processes while using the groundwater monitoring data of the basin (provided by Qinghai Provincial Department of water resources) to calibrate and validate the model parameters. Meanwhile, the simulated groundwater table depth from the MODFLOW model was transferred to the computational units of surface water through the abovementioned mapping relationship to impose boundary conditions on the simulation of irrigation groundwater extraction, crop growth, and vegetation transpiration, as well as to test the simulation results [16].

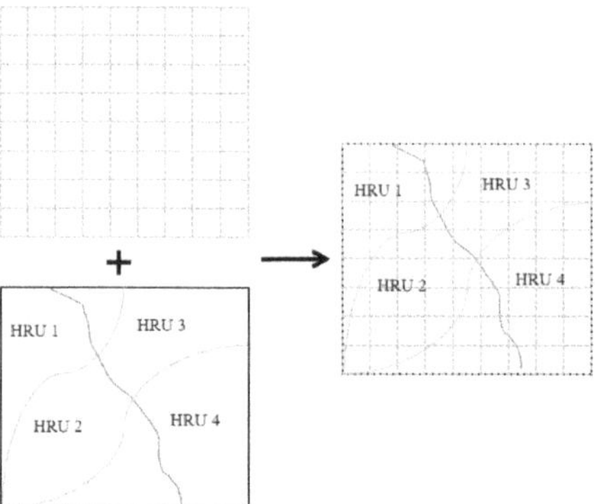

Figure 2. Integration of Soil and Water Assessment Tool (SWAT) and MODFLOW computing units.

2.3. SWAT-MODFLOW Model Coupled with Dynamic HRUs

In view of the inability of the original SWAT model to effectively reflect complete or partial land cover type conversions within the same HRU, this study transformed the HRUs of the original SWAT model to dynamic HRUs to improve the original model. The generation process of dynamic HRUs is illustrated in Figure 3. In contrast to the original HRUs, the generation process of dynamic HRUs involved the defining of spatial units where there were land use/cover changes, i.e., it incorporated the concept of dynamic land use/cover. Such spatial units were combined with soil type and slope data to generate dynamic HRUs such that each had a specific and invariant location, area, and shape with variable attributes. Such dynamic HRUs can more truly reflect the land use/cover changes in the basin.

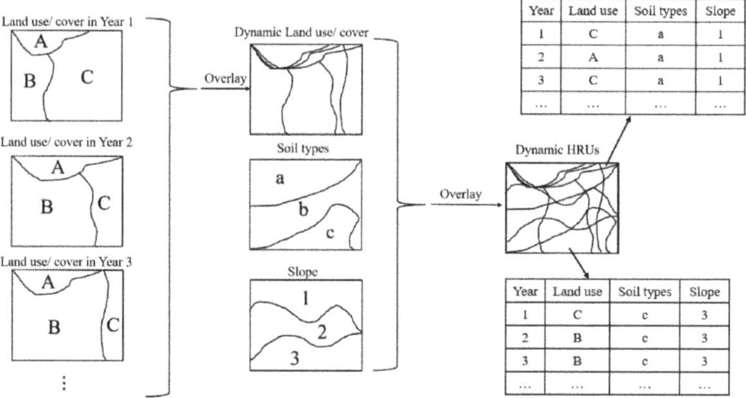

Figure 3. Production of dynamic hydrological response units (HRUs).

The operational flow chart of the coupled SWAT-MODFLOW model based on dynamic HRUs is shown in Figure 4. First, annual cycle simulations were performed by using corresponding HRUs (land cover). Second, daily cycle simulations were performed in which hydrological processes were simulated by SWAT. The simulation results on each HRU were mapped to the computational grid cells of MODFLOW, where these were taken as boundary conditions for groundwater flow simulations. Third, the simulated groundwater data within the grid cells were mapped to the HRUs of SWAT for subsequent SWAT computations. The above process was conducted in nested loops until the end of the simulation. Considering that the dynamic HRU-based SWAT-MODFLOW coupled model could reflect the dynamic changes in land use/cover, the model was referred to as LU-SWAT-MODFLOW.

Start
Yearly loop
Last year of simulation??
Yes
End
No
SWAT-call for Dynamic HRUs
Daily loop
Print the final result
Yes
Last day of the year?
No
SWAT-Surface water simulation
SWAT HRU-MODFLOW cell conversion
MODFLOW-Groundwater Simulation
MODFLOW cell-SWAT HRU conversion
SWAT simulation

Figure 4. Flowchart of the SWAT-MODFLOW coupled model based on dynamic HRUs.

2.4. Model Establishment

Meteorological data required for establishing the SWAT model included the daily monitoring data for precipitation, temperature, relative humidity, wind speed, and solar radiation at the Delingha weather station, which is located at the inlet shown in Figure 1. Agricultural, forestland, and grassland irrigation data were obtained from the Delingha Municipal Water Affairs Bureau (Table 1). The newest digital elevation model data (Figure 1) of 30 m resolution Shuttle Radar Topography Mission data [28] were used. Soil type data (Figure 5a) at the scale 1:1,000,000 from China were used [29], and the corresponding soil hydrological attributes were retrieved from the Qinghai Soil Record [30]. Figure 5b presents the sub-basin division map of the study region.

Table 1. Irrigation volume in the study region.

Irrigation Type	Annual Irrigation Rate (m^3/hm^2)	Number of Times of Irrigation	Irrigation Duration
Agricultural irrigation	5800	6	March–October
Forest irrigation	5400	6	April–November
Grassland irrigation	3600	5	April–November

Land use type data for the simulated period (2000–2018) were derived from 30 m Landsat images. Remote sensing interpretation marks were created according to spectral features combined with field survey data and relevant geographic maps [31]. Data quality was examined by comparatively analysing field survey patches versus randomly selected patches, and the classification accuracy was determined to be over 90%. Figure 6 presents the regional spatial distribution of land use/cover in 2000 (Figure 6a) versus 2018 (Figure 6b). There were six types of land use/cover in 2000, namely, spring wheat, forest, grassland, water, residences, and barren land. There were seven types in 2018, including 'Chinese wolfberry' as a new type in addition to the existing six types. In the study region,

Chinese wolfberry was the main tree species used for artificial revegetation. Considering the absence of Chinese wolfberry-related parameters in the built-in land use and vegetation database of SWAT, relevant initial parameters for apple trees in the SWAT model were taken as default parameters for Chinese wolfberry, and these were calibrated using the LAI data to simulate the growth process of Chinese wolfberry. Other relevant parameters of land use/cover types were either set to the default values in the built-in database of the model or obtained by calibration. Revegetation in the study region was mainly characterised by the conversion of farmland to forestland and the conversion of bare land to forestland and grassland. From 2000 to 2018, the years when evident vegetation changes (restoration) occurred in the study region were 2005, 2008, 2015, and 2018. Accordingly, the land use data for 2000, 2005, 2008, 2015, and 2018 were used to generate dynamic HRUs. In contrast, land cover types in the other years were only weakly altered and therefore ignored.

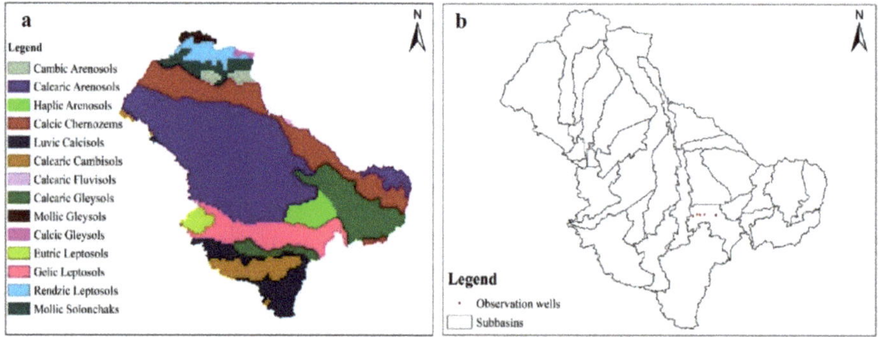

Figure 5. (**a**) Soil types in the study area and (**b**) sub-basins input to the SWAT model.

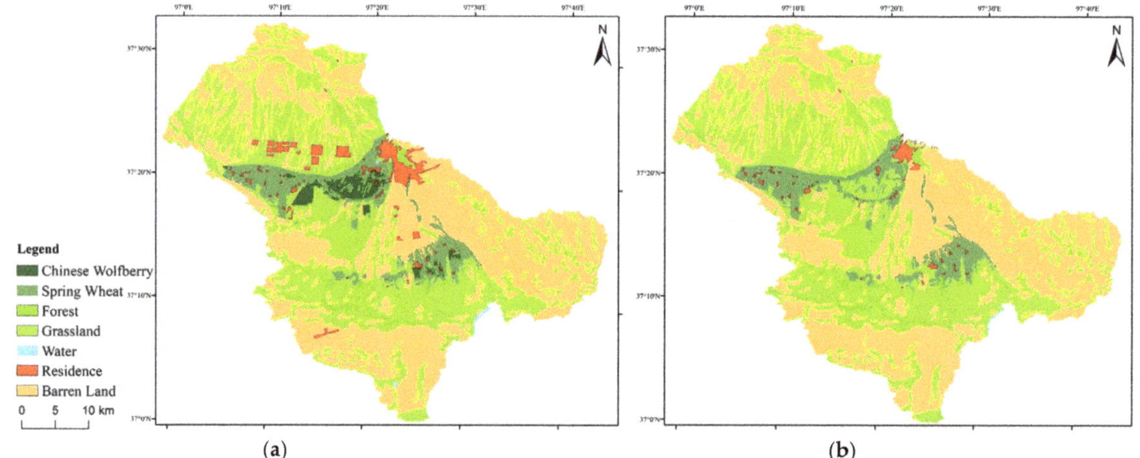

Figure 6. Revegetation status of the study region in (**a**) 2000 and (**b**) 2018.

Basin boundaries delineated by the SWAT model were considered as impermeable boundaries to groundwater flow in the MODFLOW model, where the western and eastern outlets of the basin were considered as constant flow boundaries. The river network extracted by the SWAT model was considered to constitute river boundaries in the MODFLOW model. The simulated steady-state groundwater head (Figure 7) was used as the initial head for simulations of transient flow [32]. In addition, the shallow aquifers in the study region were conceptualised as being non-homogeneous and anisotropic according to relevant studies [32], and the groundwater flow was conceptualised as a two-dimensional transient flow.

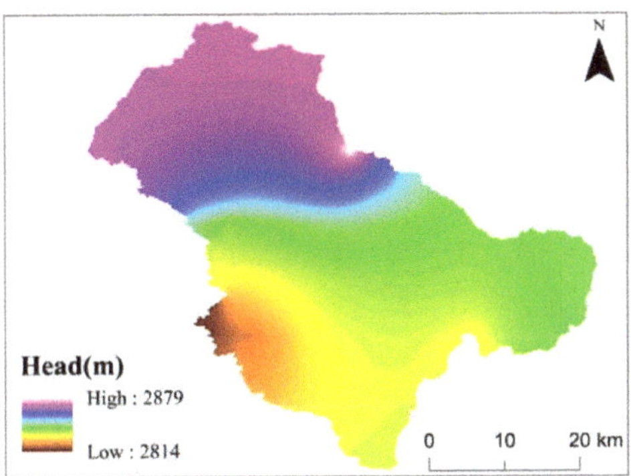

Figure 7. Initial groundwater head for MODFLOW.

2.5. Model Calibration

The years 2000–2001 were used as the model warm-up period, 2002–2011 as the model calibration period, and 2012–2018 as the model validation period. Vegetation growth parameters in the SWAT model were calibrated at HRU scales against the monthly 30 m resolution LAI data of 2002–2018 provided by the National Tibetan Plateau/Third Pole Environment Data Centre [29]. The dataset is the fusion of MODIS LAI and observed LAI in Qilian mountainous area (including the Qaidam Basin).

ET simulated by the LU-SWAT-MODFLOW model was calibrated at sub-basin scales against the ET data at a $0.1° \times 0.1°$ resolution (Figure 8) provided by the National Tibetan Plateau/Third Pole Environment Data Centre [29]. The dataset is derived by employing a calibration-free nonlinear complementary relationship model with inputs of air and dew-point temperature, wind speed, precipitation, and net radiation from the China Meteorological Forcing Dataset. The dataset is validated in Northwest China and is proved to have good spatial and temporal performance [24]. Furthermore, relevant parameters (conductivity and storage) of the MODFLOW model were calibrated against monthly recorded groundwater table depth at numerous observation wells (Figure 5b) in the basin.

In this study, land cover types in the LU-SWAT-MODFLOW model varied over the years within some of the dynamic HRUs or remained invariant throughout a relatively long period within some of the HRUs. The HRUs in the latter scenario were chosen to calibrate the relevant vegetation growth parameters (Table 2) against monthly 30 m resolution LAI data [26]. The calibrated parameters were stored in a separate file, which could be visited during SWAT operations to directly tune parameters in HRUs where changes in land cover type were detected. ET-related parameter (Table 2) calibration at sub-basin scales in the present study was mainly based on the aforementioned remote sensing-derived ET data in accordance with the method of White and Chaubey [33] and Immerzeel and Droogers [25].

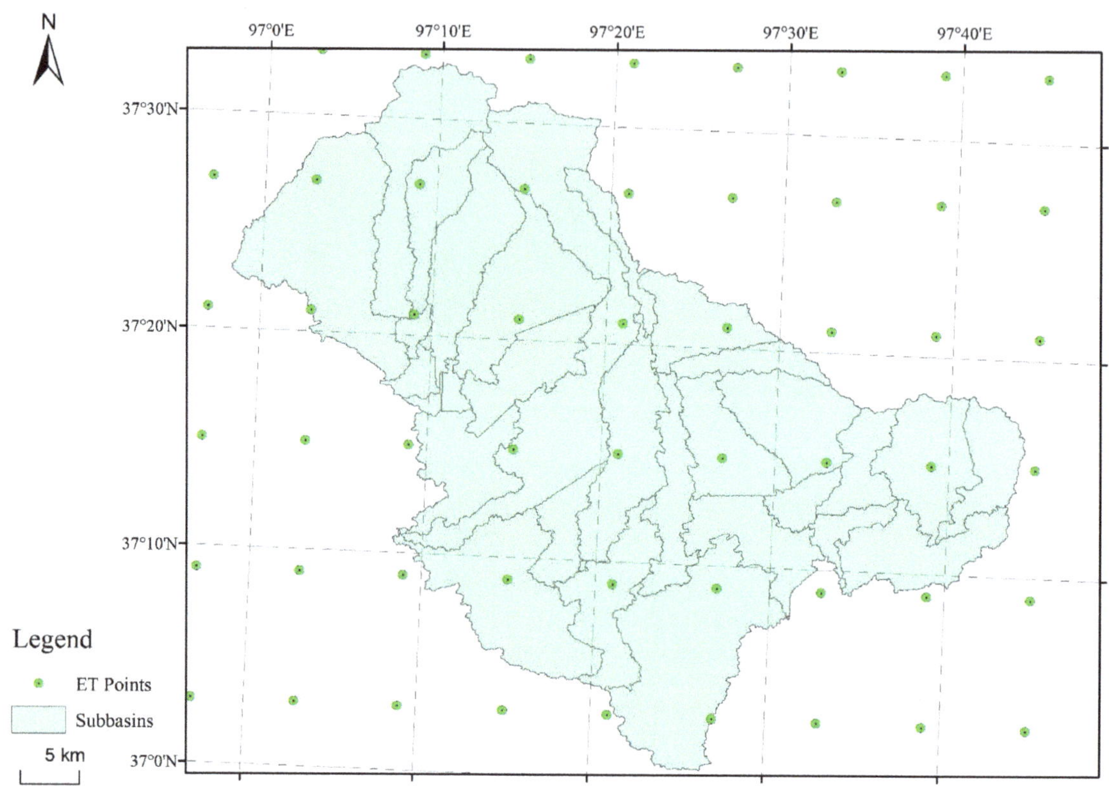

Figure 8. Locations of ET data points.

Table 2. Vegetation growth and ET-related parameters in SWAT.

Processes	Parameter	Description
Vegetation growth-related parameters	BLAI	Maximum potential leaf area index
	LAIMX_1	Fraction of the maximum leaf area index corresponding to the 1st point on the optimal leaf area development curve
	FRGRW1	Fraction of the plant growing season corresponding to the 1st point on the optimal leaf area development curve
	LAIMX_2	Fraction of the maximum leaf area index corresponding to the 2nd point on the optimal leaf area development curve
	FRGRW2	Fraction of the plant growing season corresponding to the 2nd point on the optimal leaf area development curve
	DLAI	Fraction of growing season when leaf area begins to decline
	BIO_E	Radiation use efficiency
	EXT_COEF	Light extinction coefficient
	GSI	Maximum canopy stomatal conductance
	HVSTI	Harvest index for the optimal growing condition
	T-BASE	Minimum (base) temperature for plant growth
ET-related parameters	SOL_AWC	Available water content
	RFINC	Monthly rainfall increment
	GWREVAP	Groundwater revap coefficient
	BLAI	Maximum potential leaf area index
	ESCO	Soil evaporation compensation factor
	EPCO	Plant uptake compensation factor

2.6. Model Performance Metrics

The applicability of SWAT was evaluated in terms of Nash–Sutcliffe Efficiency (NSE), percent bias (PBIAS), and squared correlation coefficient (R^2). The NSE can range between $-\infty$ and 1. If the value of the NSE was closer to 1, it was indicative of a better simulation performance and high reliability of the SWAT model. When the NSE was closer to 0.5, the model simulation results were similar to the mean of observations, that is, the model results in general were reliable. A PBIAS between −10% and 10% indicated a good simulation performance of the model. Additionally, larger values of R^2 were indicative of a better simulation performance of the model. The calculation process and significance of the three metrics have been elaborated elsewhere [34]. Model performance during the simulations of the groundwater table depth was evaluated mainly in terms of the absolute error and R^2 in this study.

$$NSE = 1 - \left[\frac{\sum_{i=1}^{n}\left(V_i^{obs} - V_i^{sim}\right)^2}{\sum_{i=1}^{n}\left(V_i^{obs} - V^{mean}\right)^2}\right] \tag{1}$$

$$PBIAS = \left[\frac{\sum_{i=1}^{n}\left(V_i{}^{obs} - V_i{}^{sim}\right) \times 100}{\sum_{i=1}^{n} V_i{}^{obs}}\right] \tag{2}$$

$$R^2 = \frac{\left[\sum_{i=1}^{N}\left(V_i^{sim} - \overline{V}^{sim}\right)\left(V_i^{obs} - \overline{V_i}^{obs}\right)\right]^2}{\sum_{i=1}^{N}\left(V_i^{sim} - \overline{V}^{sim}\right)^2 \sum_{i=1}^{N}\left(V_i^{obs} - \overline{V}^{obs}\right)^2} \tag{3}$$

3. Results

3.1. Regional Vegetation Change

The main types of vegetation change in the study region from 2002 to 2018 are summarised in Table 3. Specifically, the main types of vegetation change pertained to the conversion of low-coverage grassland, farmland, and bare land to forestland, in which the converted areas amounted to 2,877,518 and 321 hm^2, respectively. Correspondingly, the irrigation water volume of each revegetation plot underwent dramatic changes. In addition, the conversion of farmland to forestland mainly occurred in the irrigated area of the Gahai Lake in the south-eastern part of the basin, while the conversion of low-coverage grassland and bare land to forestland mainly occurred in the irrigated area of Delingha, which is situated in the north-western part of the basin (Figure 6).

Table 3. Main types of revegetation in the study region.

Main Type of Revegetation	Revegetation Area (hm^2)	Change in the Annual Irrigation Rate ($m^3/hm^2 \cdot a$)
Conversion of low-coverage grassland to forestland	2877	0→5400
Conversion of farmland to forestland	518	5800→5400
Conversion of bare land to forestland	321	0→5400

3.2. Comparison of the Original SWAT-MODFLOW and LU-SWAT-MODFLOW

3.2.1. Difference in HRUs

Figure 9 shows the HRUs generated by the original SWAT-MODFLOW model versus those from the LU-SWAT-MODFLOW model. The original SWAT-MODFLOW model generated 1304 HRUs, and the LU-SWAT-MODFLOW model generated 2978 HRUs. The higher number of HRUs in the new model was due to the land use/cover data of LU-SWAT-MODFLOW being a superposition of years-long data and thereby covering a higher number of patches. The higher number of HRUs also implied that the operation and parameter tuning of the LU-SWAT-MODFLOW model would be more complicated [35].

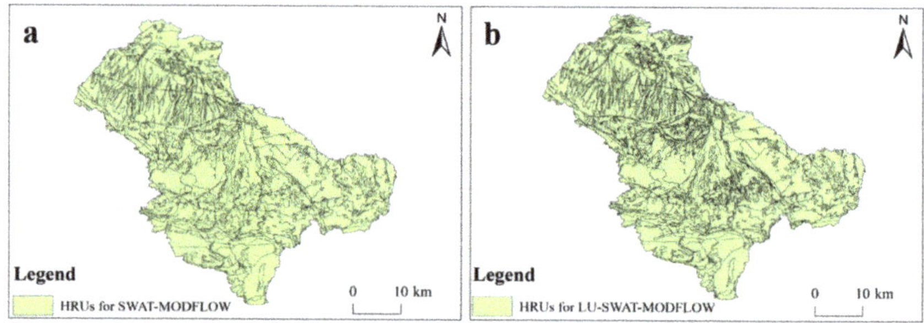

Figure 9. HRUs generated by (**a**) SWAT-MODFLOW and (**b**) LU-SWAT-MODFLOW.

3.2.2. Comparison of the LAI Simulation Results

Figure 10 shows the simulated LAI from the calibrated original SWAT-MODFLOW model and the calibrated LU-SWAT-MODFLOW model for July 2005, 2010, 2015, and 2018. Compared to the remote-sensed LAI, the calibrated LU-SWAT-MODFLOW model has better performance in expressing the spatial variation of the LAI than SWAT-MODFLOW since the LU-SWAT-MODFLOW model has more HRUs. In addition, we randomly chose 10 positions in different parts of the study area and calculated the performance metrics of LAI in corresponding HRUs of original SWAT-MODFLOW and LU-SWAT-MODFLOW model, respectively. The performance metrics for the calibrated and validated original SWAT-MODFLOW model were NSE > 0.75, PBIAS of −25–25%, and $R^2 > 0.73$, and the counterparts for the calibrated and validated LU-SWAT-MODFLOW were NSE > 0.83, PBIAS of −20–20%, and $R^2 > 0.83$ (Table 4). This indicates that the LU-SWAT-MODFLOW model was more accurate than the original SWAT-MODFLOW model in simulations of the monthly LAI after both models were calibrated and validated.

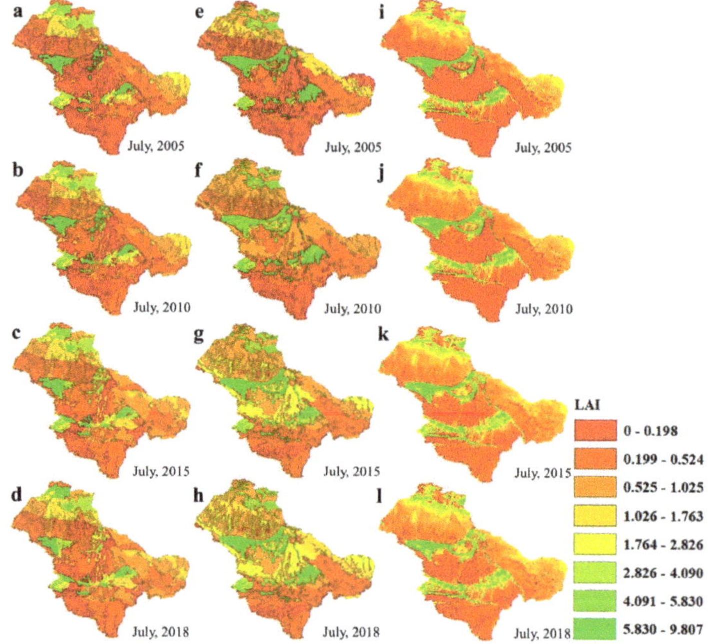

Figure 10. Simulated leaf area index (LAI) from the (**a–d**) calibrated original SWAT-MODFLOW model, (**e–h**) calibrated LU-SWAT-MODFLOW model, and (**i–l**) remote-sensed LAI.

Table 4. Performance of the original SWAT-MODFLOW model versus the LU-SWAT-MODFLOW model in simulating LAI.

HRU	SWAT-MODFLOW						LU-SWAT-MODFLOW					
	Calibration Period			Validation Period			Calibration Period			Validation Period		
	NSE	PBIAS	R^2	NSE	PBIAS	R^2	NSE	PBIAS	R^2	NSE	PBIAS	R^2
E1	0.85	20.84	0.83	0.81	17.63	0.79	0.92	14.11	0.88	0.87	16.54	0.85
E2	0.86	23.71	0.85	0.80	21.25	0.76	0.91	11.13	0.87	0.86	14.52	0.83
W1	0.82	19.89	0.82	0.78	15.65	0.75	0.88	13.45	0.90	0.84	17.89	0.88
W2	0.83	17.41	0.81	0.75	19.32	0.73	0.89	12.65	0.89	0.83	19.21	0.84
S1	0.85	13.72	0.84	0.81	17.29	0.77	0.90	10.99	0.90	0.86	16.54	0.86
S2	0.83	19.24	0.81	0.79	14.23	0.78	0.91	11.11	0.90	0.85	17.32	0.87
N1	0.87	15.36	0.85	0.77	13.22	0.78	0.93	14.65	0.92	0.88	15.21	0.89
N2	0.86	17.52	0.84	0.76	11.21	0.77	0.90	17.53	0.89	0.84	18.56	0.90
C1	0.84	21.49	0.83	0.78	22.17	0.76	0.92	16.46	0.90	0.87	14.12	0.87
C2	0.85	20.76	0.83	0.81	19.78	0.74	0.91	13.02	0.92	0.83	11.76	0.89

3.2.3. Comparison of the ET Simulation Results

During the calibration and validation period, the performance metrics of the original SWAT-MODFLOW model were NSE > 0.65, PBIAS of −20%–20%, and $R^2 > 0.63$ in simulations of the monthly mean ET for each sub-basin, while the counterparts for the LU-SWAT-MODFLOW model were NSE > 0.72, PBIAS of −20%–20%, and $R^2 > 0.73$ (Figure 11). This indicates that the LU-SWAT-MODFLOW model was more accurate than the original SWAT-MODFLOW in simulations of the monthly mean ET for most of the sub-basins. Figure 12 shows the multi-year mean of the simulated ET from the original SWAT-MODFLOW model (Figure 12a) versus that of the LU-SWAT-MODFLOW model (Figure 12b) in comparison with the multi-year mean of remote sensing-derived ET (Figure 12c). The multi-year mean of remote sensing-derived ET exhibited a spatial distribution pattern of high values in the north-eastern mountains and low values in the southwestern plains, and such a distribution pattern existed for both calibrated models.

3.2.4. Comparison of the Simulation Results for Groundwater Table Depth

Groundwater table depth data were scarce within the study region. Observation wells 1, 2, and 3 only provided monthly data for 2009–2011, and observation well 4 only provided monthly data for 2013–2015; meanwhile, observation well 5 only provided monthly data for 2014–2015. Thus, the observed groundwater table depth of wells 1, 2, and 3 were used to calibrate the two models, and the rest were used to validate the models. Linear regression results of the simulated groundwater table depth on observed groundwater table depth were compared between the original SWAT-MODFLOW model and the LU-SWAT-MODFLOW model (Figure 13). Both models performed well in simulating the changes in the groundwater table depth of the study region, with an $R^2 > 0.95$ and absolute error within 0.5 m. In addition, the simulation performance of LU-SWAT-MODFLOW was slightly better than that of SWAT-MODFLOW, which was likely attributed to the detailed consideration of the spatiotemporal changes in irrigation and land cover by the former model versus the latter model.

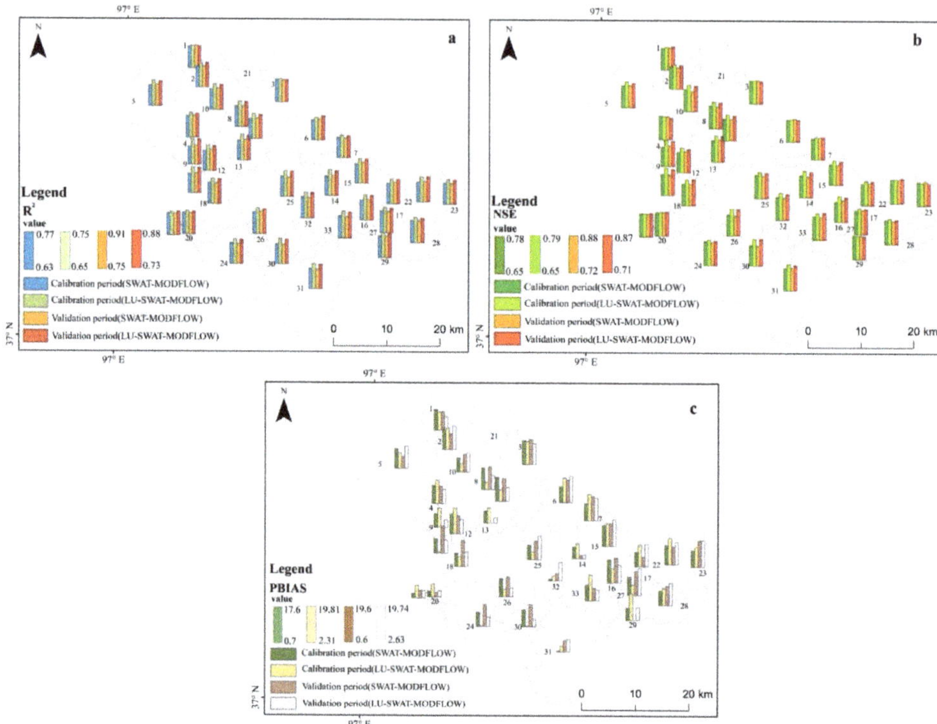

Figure 11. NSE (**a**), R^2 (**b**), PBIAS (**c**) of the calibrated original SWAT-MODFLOW model versus the calibrated LU-SWAT-MODFLOW model in simulating evapotranspiration in each sub-basin.

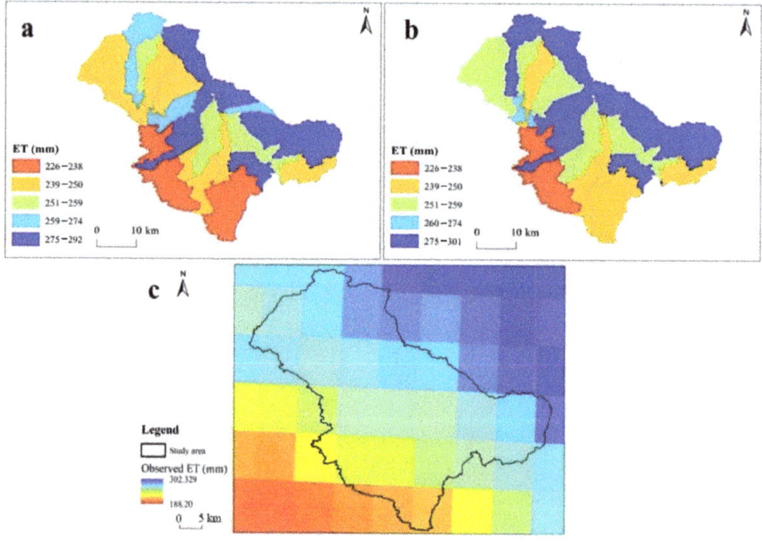

Figure 12. Multi-year mean ET simulated by (**a**) SWAT-MODFLOW, (**b**) LU-SWAT-MODFLOW, and (**c**) retrieved from remote sensing images.

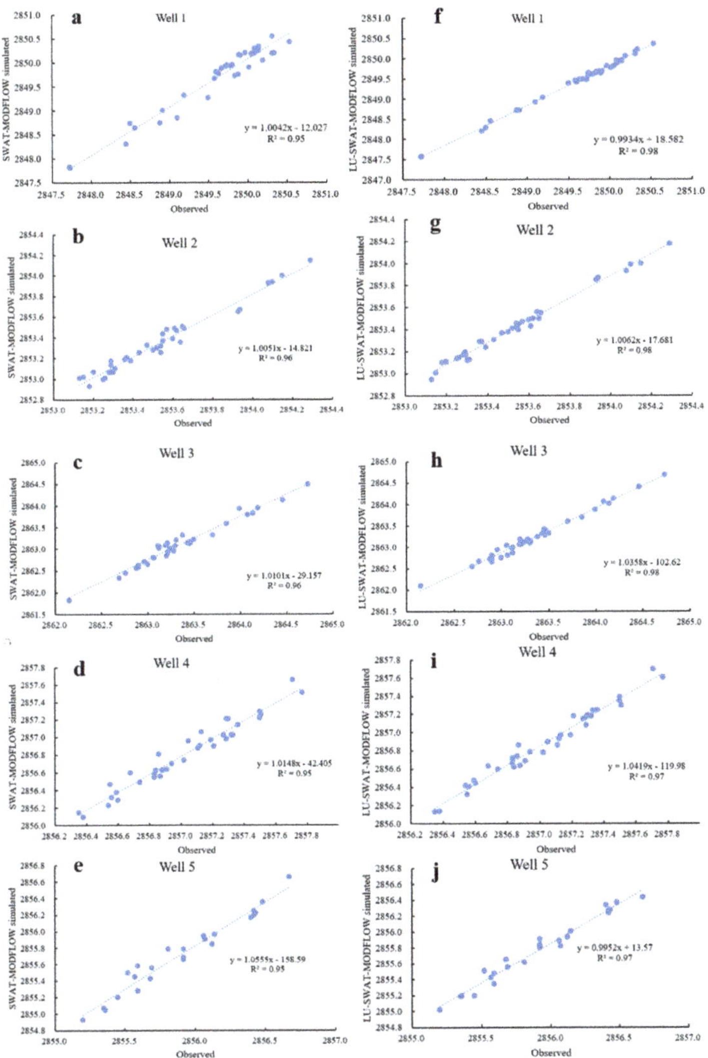

Figure 13. Monthly groundwater table depth simulated by the (**a–e**) SWAT-MODFLOW model versus the (**f–j**) LU-SWAT-MODFLOW model.

3.3. Impacts of Vegetation Change on Hydrological Processes

The case study area is located in a water consumption area of an inland river and almost never generates runoff. Thus, here we focus on the analysis of the vegetation change impacts on ET and groundwater processes.

3.3.1. Impacts on ET

The LU-SWAT-MODFLOW model was run in the following two scenarios to accurately analyse the impacts of revegetation and the related extensive irrigation on ET: (1) revegetation was assumed absent while considering the actual changes in other types of land use/cover; and (2) the actual changes in land use/cover were considered, including those pertinent to revegetation (irrigation) and other types of land use/cover.

Figure 14a shows the simulated monthly ET in the revegetation-absent scenario versus the revegetation-present scenario from 2002 to 2018. The results indicate that revegetation and related irrigation did not change the trend of monthly ET in the basin, in which the monthly ET in the revegetation-present scenario was only 1.5 mm higher than that in the revegetation-absent scenario for most months. Similarly, the trend of annual ET was almost the same in both scenarios. In 2004 and later years, ET showed weakly higher values in the revegetation-present scenario than in the revegetation-absent scenario (Figure 14b). Figure 13c illustrates the difference in the multi-year mean ET between the two scenarios in each sub-basin. Such a difference was greater than 10 mm in sub-basins 4, 12, 13, 14, 26, and 33, that is, the ET increase was most obvious in these sub-basins. Comprehensive comparisons of the land use/cover map (Figure 6) with the LAI map for the study region during the study period further confirmed that relatively obvious revegetation had been achieved in these sub-basins.

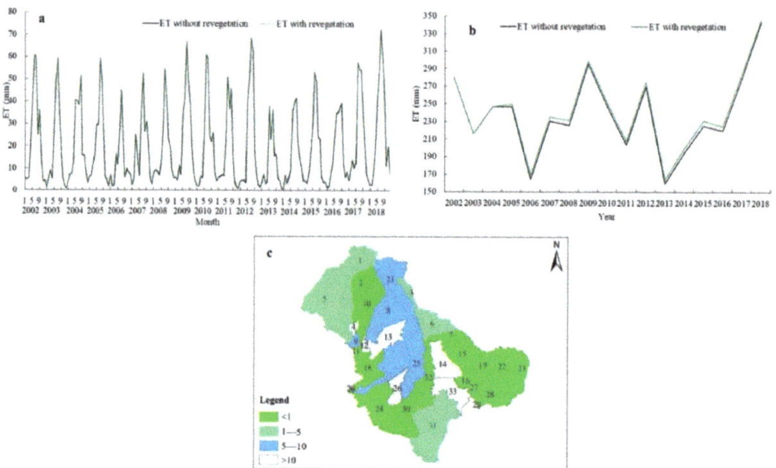

Figure 14. (**a**) Monthly and (**b**) yearly ET with revegetation and without revegetation; (**c**) yearly average ET change in different sub-basins after revegetation.

3.3.2. Impacts on Groundwater Recharge

Groundwater is the most important water resource in the arid endorheic river watershed. Changes in groundwater recharge may affect the groundwater storage and further impact the ecological environment. Figure 15 shows the monthly (Figure 15a) and yearly (Figure 15b) groundwater recharge in the entire study area. After revegetation, the groundwater recharge increased by approximately 1.27 mm on average per month and 14.02 mm on average per year. Fan et al. [36], Yang and Lu [37], and Qubaja et al. [30] showed that canopy interception and root water absorption would lead to reduction of soil water, surface runoff, and groundwater recharge in woodland. However, here, although considerable areas of low-coverage grassland, farmland, and bare land were converted to forestland, the groundwater recharge with revegetation was evidently higher than that without revegetation. We reported the yearly average groundwater recharge after revegetation in the entire study area (Figure 14c). The groundwater recharge in the irrigation district where the revegetation was applied was the highest (>14.51 m^3/day); that is, the irrigation for the recovered vegetation strongly affected the groundwater recharge.

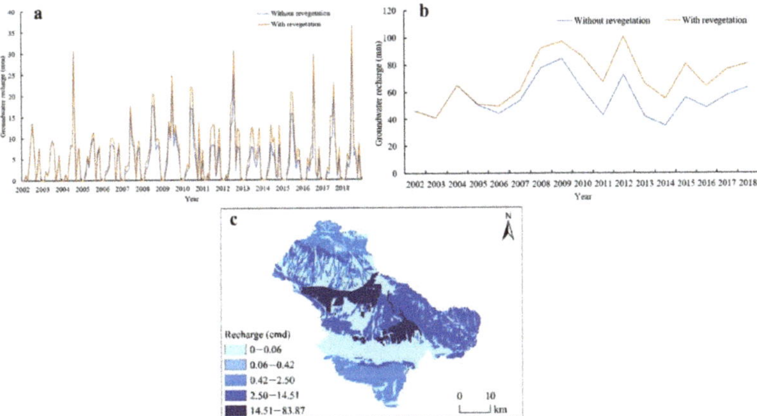

Figure 15. (**a**) Monthly and (**b**) yearly groundwater recharge with and without revegetation; (**c**) yearly average groundwater recharge after revegetation in space.

3.3.3. Impacts on Surface Water and Groundwater Exchange

There was frequent surface-water–groundwater exchange in the study region, which dominated the regional hydrological processes. We analysed the surface water and groundwater exchange affected by revegetation. Figure 16 shows the amount of groundwater recharge and discharge in the revegetation-absent scenario minus the value in the revegetation-present scenario. Specifically, river reach I was in the upper study region, where groundwater was recharged by river water. River reach II was situated in the lower study region, where significant amounts of groundwater were discharged to the river. In river reach III, both surface water recharge to groundwater and groundwater discharge to surface water were present. River reach III was situated in the irrigated area of Delingha, where it was greatly affected by agricultural, forestland, and grassland irrigation, which led to a relatively complex pattern of surface-water–groundwater exchange. The area where river reach III was situated was also the main revegetation area of the study region. Comparisons of Figure 16 revealed that the direction of surface-water–groundwater exchange was reversed in six grid cells, which was attributed to the changes in the irrigation volume within these grid cells after revegetation.

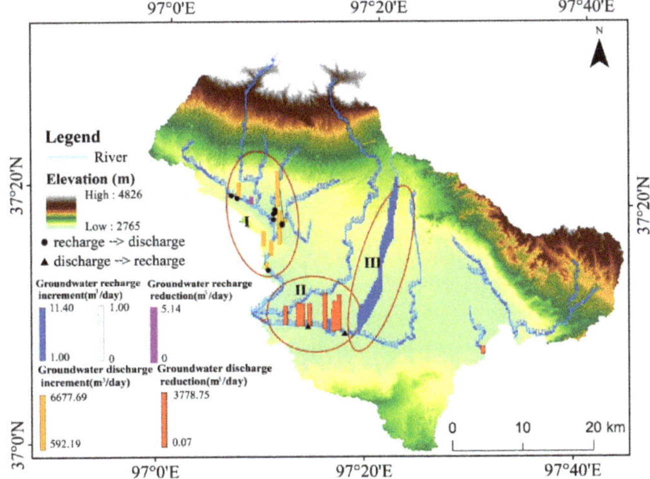

Figure 16. Revegetation impacts on groundwater recharge and discharge.

4. Discussion

The LU-SWAT-MODFLOW model was more accurate than the original SWAT-MODFLOW model in simulating the monthly LAI after both models were calibrated and validated. LAI plays a key role in SWAT for estimating ET, canopy interception, and biomass accumulation [35]. The enhanced modelling of LAI could improve the performance of the SWAT model in eco-hydrological processes [26,38]. However, accurate simulation of LAI relies on many parameters which are difficult to calibrate. Generally, parameters of SWAT-MODFLOW are calibrated with observed data in watershed outlets or sub-basins [23,27]. Only few studies have calibrated the parameters at the HRU level because of its difficulty and complexity. In this study, the remote-sensed monthly LAI data were used to calibrate the SWAT-MODFLOW and LU-SWAT-MODFLOW at the HRU level using the SWAT-CUP software (https://swat.tamu.edu/software/swat-cup/) (accessed on 1 October 2021) with a satisfactory result. This suggests that the model calibration at the HRU level is possible and effective if the related observation data exist.

The LU-SWAT-MODFLOW model was more accurate than the original SWAT-MODFLOW in simulating the monthly mean ET for most sub-basins. Ma et al. [26] reported that canopy interception and soil water content would be seriously affected by LAI in SWAT, which would further affect ET. Therefore, the enhancements of LU-SWAT-MODFLOW in modelling the monthly ET can be attributed to the more accurate simulation of the LAI.

Revegetation projects have been conducted in both the Gahai Lake irrigated area and the irrigated area of Delingha, but the revegetation had a relatively high impact on the direction and amount of surface-water–groundwater exchange in the latter area; in the former area, there was an almost negligible impact. This discrepancy was attributed to the fact that revegetation in the Gahai Lake irrigated area was mainly characterised by the conversion of farmland to forestland, and the irrigation volume did not differ significantly between the two land cover types [38]. In contrast, the irrigated area of Delingha was dominated by the conversion of low-coverage grassland to bare land and forestland, and the former two land cover types required no irrigation, while the latter land cover type required a large irrigation volume.

This study is subjected to some limitations. On the one hand, we used land use/cover map to analyse the revegetation process in our study area. In fact, plant density, age, and growth status were not considered because of the limitations in the SWAT model. Moreover, these factors may affect the eco-hydrological processes in such an arid area [26]. On the other hand, meteorological data were scarce in both original SWAT-MODFLOW and LU-SWAT-MODFLOW models. This may impact the model performance in formulating the water budget [39–41]. Nonetheless, these limitations should be addressed in future studies by using and analysing different datasets.

5. Conclusions

This study was carried out in the middle and lower reaches of the Bayin River basin in the north-eastern part of the Qaidam Basin, China, where there is frequent surface-water–groundwater interaction and evident vegetation change. A LU-SWAT-MODFLOW model was developed by integrating a coupled SWAT-MODFLOW model with dynamic HRUs in view of their ability to reflect the actual land cover changes in the basin. The impacts of revegetation and related irrigation on the main hydrological processes in the basin were more accurately simulated and analysed by the LU-SWAT-MODFLOW model than by the original SWAT-MODFLOW model.

The LU-SWAT-MODFLOW model generated dynamic HRUs by pre-defining spatial units where land use/cover changes occurred during the simulated period, thereby overcoming the inability of the original SWAT model to effectively reflect the complete or partial land cover type conversion within the same HRU. This new model outperformed the original SWAT-MODFLOW model in simulating the LAI. The LAI is an important parameter of SWAT as it affects a series of processes, such as ET and infiltration; therefore, accurate simulations of the LAI are a key to accurate hydrological simulations. Moreover, the LU-SWAT-MODFLOW model outperformed the original SWAT-MODFLOW model in simulating the ET and groundwater table depth of the basin.

The LU-SWAT-MODFLOW model was run in two different scenarios, one with revegetation and the other without it, to assess the impacts of revegetation and related irrigation on the main hydrological processes in the study region. The results showed that after regional revegetation, ET in the different sub-basins increased by approximately 1.5 mm per month and by 6 mm per year. After revegetation, the groundwater recharge increased by approximately 1.27 mm on average per month and 14.02 mm on average per year. Irrigation for the recovered vegetation strongly affected the groundwater recharge. Meanwhile, the direction and amount of surface-water–groundwater exchange underwent evident changes in areas where revegetation was characterised by the conversion of low-coverage grassland and bare land to forestland. In areas where revegetation was characterised by the conversion of farmland to forestland, the irrigation volume was not greatly altered; thus, this transition had a weak impact on the direction and amount of surface-water–groundwater exchange. Changes in the direction and amount of surface-water–groundwater exchange may lead to a series of ecological and environmental issues. To avoid problems in the future, water-saving irrigation techniques should be advocated when conducting revegetation in arid inland river basins. In addition, our findings indicate that it would be advantageous to preferentially apply revegetation measures that promote the conversion of farmland to forestland/grassland provided that they do not adversely affect regional economic development.

Author Contributions: Conceptualization, X.J. and X.M.; methodology, X.J.; formal analysis, X.J.; investigation, X.J.; resources, Y.J.; data curation J.Z. and D.F.; writing—original draft preparation, X.J.; writing—review and editing, Y.J.; project administration, X.J.; funding acquisition, X.J. and X.M. All authors have read and agreed to the published version of the manuscript.

Funding: This research was funded by the National Natural Science Foundation of China, grant number 41801094 and grants from the Natural Science Foundation of Qinghai Province, China, grant number 2021-ZJ-705.

Data Availability Statement: The processed data required to reproduce these findings cannot be shared at this time as the data is also a part of an ongoing study.

Conflicts of Interest: The authors declare no conflict of interest.

References

1. Feng, X.M.; Fu, B.J.; Piao, S.; Wang, S.; Ciais, P.; Zeng, Z.; Lü, Y.; Zeng, Y.; Li, Y.; Jiang, X.; et al. Revegetation in China's Loess Plateau is approaching sustainable water resource limits. *Nat. Clim. Chang.* **2016**, *6*, 1019–1022. [CrossRef]
2. Li, T.; Xia, J.; Zhang, L.; She, D.; Wang, G.; Cheng, L. An improved complementary relationship for estimating evapotranspiration attributed to climate change and revegetation in the Loess Plateau, China. *J. Hydrol.* **2021**, *592*, 125516. [CrossRef]
3. Paul, M.; Rajib, A.; Negahban-Azar, M.; Shirmohammadi, A.; Srivastava, P. Improved agricultural Water management in data-scarce semi-arid watersheds: Value of integrating remotely sensed leaf area index in hydrological modelling. *Sci. Total Environ.* **2021**, *791*, 148177. [CrossRef]
4. Reynolds, J.F.; Smith, D.M.S.; Lambin, E.F.; Turner, B.L., II; Mortimore, M.; Batterbury, S.P.J.; Downing, T.E.; Dowlatabadi, H.; Fernández, R.J.; Herrick, J.E.; et al. Global desertification: Building a science for dryland development. *Science* **2007**, *316*, 847–851. [CrossRef] [PubMed]
5. Jackson, R.B.; Jobbágy, E.G.; Avissar, R.; Roy, S.B.; Barrett, D.J.; Cook, C.W.; Farley, K.A.; le Maitre, D.C.; McCarl, B.A.; Murray, B.C. Trading water for carbon with biological carbon sequestration. *Science* **2005**, *310*, 1944–1947. [CrossRef]
6. Menz, M.H.M.; Dixon, K.W.; Hobbs, R.J. Hurdles and opportunities for landscape-scale restoration. *Science* **2013**, *339*, 526–527. [CrossRef]
7. Quinn, P.; Beven, K.; Chevallier, P.; Planchon, O. The prediction of hillslope flow paths for distributed hydrological modelling using digital terrain models. *Hydrol. Process* **1991**, *5*, 59–79. [CrossRef]
8. Minacapilli, M.; Iovino, M.; D'Urso, G. A distributed agro-hydrological model for irrigation water demand assessment. *Agric. Water Manag.* **2008**, *95*, 123–132. [CrossRef]
9. Xu, X.; Jiang, Y.; Liu, M.; Huang, Q.; Huang, G. Modeling and assessing agro-hydrological processes and irrigation water saving in the middle Heihe River basin. *Agric. Water Manag.* **2019**, *211*, 152–164. [CrossRef]

10. Arnold, J.G.; Moriasi, D.N.; Gassman, P.W.; Abbaspour, K.C.; White, M.J.; Srinivasan, R.; Santhi, C.; Harmel, R.D.; van Griensven, A.; Van Liew, M.W.; et al. SWAT: Model use, calibration, and validation. *Trans. ASABE* **2012**, *55*, 1491–1508. [CrossRef]
11. Panagopoulos, Y.; Makropoulos, C.; Baltas, E.; Mimikou, M. SWAT parameterization for the identification of critical diffuse pollution source areas under data limitations. *Ecol. Model.* **2011**, *222*, 3500–3512. [CrossRef]
12. Nyeko, M. Hydrologic modelling of data scarce basin with SWAT model: Capabilities and limitations. *Water Resour. Manag.* **2015**, *29*, 81–94. [CrossRef]
13. Alemayehu, T.; van Griensven, A.; Bauwens, W. Evaluating CFSR and Watch data as input to SWAT for the estimation of the potential evapotranspiration in a data-scarce Eastern-African Catchment. *J. Hydrol. Eng.* **2016**, *21*, 12–18. [CrossRef]
14. Yang, Q.; Zhang, X. Improving SWAT for simulating water and carbon fluxes of forest ecosystems. *Sci. Total Environ.* **2016**, *569–570*, 1478–1488. [CrossRef]
15. Kim, N.W.; Chung, I.M.; Won, Y.S.; Arnold, J.G. Development and application of the integrated SWAT-MODFLOW model. *J. Hydrol.* **2008**, *356*, 1–16. [CrossRef]
16. Bailey, R.T.; Wible, T.C.; Arabi, M.; Records, R.M.; Ditty, J. Assessing regional-scale spatio-temporal patterns of groundwater-surface water interactions using a coupled SWAT-MODFLOW model. *Hydrol. Process* **2016**, *30*, 4420–4433. [CrossRef]
17. Jin, X.; He, C.; Zhang, L.; Zhang, B. A modified groundwater module in SWAT for improved streamflow simulation in a large, arid endorheic river watershed in Northwest China. *Chin. Geogr. Sci.* **2018**, *28*, 47–60. [CrossRef]
18. Liu, W.; Park, S.; Bailey, R.T.; Molina-Navarro, E.; Andersen, H.E.; Thodsen, H.; Nielsen, A.; Jeppesen, E.; Jensen, J.S.; Jensen, J.B.; et al. Comparing SWAT with SWAT-MODFLOW hydrological simulations when assessing the impacts of groundwater abstractions for irrigation and drinking water. *Hydrol. Earth Syst. Sci.* **2019**. [CrossRef]
19. Semiromi, M.T.; Koch, M. Analysis of spatio-temporal variability of surface–groundwater interactions in the Gharehsoo river basin, Iran, using a coupled SWAT-MODFLOW model. *Environ. Earth Sci.* **2019**, *78*, 1–21. [CrossRef]
20. Mosase, E.; Ahiablame, L.; Park, S.; Bailey, R. Modelling potential groundwater recharge in the Limpopo River Basin with SWAT-MODFLOW. *Groundw. Sustain. Dev.* **2019**, *9*, 100260. [CrossRef]
21. Jafari, F.; Kiem, A.S.; Javadi, S.; Nakamura, T.; Nishida, K. Fully integrated numerical simulation of surface water-groundwater interactions using SWAT-MODFLOW with an improved calibration tool. *J. Hydrol.* **2021**, *35*, 100822. [CrossRef]
22. Zhang, C.; Shoemaker, C.A.; Woodbury, J.D.; Cao, M.; Zhu, X. Impact of human activities on stream flow in the Biliu River basin, China. *Hydrol. Process* **2013**, *27*, 2509–2523. [CrossRef]
23. Wang, Q.; Liu, R.; Men, C.; Guo, L.; Miao, Y. Effects of dynamic land use inputs on improvement of SWAT model performance and uncertainty analysis of outputs. *J. Hydrol.* **2018**, *563*, 874–886. [CrossRef]
24. Ma, N.; Szilagyi, J.; Zhang, Y.; Liu, W. Complementary-relationship-based modeling of terrestrial evapotranspiration across China during 1982–2012: Validations and spatiotemporal analyses. *J. Geophys. Res. Atmos.* **2019**, *124*, 4326–4351. [CrossRef]
25. Immerzeel, W.W.; Droogers, P. Calibration of a distributed hydrological model based on satellite evapotranspiration. *J. Hydrol.* **2008**, *349*, 411–424. [CrossRef]
26. Ma, T.; Duan, Z.; Li, R.; Song, X. Enhancing SWAT with remotely sensed LAI for improved modelling of ecohydrological process in subtropics. *J. Hydrol.* **2019**, *570*, 802–815. [CrossRef]
27. Jin, X.; Jin, Y.X. Calibration of a distributed hydrological model in a data-scarce basin based on GLEAM datasets. *Water* **2020**, *12*, 897. [CrossRef]
28. 30 m Resolution Shuttle Radar Topography Mission Data. Available online: http://gdex.cr.usgs.gov/gdex/ (accessed on 1 October 2021).
29. Available online: https://data.tpdc.ac.cn/ (accessed on 1 October 2021).
30. Qubaja, R.; Amer, M.; Tatarinov, F.; Rotenberg, E. Partitioning evapotranspiration and its long-term evolution in a dry pine forest using measurement-based estimates of soil evaporation. *Agric. For. Meteorol.* **2019**, *281*, 107831. [CrossRef]
31. Liu, J.; Liu, M.; Zhuang, D.; Zhang, A.; Deng, X. Study on spatial pattern of land-use change in China during 1995–2000. *Sci. China Earth Sci.* **2003**, *46*, 373–384. [CrossRef]
32. Yang, N.; Zhou, P.; Wang, G.; Zhang, B.; Shi, Z.; Liao, F.; Li, B.; Chen, X.; Guo, L.; Dang, X.; et al. Hydrochemical and isotopic interpretation of interactions between surface water and groundwater in Delingha, Northwest China. *J. Hydrol.* **2021**, 126243. [CrossRef]
33. White, K.L.; Chaubey, I. Sensitivity analysis, calibration, and validations for a multisite and multivariable SWAT model. *J. Am. Water Resour. Assoc.* **2005**, *41*, 1077–1089. [CrossRef]
34. Moriasi, D.N.; Arnold, J.G.; Van Liew, M.W.; Bingner, R.L.; Harmel, R.D.; Veith, T.L. Model evaluation guidelines for systematic quantification of accuracy in watershed simulations. *Trans. ASABE* **2007**, *50*, 885–900. [CrossRef]
35. Alemayehu, T.; Griensven, A.; Woldegiorgis, B.T.; Bauwens, W. An improved SWAT vegetation growth module and its evaluation for four tropical ecosystems. Hydrol. *Earth Syst. Sci.* **2017**, *21*, 4449–4467. [CrossRef]
36. Fan, J.; Oestergaard, K.T.; Guyot, A.; Lockington, D.A. Estimating groundwater recharge and evapotranspiration from water table fluctuations under three vegetation covers in a coastal sandy aquifer of subtropical Australia. *J. Hydrol.* **2014**, *519*, 1120–1129. [CrossRef]
37. Yang, K.; Lu, C. Evaluation of land-use change effects on runoff and soil erosion of a hilly basin—The Yanhe River in the Chinese Loess Plateau. *Land Degrad. Dev.* **2018**, *29*, 1211–1221. [CrossRef]

38. Samimi, M.; Mirchi, A.; Moriasi, D.; Ahn, S.; Alian, S.; Taghvaeian, S.; Sheng, Z. Modeling arid/semi-arid irrigated agricultural watersheds with SWAT: Applications, challenges, and solution strategies. *J. Hydrol.* **2020**, 125418. [CrossRef]
39. Earls, J.; Dixon, B. A Comparison of SWAT Model-Predicted Potential Evapotranspiration Using Real and Modelled Meteorological Data. *Vadose Zone J.* **2008**, *7*, 570–580. [CrossRef]
40. Ji, L.; Duan, K. What is the main driving force of hydrological cycle variations in the semiarid and semi-humid Weihe River Basin, China? *Sci. Total Environ.* **2019**, *684*, 254–264. [CrossRef]
41. Li, D.C.; Zhao, X.; Zhang, G.L. *Chinese Earth Series—Qinghai Volume [M]*; Science Press: Beijing, China, 2019. (In Chinese)

Article

Predicting Snowmelt Runoff at the Source of the Mountainous Euphrates River Basin in Turkey for Water Supply and Flood Control Issues Using HEC-HMS Modeling

Selim Şengül * and Muhammet Nuri İspirli

Department of Civil Engineering, Faculty of Engineering, Atatürk University, Erzurum 25100, Turkey; m.nurii@hotmail.com
* Correspondence: ssengul@atauni.edu.tr; Tel.: +90-442-2314569

Abstract: Predicting the runoff from snowpack accumulated in mountainous basins during the melting periods is very important in terms of assessing issues such as water supply and flood control. In this study, the Hydrological Engineering Center–Hydrological Modeling System (HEC-HMS) was used to simulate snowmelt runoff in the Kırkgöze–Çipak Basin that has a complex topography where altitude differences range from 1823 m to 3140 m above the sea level. The Kırkgöze–Çipak Basin, located in eastern Turkey, is a basin where snowfall is highly effective during the cold season. There are three automatic meteorology and snow observation stations and three stream gauge stations in the basin, which are operated especially for the calibration and validation of hydrological parameters at different altitudes and exposures. In this study, the parameters affecting snow accumulation–melting and runoff were investigated using the simulations on an hourly basis carried out over a three-year period for temporal and spatial distribution at the basin scale. Different from previous studies focusing on the rate of snowmelt, the temperature index method, which is calculated with physically-based parameters (R^2 = 0.77~0.99), was integrated into the runoff simulations (R^2 = 0.84) in the basin. The snowmelt-dominated basin is considered to be the source of the headwaters of the Euphrates River.

Keywords: snowmelt; hydrologic modeling; ATIMR; HEC-HMS; Euphrates River; Kırkgöze–Çipak Basin

Citation: Şengül, S.; İspirli, M.N. Predicting Snowmelt Runoff at the Source of the Mountainous Euphrates River Basin in Turkey for Water Supply and Flood Control Issues Using HEC-HMS Modeling. *Water* **2022**, *14*, 284. https://doi.org/10.3390/w14030284

Academic Editors: Dengfeng Liu, Hui Liu and Xianmeng Meng

Received: 15 December 2021
Accepted: 15 January 2022
Published: 18 January 2022

1. Introduction

Water is the source of life and is probably the most valuable natural asset in the Middle East. Within this perspective, the history of water management is nothing less than the history of humankind. From the inception of our species, coping with the availability—or unavailability—of water resources has been an essential element of human beings' strategies for survival and wellbeing [1]. The two largest rivers in Western Asia, the Euphrates and Tigris, flow in Turkey, Syria, Iran, Iraq, and Saudi Arabia. The Euphrates and Tigris basins are fed predominantly by snow precipitation. Approximately two-thirds of this occurs in winter, and the snow may remain for half a year [2]. Consequently, where water supplies are under stress, such as the semiarid regions of the Mediterranean basins, the activity of snowmelt-derived streamflows are extremely important [3].

The mountain snowfall acts as a natural reservoir for storing precipitation during the cold season, and during the spring months it melts and flows to the rivers. Understanding when the snow melts and the resulting streamflow occurs is essential to be able to effectively manage water resources. Analyses of how the amount and timing of these hydrological quantities vary are crucial to the water supply systems in mountain regions [4]. It is particularly important in the Euphrates and Tigris basins where there are large reservoirs. Results obtained from the hydrological modeling system algorithms of the snowmelt-dominated mountainous Kırkgöze–Çipak Basin improve the accuracy of water resource simulations and help in the planning and operation of the Euphrates River flows.

To date, researchers have introduced a wide variety of modeling frameworks to model the hydrological process [5–7]. In general, these modeling frameworks can be divided into three main groups: conceptual, physically-based, and machine learning models. Conceptual and physically-based models can be used for research purposes to improve knowledge and understanding of the hydrological processes that govern the real-world system. On the other hand, machine learning models create a direct mapping between precipitation and runoff variables and infer their relationships based on historical observations with machine learning algorithms without prior knowledge of internal hydrological processes [8]. Hydrological models are also developed and used for simulation and forecasting tools that allow decision-makers to make the most effective decisions for planning and operations, taking into account the interactions of the physical, ecological, economic, and social aspects of the real-world system. In addition, real-time flood forecasting and warning, flood frequency forecasting, flood route and overflow forecasting, climate and land-use change, and impact assessments of integrated basin management are examples of other applications in which the hydrological models are used [7,9].

In regions where most of the precipitation falls as snow during the winter months as the altitude increases, the snowmelt component of the hydrological models is vital for water resources management [10]. From a hydrological perspective, two main methods are generally used to simulate snowmelt: energy budget and temperature index methods. The energy budget method needs detailed observation data and a wide range of model parameters. The distribution of meteorological and hydrological stations in mountain basins is often limited, making it difficult to obtain and process the detailed information required for model study [11]. In contrast, the temperature index method uses air temperature as the only index of energy exchange at the snow surface [12]. The latter approach is commonly used in real-time hydrological forecasts. Examples of numerical models using the temperature index method include the National Weather Service River Forecast System model (1995), Streamflow Synthesis and Reservoir Regulation (SSARR) model, Hydrologic Engineering Center (HEC-1) model, Snowmelt Runoff Model for Windows (WinSRM), Cold Regions Hydrological Model (CRHM), Mesoscale Hydrologic Model (mHM), and the HEC–Hydrologic Modeling System (HEC-HMS) [13–17]. HEC-HMS model is a flexible hydrological model with particular physical significance designed to simulate a comprehensive range of hydrological processes coupled with a very sophisticated graphical user interface [18]. Modified melting rates have been used by many studies, using the hypothetical ATIMR (antecedent temperature index—melt rate) function used in the snowmelt module of HEC-HMS during calibration [19–22]; however, a commonly observed shortcoming in published literature is that no particular data is used to directly estimate the ATIMR curve. Therefore, its estimation and application to a mountainous basin with flow sources of complex composition is noteworthy here [23]. The method provided by Fazel et al. (2014) for one snowmelt period at distinct station locations was subsequently developed and applied by Şengül and İspirli (2021) to create ATIMR curves specific to the Kırkgöze–Çipak Basin using hourly temperatures and snow–water equivalent (SWE) data using error analysis methods recommended by Bombardelli and García (2003) obtained from the three meteorology and snow observation stations [24–26]. Their results showed that the application of the ATIMR function using the observed data significantly improves the snowpack simulations, and it is quite useful for runoff simulations.

Although Turkey is a peninsula, it has a geography with an average altitude of over 1100 meters. Snowmelt runoff in the mountainous eastern part of Turkey is of great importance as it constitutes 60 to 70% in volume of the total yearly runoff during the spring and the early summer months [27]. Most of the annual water volumes in the dam reservoirs built in this region come from the precipitation in the winter months, snowmelt, and the rain falling on the snow cover in the spring. For this reason, conducting hydrological model studies based on snowmelt in the Eastern Anatolia Region of Turkey, where the snow potential is quite high, are of great importance both on a regional, national, and international scale in terms of the planning and economic management of water resources [25,27,28].

Advances in Geographic Information Systems and availability of geospatial databases have paved the way for estimation of several hydroclimatic variables. Reducing the uncertainties in these estimations made at various scales provides a better description of hydrological regimes [29,30].

In this study, which uses these advances in the availability of geospatial data, a continuous hydrological modeling approach is discussed by incorporating the soil moisture account (SMA) algorithm [31] with the snow accumulation and melting algorithm. The Hydrologic Engineering Center's Hydrologic Modeling System [18] was applied to the Kırkgöze–Çipak basin (Figure 1), considering the characteristic behaviors of point and area-based snow–water equivalent simulations by using the most sensitive ATIMR functions calculated on a physical basis [25], and the precipitation distribution algorithms embedded in the model were modified for depicting the actual watershed conditions. The development stages of the model, the determination of the parameters, and the calibration process are explained, and the model results are discussed.

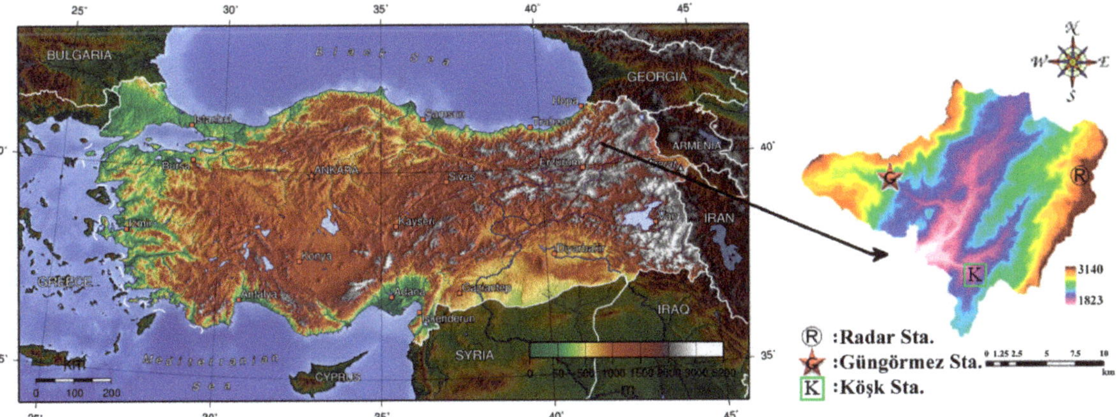

Figure 1. Study area and station locations.

2. Materials and Methods

2.1. Area of Study

This study chose the Kırkgöze–Çipak Basin, located near the source of the Karasu Basin (Upper Euphrates Basin)—which is itself a sub-basin of the Euphrates River—as the test area for this research. With an area of 242 km^2 and altitude ranging from 1823 to 3140 m, the Kırkgöze–Çipak Basin is shown on the digital elevation model (DEM) in Figure 1. The median elevation of the basin is 2325 m, while the mean total basin slope is 15.3 degrees. The geography comprises a rugged mountainous area with the main area being pasture and bare land. The characteristic climatological conditions are those of a cold, dry, and windy region. The region is covered by snow at least 150 days per year, and a significant part of the precipitation falls in the form of snow. The catchment area is not affected by urbanization or by reservoir regulation. Although the basin can be considered small in terms of scale, it has a large elevation difference that makes it possible to conduct snow modeling of major basins such as the Euphrates Basin. Previous snow studies in the area have shown how important snow dynamics and snow modeling are for this region [2,10,27,32–41]. The study area is located within the city center limits of Erzurum in Turkey, which is located at the intersection of Turkey's three major basins: the Çoruh, Aras, and Euphrates basins, and the snowmelt of the mountains in this region is the main source of water for these basins [3,42]. Therefore, the input parameters of the snowmelt model applied in this study will also be a good starting point for hydrological modeling studies of other mainstream resources in the vicinity.

The Kırkgöze–Çipak Basin includes a few state-built stations in its vicinity; however, these stations cannot provide enough information to effectively represent the pertinent spatial and temporal quality of the snowmelt-dominated basin. To compensate for this, three different automatic meteorology and snow observation stations that had been established in the Kırkgöze–Çipak Basin at the villages of Güngörmez and Köşk, inside the grounds of a military radar location under a prior project numbered TÜBİTAK 106Y293, were developed over time. Station information is provided in Table 1 for each of the locations that are shown in Figure 1. This allowed climate data from stations in a mountainous basin with high snow potential to be collected in real time and of sufficient quality [35,36].

Table 1. Parameters of the meteorology stations.

Automatic Meteorology and Snow Observation Station	Altitude (m)	Aspect	Land Use	Average Slope of Land Close to the Station (Degrees)	Simulation Time Interval
KÖŞK	2019	Northwest	Dry Farming	9.90	10/22/2008–9/30/2011
GÜNGÖRMEZ	2454	Southeast	Transition Area	24.10	10/22/2008–9/30/2011
RADAR	2891	Northwest	Transition Area	12.06	10/22/2008–9/30/2011

The upper levels of the study area are surrounded by basalts. These structures were formed as a result of numerous volcanic activities, so they show a complex structure that includes other volcanic rocks. The accumulated groundwater either discharges as small seasonal springs or is channeled to the adjacent formation comprising tuff and agglomerate (Figure 2). Tuff and agglomerate are common under basalts in this region. They were formed as a result of the cementation of angular pebbles of different size and blocks containing basalt, andesite, and tuff with fine-grained volcanic rocks.

The agglomerates, which are faulted and fractured in several directions, carry a small amount of groundwater in the fracture zones. In the region, tuff and agglomerate-inclusive claystone and marl layers are located due to the unconformity under the agglomerate. Many small seasonal springs are observed at the boundary of the clay and marl layer, which has a more impermeable structure than the formations above it [43].

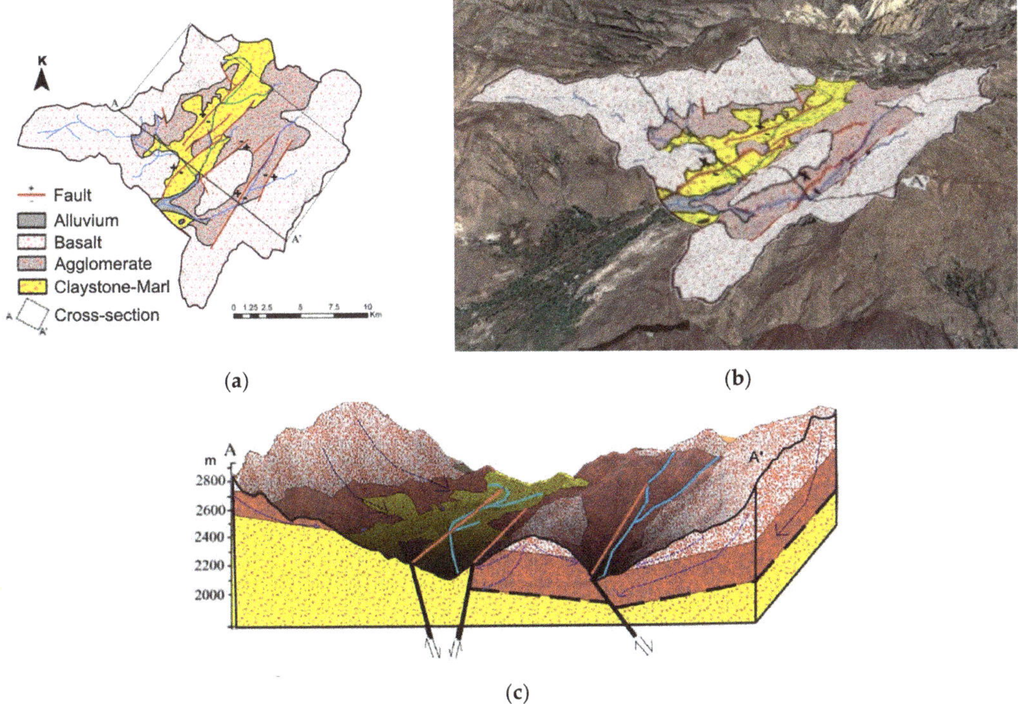

Figure 2. (**a**) Geological map. (**b**) Land survey. (**c**) A–A′ geological cross-section of Kırkgöze–Çipak basin [44].

2.2. Hydrological Model

The Hydrologic Engineering Center's Hydrologic Modeling System (HEC-HMS) is a hydrological model developed to simulate the precipitation–runoff processes of dendritic drainage basin systems [18]. The model is designed for both continuous and event-based hydrological modeling and offers several different options for modeling the various components of the hydrological cycle. In event-based modeling, storm precipitation is simulated during the simulation time interval ranging from a few hours to several days, depending on the basin size [45]. In continuous modeling, a continuous historical record of hydrological events, including dry and wet periods over several years, is simulated [46]. The main difference is that evapotranspiration and groundwater seepage can be neglected in event-based modeling, while they cannot be ignored in continuous simulation [47]. HEC-HMS can conduct hydrological simulation over a wide range with various simple modules to represent different components of the hydrological cycle. The selection of the appropriate model for each component depends on the experience of the modeler, the purpose of the modeling, and the usability of the input data [48].

The HEC-HMS modeling system has three main components: the basin model in which the topographic and physical characteristics of the basin are determined, the meteorological model in which the meteorological data are processed, and the control manager.

2.3. Basin Model

The HEC-HMS basin model (Figure 3) simulates the process of the water falling to Earth by precipitation from the canopy to become groundwater, excluding bottom percolation. HEC-HMS uses the soil moisture accounting (SMA) [31] algorithm to simulate the movement of water in soil under continuous simulations. This algorithm takes precipitation and evapotranspiration as inputs and computes surface runoff, groundwater runoff, evapotranspiration, and losses from bottom percolation (Figure 4; see USACE (2016) for further detail). The Clark Unit Hydrograph was chosen for the transformation method (or hydrograph simulation) and the monthly constant baseflow method was chosen for the baseflow calculation [49,50]. Initial parameters for the Clark method were obtained using the Kerby equation (Tc = G(L*r/S0.5)0.467). The physical parameters of the sub-basins at the exit of the three selected meteorological stations (for example, river length, drainage area, slope, etc.) were computed with Geographic Information Systems (GIS) by using the digital elevation model (DEM) obtained from the 10 m contour maps (Table 2). The initial values for the baseflow were taken as the current river flows because the beginning of the simulation was in the dry period, and they were distributed on the basis of the average area-based distribution in the sub-basins.

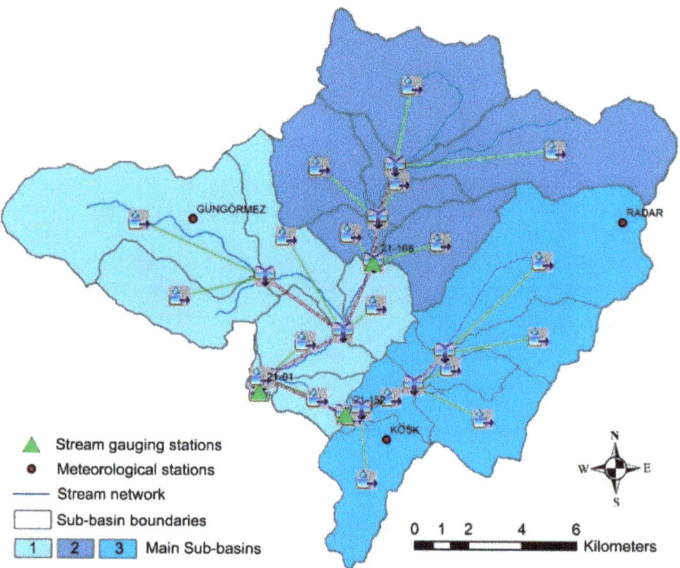

Figure 3. Conceptual model of the Kırkgöze–Çipak Basin in HMS, showing junctions, reaches, flow direction, and sub-basins.

The basin model includes many parameters used for baseflow, hydrograph simulation, and SMA. For the estimation of these parameters in previous studies [10,45,51–53], it was found appropriate to use geodatabases, reducing the number of free parameters by starting the simulation during periods when initial conditions are easier to predict (i.e., the start of the water year), and use empirical equations or reliable sources. A combination of these methods was used in this study. As there was no map from which soil texture information of the study area could be obtained, the initial values of SMA were obtained from previous studies and then calibrated to match the observed streamflow. Canopy maximum retention and soil surface deposition were estimated by vegetation type and percentage of land slope, respectively [54,55]. The rate and amount of seepage in the soil profile and groundwater were estimated based on hydraulic conductivity [52]. Active soil depth was assumed to be 60 cm, considering the land cover. Fleming and Neary (2004) predicted HEC-HMS groundwater storage (groundwater 1 and groundwater 2), and seepage parameters [18] were based on recession analysis. These estimates from published literature were taken as initial values and they were calibrated during the simulations.

2.4. Meteorological Model

The Kırkgöze–Çipak basin is divided into a few sub-basins, as shown in Figure 3. The data obtained from three automatic meteorology and snow observation stations in the basin at altitudes of 2019 m (Köşk), 2454 m (Güngörmez), and 2891 m (Radar) were used for the meteorological data required for the different parameter methods selected in the basin and for the meteorological model simulating the precipitation–runoff process. These stations provided time series of the maximum wind speed (m/s), wind direction, average air temperature (°C), average humidity (% rh), air pressure (mbar), average soil temperature (°C), solar radiation (W/m^2), average albedo, precipitation (mm), snow height (cm), snow density (gr/cm^3), and snow–water equivalent (cm) parameters over 15 min periods. References [35,36] showed that the climate data from the stations in the basin was sufficient, of good quality, and could be collected in real time. Measurements from the years 2008 to 2011 obtained from the Köşk, Güngörmez, and Radar meteorology stations were used for the hydrological simulations to be conducted with HEC-HMS.

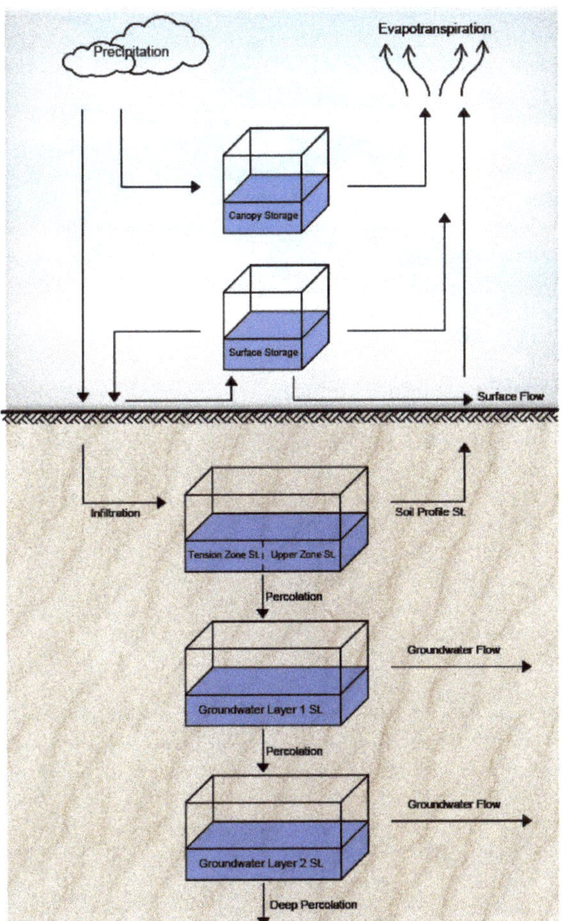

Figure 4. Schematic of basin model in HEC-HMS and its principal components.

Table 2. Physical properties of the main sub-basins.

	MS1	MS2	MS3
Basin Slope (%)	0.173	0.167	0.215
Elevation (m)	2125.02	2175.37	2204.53
River Length (m)	21,917.56	12,094.25	11,208.97
Area (km^2)	91.53	75.50	74.51

2.5. Precipitation Model

A variety of different statistical techniques to distribute point observations over complex topography is given in published literature [56–64]. Although these studies improved high-resolution grid-type climate data estimations, uncertainties remained. In particular, it is more difficult to estimate the spatial distribution and the intensity of precipitation compared to other variables such as temperature, due to the regional, seasonal, and topographic characteristics [65]. The Kırkgöze–Çipak basin study area has a very large altitude range and other variable aspects, even though it is small in terms of scale. As a result of observations in the basin over a long time, it was determined that some convective precipitations took place independently from each other as in the northern aspects with quite high land altitudes where the Radar station is located and in the southern aspects where the Güngörmez station is located. Therefore, while snowmelt runoff simulations are performed throughout the basin, the emphasis is on how the precipitation is distributed regionally rather than how the precipitation may be distributed in the basin. The HEC-HMS program offers grid-based and polygonal-based solution alternatives to determine the precipitation distribution over the basin. This study was carried out on a polygonal basis, and the gage weights method was chosen for modeling the precipitation processes. The gage weights method is based on the Thiessen polygon method. The Thiessen polygon method, which is usually recommended for use in vast areas, does not distribute precipitation with respect to topographical effects and precipitation characteristics; instead, it performs it only over polygonal areas determined by the positions of the stations [66]. Therefore, the gage weights method used in HEC-HMS was modified for the study basin, which is heterogeneous in terms of altitude and exposure. While developing this polygonal area-based algorithm, in addition to the general behavior that is dependent on the topography of the region—the barrier effect (Figure 5, 4th elevation zone), the measurement data at the stream gauge stations and the ambient temperature, humidity, atmospheric pressure, rate of increase in cloudiness (observed), albedo, wind speed, and SWE values were all simultaneously examined. The area-based distribution of precipitation was simulated by six different zonal polygons shown in Figure 5.

In basins where there is a large altitude range, the use of data obtained from stations representing low levels may cause the precipitation input calculated for the entire basin to be lower than the actual value. It is recommended to extrapolate precipitation data to average hypsometric elevations for zones with elevation gradients [67], so that the point-based input values used in the modeling procedure can better represent a specific area. It is important for the precipitation data to align with other meteorological data with respect to time, so that the model can perform the necessary iterations accurately and reliably. For this reason, while making the area-based distribution of meteorological data, a general grouping based on altitude and exposure, taking into account station locations, was deemed appropriate so that simultaneous atmospheric homogeneity could be assured. For this reason, the meteorological variables in altitude zones 1 and 2 were based on the Güngörmez station, the meteorological variables in altitude zones 3 and 4 were based on the Köşk station, and the meteorological variables in altitude zones 5 and 6 were based on the Radar station variables (Figure 5). In Table 3, the hypsometric elevations for each zone and the altitudes of the meteorological stations in these zones are given.

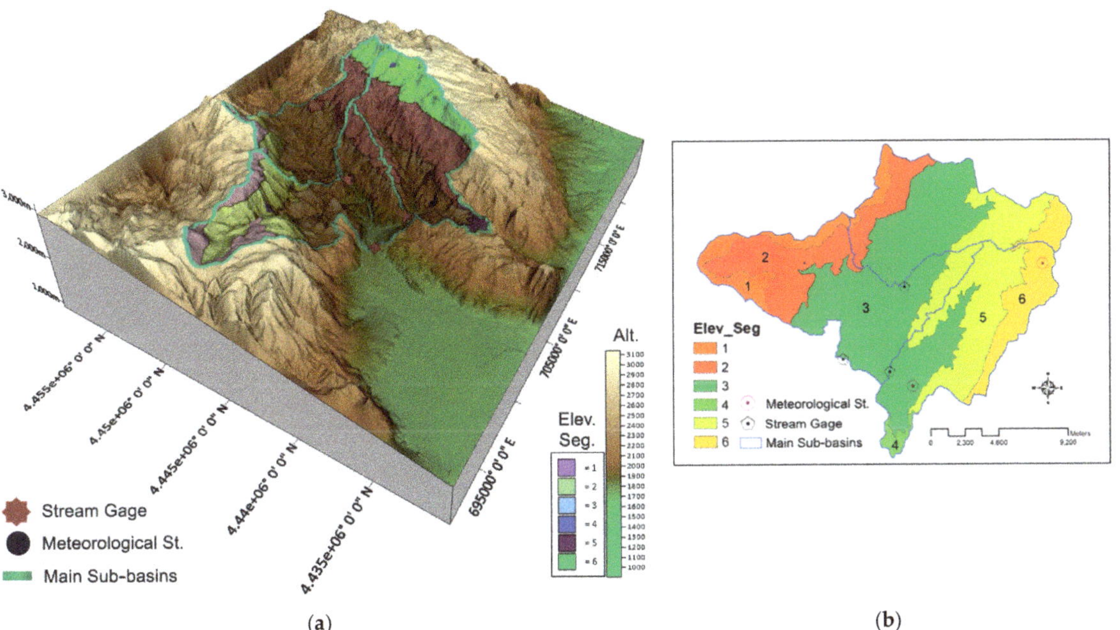

(a) (b)

Figure 5. (**a**) Land survey of the elevation zones and (**b**) polygonal representation of the Kırkgöze–Çipak basin.

Table 3. Hypsometric values for the Kırkgöze–Çipak basin.

Zone	Hypsometric Elevation (m)	Weather Station	Weather Station Altitude (m)
1	2777.236		
2	2509.433	Güngörmez	2444
3	2098.651	Köşk	2042
4	2297.814		
5	2454.838		
6	2856.015	Radar	2887

As the hypsometric averages of zones 2, 3, and 6 are close to the altitudes of Güngörmez, Köşk, and the Radar stations, respectively, the average area-based precipitation calculations in these zones were taken directly at the respective stations. Based on the hypsometric elevations in other zones, a series of algorithms were run to obtain values in the direction of increasing or decreasing precipitation.

Firstly, the 15 min precipitation series recorded at each station were converted into daily total precipitation series while removing possible measurement errors. This daily sum is a precaution for the following algorithm (Table 4), especially for modeling the natural distribution of the interpolated or extrapolated zonal values of the convective characteristic heavy snowfalls observed at the station points. Otherwise, if the precipitation transition between stations exceeds the simulation time interval of 1 h, the precipitation is only distributed on the station's zone. Therefore, the predictive values may take zero values mathematically on the transition zones noticed on field trips. The daily total precipitation values in zones 1, 4, and 5 were analyzed according to the flow chart in Table 4, which was prepared by considering the station locations given in Figure 5 and calculated within the designated rules.

After calculating the daily precipitation altitudes for zones 1, 4, and 5, these altitudes were proportioned to the daily total precipitation altitude of zones 2, 3, and 6, respectively, and precipitation coefficients were obtained. These coefficients were used for the conversion from the daily total precipitation altitude to the 15 min time interval. The first precipitation coefficient calculated to obtain the 15 min precipitation values for the 1st zone was multiplied by the 15 min precipitation series of the 2nd zone. This procedure was also performed for zones 4 and 3 and zones 5 and 6, respectively. Thus, while maintaining the atmospheric homogeneity, the precipitation altitudes and timings of the 1st, 4th, and 5th zones were adjusted with reference to the measurements taken from the 2nd, 3rd, and 5th zones, respectively. The results obtained were increased to a one-hour time interval selected as the HEC-HMS simulation time interval and entered into the program.

Complete reliable data could not be obtained from the pluviographs during the winter months because the diluted antifreeze in the rain gauge froze after a certain period of time under the effect of the cold weather and excessive precipitation at both the Radar station and the Güngörmez station; the movable scale shaft which measures the amount of precipitation discharged from the reservoir did not work due to freezing and jamming, even when the antifreeze did not freeze. The data from the Köşk station showed that the freezing did not occur there due to the fact that the temperature was relatively higher than the other stations due to its lower altitude. As a result, much more reliable precipitation data were obtained there compared to the other stations during the winter months. Due to the problems encountered, especially at the Radar and Güngörmez stations, it was not found appropriate to use the data obtained from the rain gauges as direct precipitation data.

Table 4. The algorithm used for determining precipitation altitudes in the elevation zones where there was no station.

1st Zone Calculated by extrapolating Köşk and Güngörmez station data	***1st Rule:*** If precipitation was observed at both stations, the total daily precipitation at the Köşk and Güngörmez stations was linearly extrapolated from the station altitudes to the average zone altitudes. If the extrapolation result was negative, the zonal precipitation altitude was taken as zero. (If Köşk > 0 and Güngörmez > 0, then the trend was applied. If the trend < 0, then 1st zone = 0). ***2nd Rule:*** If there was no precipitation at Köşk but there was precipitation at the Güngörmez station, then precipitation altitude equals the Güngörmez station. (If Köşk = 0 and Güngörmez > 0, then 1st zone = Güngörmez).
4th Zone Calculated by interpolating the Köşk and Radar station data	***1st Rule:*** If Köşk = 0, then 4th zone = 0. ***2nd Rule:*** If Radar = 0, then 4th zone = Köşk. ***3rd Rule:*** If Köşk > 0 and Radar > 0, then the trend is applied. If Trend < 0, then 4th zone = Köşk.
5th Zone Calculated by extrapolating the Köşk and Radar station data	***1st Rule:*** If Köşk = 0 and Radar ≥ 5 mm, then the trend is applied. ***2nd Rule:*** If Köşk = 0 and Radar < 5 mm, then 5th zone = 0. ***3rd Rule:*** If Köşk < 5 mm and Radar = 0, then 5th zone = 0. ***4th Rule:*** If Köşk ≥ 5 mm and Radar = 0, then the trend is applied by checking meteorological data from the Güngörmez and Radar stations. ***5th Rule:*** If Köşk > 0 and Radar > 0, then the trend is applied.

In winter, while the precipitation series were formed during the snow accumulation period, the differences in the 24-h averages of the snow–water equivalent altitudes (SWE) obtained from the snow pillows were taken. If the difference between these daily averages was positive, the SWE difference for that day was added to the station as precipitation. The timing of precipitation was adjusted in correlation with simultaneous albedo and humidity data, taking into account the effect of snow drift, while the distribution of precipitation during the day was determined by the amount of increase in the measured SWE during the day.

2.6. Snowmelt Model

In HEC-HMS (Version 4.2.1), there are two snowmelt modeling options. One of them is the gridded temperature index method and the second is the temperature index method, which is the method that was used in this study. The temperature index method is an extension of the degree-day approach to modeling a snowpack. A typical approach to the degree day is to have a fixed amount of snowmelt for each degree above freezing. This method includes a conceptual representation of the cold energy stored in the pack along with a limited memory of past conditions and other factors to compute the amount of melt for each degree above freezing. As the snowpack internal conditions and atmospheric conditions change, the melt coefficient also changes [18].

If the main source of energy in the spring is not the solar irradiance, snowmelt can be more effectively and simply computed using a temperature index model [10,68–73]. In hydrologics, an index is a meteorological or hydrological variable. Changes in the variable are associated with those in the parameter it is estimating, and which are more easily measured than the actual parameter. Either a coefficient (such as a degree-day factor) or a formula for more complex linear or curvilinear functions (such as the antecedent temperature index—melt rate function) may be used to describe this index relationship. Depending upon changing associated factors, it may be either constant or variable. Spatial and temporal basin value point measurements are represented by the index where average fixed relationships are known to exist between the measured values and basin values. However, snow accumulation and melting topics are complex, and the data required for physically-based energy budget calculations are comprehensive and challenging to obtain [68].

Some temperature index models require the snowpack's melt rate to be characterized [74,75]. This melt rate can be stated differently. One example is to express changes in the melting rate as a function of air temperature accumulation over several warm days for melting snow. This is achieved by using the ATIMR (antecedent temperature index—melt rate) function to determine the melt rate for a certain antecedent temperature index. Snow physics indicate that melting rates increase throughout the season due to both metamorphic processes causing ice crystal consolidation and the snowpack producing more water over time [24].

Past modeling studies have generally been based on a theoretical constant ATIMR curve generated by the USACE (1991) and used for characterization of melt rates [19–22,76,77]. The theoretical curve was included in the SSARR model in 1991. The SSARR guide [78] ATIMR values are shown in Table 5.

Table 5. Tabulation of melt rate as a function of ATIMR.

ATIMR: °F-Day (°C-Day)	Melt Rate: in/°F-Day (cm/°C-Day)
0 (0)	0.025 (0.12)
100 (55.56)	0.03 (0.13)
200 (111.11)	0.05 (0.23)
300 (166.67)	0.04 (0.18)
1000 (555.56)	0.04 (0.18)

For the values in the customary U.S. system, please see the SSARR model guide in Appendix D, p. 17; the methodology used to calculate the metric system results is presented by Şengül and İspirli (2021) in detail.

Modified melting rates have been used by many studies using the hypothetical ATIMR function of Table 5 during calibration; however, a commonly observed shortcoming in published literature is that no particular data is used to directly estimate the ATIMR curve. Sometimes the hypothetical ATIMR curve is taken as a starting point for snowpack simulations and different scenarios used to modify the curve to improve simulated results during calibration [19–22]. Sometimes the theoretical ATIMR curve is not modified, but an additional rate is applied to the melting rate obtained from the ATIMR curve in proportion to the varying rate over time [77,78]. However, the physical meaning of widely used ATIMR functions is important in hydrologic modeling studies [25].

It is necessary to refer to published literature or land data to understand how a generalized hypothetical ATIMR curve was generated. The values in Table 5 are thought to be a visualization of an ATIMR function generated from the authors' information—as a resulting of engineering decisions implemented in 1991 at the start of snowmelt modeling studies—from the documented results of land data or by undocumented means. However, now a review of this parameter is necessary to determine the reliability of regional snowmelt predictions [24]. Following a comprehensive review of published literature, no study was found carrying out a formal validation of the ATIMR parameter using observed data other than the studies by Şengül and İspirli (2021) and Fazel et al. (2014). The first of these studies was a preliminary study of snowmelt modeling in this basin. The methodology determined by Fazel et al. (2014) represented only one year of data for certain single-point locations. When the HEC-HMS program performs full hydrological simulations on catchments it uses the temperature index methodology and restricts researchers to one ATIMR function for the whole basin. It is therefore necessary to develop an optimal area-based average ATIMR function later on and is hydrologically significant for modeling snowmelt-originated flows originating in complex mountainous terrains.

The HEC-HMS model program is capable of generating grid or polygonal area-based hydrological simulation models. The HEC-HMS program allows the creation of a meteorological model to represent the meteorological boundary conditions of a basin's physical behavior and some of the spatial and area-based variables distributed over that basin. However, published literature highlights a significant deficiency in the polygonal-based modeling of the HEC-HMS model program that is widely used and part of this study, in that only one meteorological model can be used for a basin model. Consequently, eighteen hydrological models must be created for eighteen sub-basins [79]. The meteorological model applies the climatic conditions represented by precipitation, evapotranspiration, and snowmelt, based upon the methods chosen. In basins where there are large differences in altitudes, it is impractical to apply one set of snowmelt parameters—such as the melt rate or snowmelt threshold temperature—over all the locations because of a range of factors that can include radiation effects, wind conditions, and others [70]. Snowmelt parameters would not be constant for a basin that exhibited a wide range of altitudes. This would include variables such as the water capacity of the snowpack and the threshold temperature at which precipitation occurs as snow or rain. In order to take this into consideration when the entire basin is modeled in one go using the polygonal-based method in HEC-HMS, it is necessary to enhance the temperature index model with area-based average ATIMR functions to cope with restricted availability of parameters and the increasing demand for accurate estimates for melt rates in both spatial and temporal terms.

The study conducted by Fazel et al. (2014) was originally the only approach to calculate the physical significance of the ATIMR curve beyond its manual calibration. As that study mentioned, although the ATI equation (the antecedent temperature index component of the ATIMR function) was provided, the SSARR guide did not describe the method used to generate the hypothetical ATIMR curve. The method provided by Fazel et al. (2014) for one snowmelt period at distinct station locations was subsequently developed and applied by Şengül and İspirli (2021) to create ATIMR curves specific to the Kırkgöze–Çipak Basin using hourly temperatures and snow–water equivalent (SWE) data using error analysis methods recommended by Bombardelli and García (2003) obtained from the three meteorology and snow observation stations. The comparisons of both characteristics and statistical information from the snowmelt component simulation results of HEC-HMS, and the observed multivariate spatial–temporal SWE values of the region, shows a very high correlation between the generated ATIMR functions and the default SSARR values used in published literature [25].

Calibration of the other parameters used in the meteorological model used in the temperature index method were performed by considering the values in published literature [48,78], namely (PX temperature = 2 °C, base temperature = 0 °C, wet melt rate = 3.2 mm/°C-day, rain rate limit = 1.3 mm/day, ATI-melt rate coefficient = 0.98, cold limit = 0 mm/day, ATI cold rate coefficient = 0, water capacity = 20%, ground melt = 0 mm/day). As a result, the SWE simulations necessary to arrive efficiently at the final water budget calculations were optimized throughout the basin [25].

The area-based common ATIMR function (Figure 6) is meant to represent all three point ATIMR functions, so point values should be examined together and in relation to their land and snow altitude. For example, at the low-altitude Köşk Station, due to the low amount of snowpack and early melting, there was a limited ATI value, and the ATI values of stations at higher altitudes were increased in proportion to the area-based ATI values exceeding that threshold. The values of the point and final area-based ATIMR functions measured in the Kırkgöze–Çipak Basin are shown in Table 6. The HEC-HMS modeled SWE results using the common area-based ATIMR function for the different stations are shown in Figure 7. The area-based ATIMR value of 125 °C-day—the last value in Table 6—is the cumulative ATI value for which the snow observed over the specified period at all station locations had completely melted. For rainfall–runoff studies to be carried out across the basin, the value had to be increased and extrapolated to account for the greater snow depths observed at higher altitudes by modifying precipitation series [25].

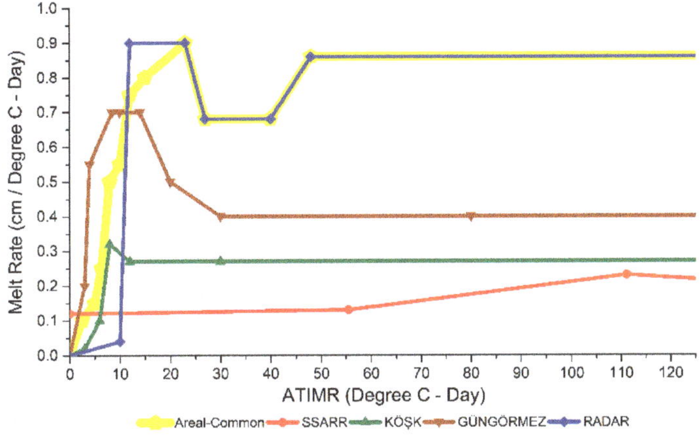

Figure 6. Point and area-based ATIMR curves.

Table 6. Point and area-based common ATIMR function for the Kırkgöze–Çipak Basin.

Point ATIMR Curves						**Areal-Common ATIMR Curve**	
Köşk Station (2019 m)		**Güngörmez Station (2454 m)**		**Radar Station (2891 m)**		**Basin-Wide**	
ATIMR (°C-Day)	**Melt Rate (cm/°C-Day)**	**ATIMR (°C-Day)**	**Melt Rate (cm/°C-Day)**	**ATIMR (°C-Day)**	**Melt Rate (cm/°C-Day)**	**ATIMR (°C-Day)**	**Melt Rate (cm/°C-Day)**
0	0.00	0	0.00	0	0.00	0	0.00
3	0.02	3	0.02	10	0.04	2.5	0.10
6	0.10	4	0.55	12	0.90	5	0.15
8	0.32	8.5	0.70	23	0.90	6.4	0.25
12	0.27	10	0.70	27	0.68	8	0.50
30+	0.27	20	0.50	40	0.68	10	0.55
		30	0.40	48	0.86	12	0.75
		80+	0.40	125+	0.86	15	0.80
						23	0.90
						27	0.68
						40	0.68
						48	0.86
						125+	0.86

2.7. Calibration Strategy

Manual parameter calibration was preferred in this study due to the karstic behavior of the basin. Manual calibration begins with an appropriate estimation of the initial parameters to run the model. The Kırkgöze–Çipak model was developed on a daily timescale over a 3-year period between the calibration (2008 to 2010) and validation (2010 to 2011) periods. In the study area, there were three stream gauge stations, namely Karasu-Çipak (DSİ, 21-01), Büyükçay-Karagöbek (DSİ, 21-168), and Köşk Dere-Köşk (DSİ, 21-152); these stations regularly performed hydrometric measurements. Calibration was carried out using the records of these stream gauge stations to simulate the flow in the simulations performed with HEC-HMS. The locations of meteorological stations and stream gauge stations are shown in Figure 3. The simulation was initiated at the beginning of autumn when the soil was almost dry. Therefore, the initial storage was assumed to be empty. Initial storage has an effect on the simulated hydrograph from a few days to a maximum of a few months [80]. However, they are insignificant for long-term water resource planning. After running the simulation, the simulated results were compared with the data observed from the stream gauge stations.

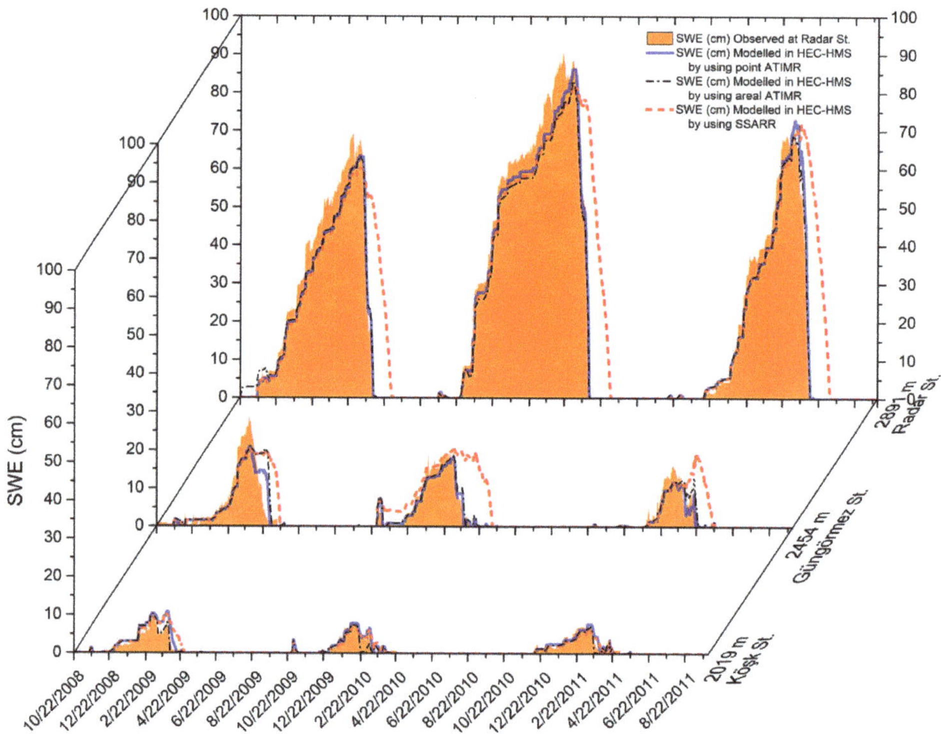

Figure 7. Snow–water equivalent (SWE) simulations computed with typical SSARR, point and common-areal ATIMR functions.

3. Results

The Kırkgöze–Çipak Basin is one where there is a lot of snowfall. In modeling the transformation process from snowfall into runoff, primarily SWE simulations were performed. For this reason, using the temperature index method, the data from the three meteorology stations at different altitudes and exposures on the basin were taken as points, and simulations were conducted. The validation criteria for these simulations are presented in Table 7. However, the fact that the basin has altitude and exposure differences due to its complex topographical structure, and because the HEC-HMS has to simulate the entire basin with a single meteorological model, SWE simulations needed to be performed spatially while normally point simulations are performed for SWE simulations. This is because if the snow–water equivalent simulations are not performed properly, the snowmelt runoff process cannot be modeled properly. For this reason, a common melting parameter was developed for the three stations and the validity of the results is shown. The typical SSARR function and the common-areal melting parameter developed to be used in this and similar studies, along with the snow–water equivalent simulations at station locations, are compared in Figure 7.

After obtaining accurate snow–water equivalent simulations (Figure 7), basin-wide snowmelt runoff simulations were performed. As the stream gauge stations Köşk Dere-Köşk (DSİ 21-152), Büyük Çay-Karagöbek (DSİ 21-168), and Karasu-Çipak (DSİ 21-01) in the basin were at the lower and main exit points of the basin, the runoff series taken from these stations were used for calibration. The calibrated parameters of the sub-basins of the Kırkgöze–Çipak Basin are summarized in Table 8 on the scale of the main sub-basins. The improved final runoff simulation results obtained from the stream gauge station points are presented in Figure 8 along with the observed values. The similarity between the runoff obtained as a result of the hydrological simulation and the runoff values obtained from the stream gauge stations in the basin where the simulation is carried out is very important in terms of validating simulation accuracy and reliability.

Table 7. The R^2, root mean square error (RMSE), Nash–Sutcliffe efficiency (NSE) and Kling–Gupta efficiency (KGE) values for modeled and actual SWE (using estimated point and area-based ATIMR functions with the default SSARR curve for the snow accumulations and melting periods from 2008 to 2011).

	Station	Point R^2	Point RMSE (cm)	Point NSE	Point KGE	Area-Based R^2	Area-Based RMSE (cm)	Area-Based NSE	Area-Based KGE	SSARR R^2	SSARR RMSE (cm)	SSARR NSE	SSARR KGE
Calibration period (2008–2010)	**Köşk**	0.891	1.351	0.735	0.624	0.802	1.276	0.763	0.861	0.757	1.601	0.628	0.678
	Güngörmez	0.887	2.610	0.876	0.830	0.770	3.591	0.766	0.857	0.385	7.312	0.030	0.322
	Radar	0.992	4.590	0.975	0.887	0.993	5.487	0.964	0.851	0.685	16.433	0.680	0.760
Validation period (2010–2011)	**Köşk**	0.781	1.096	0.752	0.849	0.840	0.939	0.818	0.845	0.761	1.100	0.751	0.773
	Güngörmez	0.922	1.834	0.874	0.730	0.814	2.235	0.812	0.871	0.230	6.207	−0.449	0.246
	Radar	0.970	4.241	0.970	0.972	0.982	3.577	0.979	0.937	0.582	17.981	0.467	0.646
3 years period (2008–2011)	**Köşk**	1.280	0.858	0.740	0.701	0.805	1.185	0.777	0.874	0.730	1.469	0.658	0.779
	Güngörmez	0.891	2.421	0.878	0.819	0.779	3.273	0.776	0.866	0.361	7.024	−0.032	0.318
	Radar	0.985	4.476	0.976	0.904	0.989	4.931	0.970	0.868	0.666	16.966	0.648	0.758

Table 8. Initial and calibrated parameters for the three main sub-basins of the Kırkgöze–Çipak Basin.

Sub-Model	**Method**	**Parameter**	**MS1 (Inc. 8 Sub-Basins)**		**MS2 (Inc. 5 Sub-Basins)**		**MS3 (Inc. 5 Sub-Basins)**	
			Initial	**Calibrated**	**Initial**	**Calibrated**	**Initial**	**Calibrated**
Canopy	Simple Canopy	Initial Storage (%)	0	0	0	0	0	0
		Max. Storage (mm)	2.5	6.77	2.5	4.23	2.5	5.97
Surface	Simple Surface	Initial Storage (%)	0	0	0	0	0	0
		Max. Storage (mm)	5	6.67	5	6.67	5	6
Loss	Soil Moisture Accounting	Max. Infiltration (mm/h)	2	1.208	2	1.95	2	1.73
		Impervious (%)	0	0	0	0	0	0
		Soil Storage (mm)	100	71.93	100	101.83	100	66.97
		Tension Storage (mm)	50	41.25	50	33.55	50	45.64
		Soil Percolation (mm/h)	0.6	0.06	0.6	0.06	0.6	0.06
		GW1 Storage (mm)	30	33.33	30	33.33	30	26.67
		GW1 Percolation (mm/h)	0.4	3.55	0.4	3.55	0.4	2.84
		GW1 Coefficient (h)	300	541.67	300	361.11	300	511.11
		GW2 Storage (mm)	40	35	40	35	40	35
		GW2 Percolation (mm/h)	0.3	2.67	0.3	2.67	0.3	2.13
		GW2 Coefficient (h)	400	433.33	400	288.88	400	408.88

Table 8. *Cont.*

Sub-Model	Method	Parameter	MS1 (Inc. 8 Sub-Basins)		MS2 (Inc. 5 Sub-Basins)		MS3 (Inc. 5 Sub-Basins)	
			Initial	Calibrated	Initial	Calibrated	Initial	Calibrated
Transform	Clark Unit Hydrograph	Time of Concentration (h)	2.386	2.386	2.528	2.529	2.176	2.175
		Storage Coefficient (h)	200	80	200	80	200	300
Base Flow	Constant Monthly	January (m^3/s)	0.048	0.048	0.065	0.065	0.056	0.056
		February (m^3/s)	0.074	0.074	0.099	0.099	0.079	0.079
		March (m^3/s)	0.121	0.121	0.158	0.158	0.143	0.143
		April (m^3/s)	0.183	0.183	0.221	0.221	0.233	0.233
		May (m^3/s)	0.226	0.226	0.205	0.205	0.272	0.272
		June (m^3/s)	0.123	0.092	0.133	0.099	0.161	0.241
		July (m^3/s)	0.084	0.063	0.097	0.072	0.112	0.126
		August (m^3/s)	0.062	0.046	0.075	0.056	0.080	0.060
		September (m^3/s)	0.054	0.041	0.067	0.050	0.071	0.053
		October (m^3/s)	0.053	0.053	0.067	0.067	0.069	0.069
		November (m^3/s)	0.050	0.050	0.065	0.065	0.064	0.064
		December (m^3/s)	0.046	0.046	0.060	0.060	0.058	0.058

It was observed that the simulations from the three-year time period simulated the real values very well both in temporal terms and statistically on the basin basis. R^2, RMSE, NSE, and KGE values of runoff simulations at stream gauge stations are presented in Table 9.

Table 9. R^2, RMSE, NSE, and KGE values of flow rate simulations at the stream gauge stations.

Station	R^2	RMSE	NSE	KGE
21-01 (Karasu-Çipak)	0.840	2.144	0.817	0.748
21-152 (Köşk Dere-Köşk)	0.656	1.967	0.431	0.262
21-168 (Büyükçay-Karagöbek)	0.586	1.359	0.406	0.265

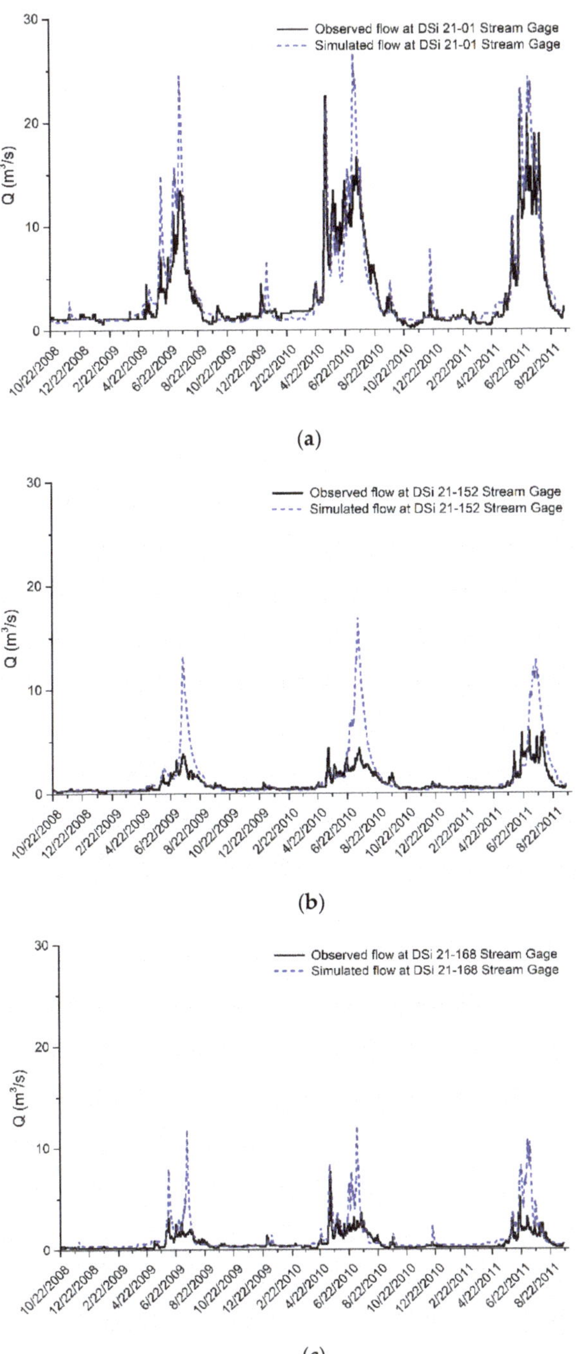

Figure 8. Observed and simulated flow rates at the (**a**) Karasu-Çipak (DSİ 21-01), (**b**) Köşk Dere-Köşk (DSİ 21-152), and (**c**) Büyük Çay-Karagöbek (DSİ 21-168) stream gauge stations.

4. Discussion

The observation of the stream hydrographs and the statistical analyses showed that the flowrate simulations at the basin outlet (DSİ 21-01) were better than at the other stream gauge stations (DSİ 21-152, DSİ 21-168). It was observed that the simulations for the peak flow rates during the melting periods at these two stream gauge stations inside the study basin were higher than the measured values. However, the fact that the water budget calculations for the main basin outlet can only be obtained with the exaggerated simulations of related hydrographs indicates the presence of karstic formations in the basin. As a matter of fact, the presence of many springs observed in the study basin due to the general hydrogeological formations in the basin, and the fact that the groundwater model does not exhibit linear behavior, confirms that the land has a karstic character [23,81]. For this reason, while performing HEC-HMS model calibrations, the automatic calibration process was initially followed up to a point but later abandoned. Still, a manual calibration process was used in the study to reveal the actual behavior of the basin in general. In hydrological modeling studies, a model that reflects the basin characteristics well is expected to have good statistical indicators such as R^2 and RMSE or metric scores such as the Nash–Sutcliffe efficiency (NSE) and the Kling–Gupta efficiency (KGE). However, the reverse may not always be accurate [82–84]. Considering the modeling limitations for karstic behavior in the HEC-HMS model, this study may not provide characteristic flow simulations at the main basin outlet by automatically optimizing the parameters while evaluating the statistical and metric results of flow rate simulations with observation gauges inside the basin. In this case, a physically meaningful manual calibration of all the natural events that may cause the change in the flow simulations for the main basin outlet is required. However, it should not be ignored that the model results obtained with a more intense effort can improve the results much more; it has been concluded that the calibration and validation period results are sufficient. Undoubtedly, many clues can be brought about the actual basin behavior with the hydrological models established in the computer environment. The studies carried out will help the development of new techniques that can fully model natural behavior at every point of the watershed and for selecting or combining the appropriate models.

Similarly, in the manual calibration process of snow–water equivalent simulations, each characteristic detail of the SWE curve, especially during the melting period, was primarily modeled in a physically meaningful way to analyze the events for future studies. For example, the R^2 values calculated on the point scale for the Radar station in Table 7 are slightly lower than the values calculated on the areal-based values, unexpectedly, because the manually selected parameters also try to better simulate the critical rain on snow events during the melting period on high elevation zones [25].

5. Conclusions

In this study, basin characterization preprocessing was conducted with GIS-based HEC-GeoHMS, and basin and meteorological models were created. The outputs obtained were used as inputs for the hydrological simulation program HEC-HMS. The simulations for the years 2008 to 2011 were carried out with the model developed for the runoff of the Kırkgöze–Çipak basin and its sub-basins, where a significant part of the annual total runoff (70 to 80%) is formed by snowmelt.

The boundaries of the chosen Kırkgöze–Çipak basin study area were determined using the HEC-GeoHMS program, and its characterization was carried out and the model inputs were obtained for the HEC-HMS application. When determining the boundaries of a basin and its sub-basins, the outer basin boundary, and the surface stream network, the longest flow path, etc., are determined and then the whole basin is divided into sub-basins. After that, the physical parameters of these sub-basins are determined. In the next stage, a meteorological model definition is created for the climate characterization of the sub-basins.

The snowmelt rate function, which is the most effective parameter for the simulation of the snow–water equivalent during the implementation of the snowmelt model with the basin temperature index method, was primarily obtained from the locations of the three meteorology stations in the study basin. Then, these curves that were originally obtained as points were reduced to a single curve representing the study basin in general. Considering the characteristic behavior of point and areal snow–water equivalent simulations, as well as R^2 and RMSE values, the parameters required for snow–water equivalent simulation—one of the most important steps in simulating the flow rate for a basin where there is a lot of snowfall—were successfully integrated into the runoff model. With this physically-based approach, it has also been shown that regional studies on snow can be carried out more reliably and quickly.

After the basin-wide snow–water equivalent simulations were successfully performed, a hydrological model was created with HEC-HMS, and the runoff outputs of this model were correlated with the observed data from Köşk Dere-Köşk (DSİ 21-152), Büyük Çay-Karagöbek (DSİ 21-168), and Karasu-Çipak (DSİ 21-01) stream gauge stations, and thus model calibration was performed.

Although the Kırkgöze–Çipak Basin is small in terms of surface area, it is a basin with a large altitude range. It is inevitable that meteorological variability will be high in such a basin. In hydrological model studies, it is very important for the accuracy and reliability of the simulations that the meteorological data distribution across the basin is in line with the real values in the field. Having the Köşk, Güngörmez, and Radar meteorology stations, which are located at the appropriate altitude and location in the basin, ensures that the meteorological variable distribution was as close to reality as possible, and it also maximized the reliability of the hydrological model parameters obtained from the HEC-HMS.

As a result, in this study, it has been shown that, with the HEC-HMS hydrological model, flow rate simulations can be performed with very good R^2 and RMSE values and also NSE and KGE scores at the outlet of the snow-dominated, mountainous Kırkgöze Basin, which has a very complex topography. It is believed that the model parameters obtained and the methodology used will be a source for hydrological model studies to be carried out in similar mountain basins and help authorities to use water resources well, not only regionally, but also nationally and internationally.

Author Contributions: All authors contributed to the study conception and design. Conceptualization: S.Ş. and M.N.İ., Methodology: S.Ş. and M.N.İ., Software: S.Ş. and M.N.İ., Validation and Formal analysis: S.Ş. and M.N.İ., Writing—Original Draft: S.Ş. and M.N.İ., Writing—Review and Editing: S.Ş., Visualization: S.Ş., Supervision: S.Ş. All authors have read and agreed to the published version of the manuscript.

Funding: This study has not received any funding from any institution/organization.

Data Availability Statement: All the data are provided in this paper, and no additional data are available to provide.

Conflicts of Interest: The authors declare no conflict of interest.

References

1. Hassan, F. Water History for Our Times. In *International Hydrological Programme*; UNESCO: Paris, France, 2011; p. 122.
2. Tekeli, A.E.; Akyürek, Z.; Arda Şorman, A.; Şensoy, A.; Ünal Şorman, A. Using MODIS Snow Cover Maps in Modeling Snowmelt Runoff Process in the Eastern Part of Turkey. *Remote Sens. Environ.* **2005**, *97*, 216–230. [CrossRef]
3. Yucel, I.; Guventurk, A.; Sen, O.L. Climate Change Impacts on Snowmelt Runoff for Mountainous Transboundary Basins in Eastern Turkey. *Int. J. Climatol.* **2015**, *35*, 215–228. [CrossRef]
4. Stewart, I.T. Changes in Snowpack and Snowmelt Runoff for Key Mountain Regions. *Hydrol. Process. Int. J.* **2009**, *23*, 78–94. [CrossRef]
5. Sood, A.; Smakhtin, V. Global Hydrological Models: A Review. *Hydrol. Sci. J.* **2015**, *60*, 549–565. [CrossRef]
6. Dong, C. Remote Sensing, Hydrological Modeling and in Situ Observations in Snow Cover Research: A Review. *J. Hydrol.* **2018**, *561*, 573–583. [CrossRef]
7. Dhami, B.S.; Pandey, A. Comparative Review of Recently Developed Hydrologic Models. *J. Indian Water Resour. Soc.* **2013**, *33*, 34–41.
8. Zhou, Y.; Cui, Z.; Lin, K.; Sheng, S.; Chen, H.; Guo, S.; Xu, C.-Y. Short-Term Flood Probability Density Forecasting Using a Conceptual Hydrological Model with Machine Learning Techniques. *J. Hydrol.* **2022**, *604*, 127255. [CrossRef]

9. Wurbs, R.A. Dissemination of Generalized Water Resources Models in the United States. *Water Int.* **1998**, *23*, 190–198. [CrossRef]
10. Yilmaz, A.G.; Imteaz, M.A.; Ogwuda, O. Accuracy of HEC-HMS and LBRM Models in Simulating Snow Runoffs in Upper Euphrates Basin. *J. Hydrol. Eng.* **2012**, *17*, 342–347. [CrossRef]
11. Chen, Y.; Li, W.; Fang, G.; Li, Z. Review Article: Hydrological Modeling in Glacierized Catchments of Central Asia—Status and Challenges. *Hydrol. Earth Syst. Sci.* **2017**, *21*, 669–684. [CrossRef]
12. Anderson, E.A. *Snow Accumulation and Ablation Model–SNOW-17*; NOAA's National Weather Service Hydrology Laboratory NWSRFS User Manual; NOAA: Washington, DC, USA, 2006; p. 61.
13. Zhou, J.; Pomeroy, J.W.; Zhang, W.; Cheng, G.; Wang, G.; Chen, C. Simulating Cold Regions Hydrological Processes Using a Modular Model in the West of China. *J. Hydrol.* **2014**, *509*, 13–24. [CrossRef]
14. Samaniego, L.; Kumar, R.; Attinger, S. Multiscale Parameter Regionalization of a Grid-Based Hydrologic Model at the Mesoscale. *Water Resour. Res.* **2010**, *46*. [CrossRef]
15. Burnash, R.J.C. The NWS River Forecast System—Catchment Modeling. In *Computer Models of Watershed Hydrology*; Water Resource Publication, LCC: Littleton, CO, USA, 1995; pp. 311–366.
16. Speers, D.D. SSARR Model. In *Computer Models of Watershed Hydrology*; Water Resource Publication, LCC: Littleton, CO, USA; pp. 367–394.
17. Scharffenberg, W.; Ely, P.; Daly, S.; Fleming, M.; Pak, J. Hydrologic Modeling System (HEC-HMS): Physically-Based Simulation Components. In Proceedings of the 2nd Joint Federal Interagency Conference, Las Vegas, NV, USA, 27 June–1 July 2010.
18. USACE. *Hydrologic Modeling System HEC-HMS User's Manual*; U.S. Army Corps of Engineers: Hanover, NH, USA, 2016.
19. OHRG. Osisko Hammond Reef Gold Project, Hydrology Technical Support Document VERSION 2, Project Number: 13-1118-0010, Document Number: DOC018. 2013. Available online: https://www.ceaa-acee.gc.ca/050/documents/P63174/123081E.Pdf (accessed on 10 November 2021).
20. Razmkhah, H.; Saghafian, B.; Ali, A.-M.A.; Radmanesh, F. Rainfall-Runoff Modeling Considnmering Soil Moisture Accounting Algorithm, Case Study: Karoon III River Basin. *Water Resour.* **2016**, *43*, 699–710. [CrossRef]
21. Dariane, A.B.; Javadianzadeh, M.M.; James, L.D. Developing an Efficient Auto-Calibration Algorithm for HEC-HMS Program. *Water Resour. Manag.* **2016**, *30*, 1923–1937. [CrossRef]
22. Darbandsari, P.; Coulibaly, P. Inter-Comparison of Lumped Hydrological Models in Data-Scarce Watersheds Using Different Precipitation Forcing Data Sets: Case Study of Northern Ontario, Canada. *J. Hydrol. Reg. Stud.* **2020**, *31*, 100730. [CrossRef]
23. Kourgialas, N.N.; Karatzas, G.P.; Nikolaidis, N.P. An Integrated Framework for the Hydrologic Simulation of a Complex Geomorphological River Basin. *J. Hydrol.* **2010**, *381*, 308–321. [CrossRef]
24. Fazel, K.; Scharffenberg, W.A.; Bombardelli, F.A. Assessment of the Melt Rate Function in a Temperature Index Snow Model Using Observed Data. *J. Hydrol. Eng.* **2014**, *19*, 1275–1282. [CrossRef]
25. Şengül, S.; İspirli, M.N. Estimation and Analysis of the Antecedent Temperature Index–Melt Rate (ATIMR) Function Using Observed Data from the Kırkgöze-Çipak Basin, Turkey. *J. Hydrol.* **2021**, *598*, 126484. [CrossRef]
26. Bombardelli, F.A.; García, M.H. Hydraulic Design of Large-Diameter Pipes. *J. Hydraul. Eng.* **2003**, *129*, 839–846. [CrossRef]
27. Şorman, A.A.; Şensoy, A.; Tekeli, A.E.; Şorman, A.Ü.; Akyürek, Z. Modelling and Forecasting Snowmelt Runoff Process Using the HBV Model in the Eastern Part of Turkey. *Hydrol. Process.* **2009**, *23*, 1031–1040. [CrossRef]
28. Yüksek, Ö. Reevaluation of Turkey's Hydropower Potential and Electric Energy Demand. *Energy Policy* **2008**, *36*, 3374–3382. [CrossRef]
29. Ogden, F.L.; Garbrecht, J.; DeBarry, P.A.; Johnson, L.E. GIS and Distributed Watershed Models. II: Modules, Interfaces, and Models. *J. Hydrol. Eng.* **2001**, *6*, 515–523. [CrossRef]
30. Emerson, C.H.; Welty, C.; Traver, R.G. Watershed-Scale Evaluation of a System of Storm Water Detention Basins. *J. Hydrol. Eng.* **2005**, *10*, 237–242. [CrossRef]
31. Leavesley, G.H.; Lichty, R.W.; Troutman, B.M.; Saindon, L.G. Precipitation-Runoff Modeling System: User's Manual. *Water-Resour. Investig. Rep.* **1983**, *83*, 207.
32. Akyürek, Z.; Şorman, A.Ü. Monitoring Snow-Covered Areas Using NOAA-AVHRR Data in the Eastern Part of Turkey. *Hydrol. Sci. J.* **2002**, *47*, 243–252. [CrossRef]
33. Yerdelen, C.; Acar, R. Study on Prediction of Snowmelt Using Energy Balance Equations and Comparing with Regression Method in the Eastern Part of Turkey. *J. Sci. Ind. Res.* **2005**, *64*, 520–528.
34. Şensoy, A.; Şorman, A.A.; Tekeli, A.E.; Şorman, A.; Garen, D.C. Point-scale Energy and Mass Balance Snowpack Simulations in the Upper Karasu Basin, Turkey. *Hydrol. Process. Int. J.* **2006**, *20*, 899–922. [CrossRef]
35. Acar, R.; Şenocak, S.; Şengül, S. Snow Hydrology Studies in the Mountainous Eastern Part of Turkey. In Proceedings of the 2009 IEEE International Conference on Industrial Engineering and Engineering Management, Hong Kong, China, 8–11 December 2009; pp. 1578–1582.
36. Acar, R.; Şenocak, S.; Şengül, S.; Coşkun, T.; Balık Şanlı, F. Erzurum Kırkgöze Havzasında Kar Erimesine Etki Eden Meteorolojik Ölçümlerin Üç İstasyonda Karşılaştırılması. In Proceedings of the III. Ulusal Kar Kongresi, Erzurum, Turkey, 17–19 February 2009.
37. Şorman, A.A.; Uysal, G.; Şensoy, A. Probabilistic Snow Cover and Ensemble Streamflow Estimations in the Upper Euphrates Basin. *J. Hydrol. Hydromech.* **2019**, *67*, 82–92. [CrossRef]
38. Şengül, S. Dağlık Havzalarda Hidrolojik Çevrime Etki Eden Parametrelerin Coğrafi Bilgi Sistemleri ve HSPF Model Programıyla İncelenmesi ve Kırkgöze Havzası Örneği. Ph.D. Thesis, Fen Bilimleri Enstitüsü, Atatürk Üniversitesi, Erzurum, Turkey, 2011.

39. Şenocak, S. Kar Erimesi Akış Modelinin (SRM), Coğrafi Bilgi Sistemleri ve Uzaktan Algılama Teknikleri de Kullanılarak Dağlık Bölgelerde Uygulaması ve Erzurum Kırkgöze Havzası Örneği. Ph.D. Thesis, Fen Bilimleri Enstitüsü, Atatürk Üniversitesi, Erzurum, Turkey, 2011.
40. İspirli, M.N. HEC-HMS Model Programı Kullanılarak Dağlık Havzalarda Kar Erimesine Etki Eden Parametrelerin Belirlenmesi ve Kırkgöze Çipak Havzası'nın Hidrolojik Modellenmesi Determination of Parameters Affecting Snow Melting in Mountain Basins Using HEC-HMS Model Program and Hydrologic Modeling of Kırkgöze Çıpak Basin. Master's Thesis, Fen Bilimleri Enstitüsü, Atatürk Üniversitesi, Erzurum, Turkey, 2019.
41. Şengül, S. Küresel İklim Değişikliğinin Yağış ve Sıcaklık Üzerindeki Etkilerinin Kırkgöze Dağlık Havzasındaki Kar Kütlesi Üzerinde 2050 Yılı İçin Beklenen Etkilerinin HSPF Model Programı İle İncelenmesi. *Avrupa Bilim Ve Teknol. Derg.* **2019**, *17*, 611–636.
42. Akbulut, N.; Bayarı, S.; Akbulut, A.; Şahin, Y. Rivers of Turkey. In *Rivers of Europe*; Academic Press: London, UK, 2009; pp. 643–672.
43. Pekkan, E. Yukarı Fırat Havzasında Kar Erimesi Süreciniuydu Görüntüsü Analizlerive Izleyici Teknikleri Ile Incelenmesi. Ph.D. Thesis, Fen Bilimleri Enstitüsü, Hacettepe Üniversitesi, Ankara, Turkey, 2009.
44. Anonymous. *Erzurum Projesi Mühendislik Jeolojisi Planlama Raporu*; DSİ Genel Müdürlüğü; Jeoteknik Hizmetler ve Yeraltı Suları Dairesi Başkanlığı: Erzurum, Turkey, 1978; p. 48.
45. Gyawali, R.; Watkins, D.W. Continuous Hydrologic Modeling of Snow-Affected Watersheds in the Great Lakes Basin Using HEC-HMS. *J. Hydrol. Eng.* **2013**, *18*, 29–39. [CrossRef]
46. Chu, X.; Steinman, A. Event and Continuous Hydrologic Modeling with HEC-HMS. *J. Irrig. Drain. Eng.* **2009**, *135*, 119–124. [CrossRef]
47. Scharffenberg, B. Introduction to HEC-HMS. In Proceedings of the Watershed Modeling with HEC-HMS, California Water and Engineering Forum, Sacramento, CA, USA, 28 May 2008.
48. Khatri, H.B.; Jain, M.K.; Jain, S.K. Modelling of Streamflow in Snow Dominated Budhigandaki Catchment in Nepal. *J. Earth Syst. Sci.* **2018**, *127*, 100. [CrossRef]
49. Viessman, W.J.; Lewis, G.L. *Introduction to Hydrology*; Addison Wesley Longman: Boston, MA, USA, 1995.
50. USACE. *Hydrologic Modeling System HEC-HMS Technical Reference Manual*; U.S. Army Corps of Engineers: Hanover, NH, USA, 2000.
51. Fortin, J.-P.; Turcotte, R.; Massicotte, S.; Moussa, R.; Fitzback, J.; Villeneuve, J.-P. Distributed Watershed Model Compatible with Remote Sensing and GIS Data. I: Description of Model. *J. Hydrol. Eng.* **2001**, *6*, 91–99. [CrossRef]
52. Fleming, M.; Neary, V. Continuous Hydrologic Modeling Study with the Hydrologic Modeling System. *J. Hydrol. Eng.* **2004**, *9*, 175–183. [CrossRef]
53. Khatami, S.; Khazaei, B. Benefits of GIS Application in Hydrological Modeling: A Brief Summary. *Vatten Tidskr. För Vattenvård/J. Water Manag. Res.* **2014**, *70*, 41–49.
54. Donigian, A.S.; Davis, H.H. *User's Manual for Agricultural Runoff Management(ARM) Model*; Environmental Protection Agency, Office of Research and Development, Environmental Research, Technology Development and Applications Branch: Washington, DC, USA, 1978.
55. Bennett, T.H. Development and Application of a Continuous Soil Moisture Accounting Algorithm for the HEC-HMS. Master's Thesis, Department of Civil and Environment Engineering, University of California, Davis, CA, USA, 1998.
56. Balk, B.; Elder, K. Combining Binary Decision Tree and Geostatistical Methods to Estimate Snow Distribution in a Mountain Watershed. *Water Resour. Res.* **2000**, *36*, 13–26. [CrossRef]
57. Erxleben, J.; Elder, K.; Davis, R. Comparison of Spatial Interpolation Methods for Estimating Snow Distribution in the Colorado Rocky Mountains. *Hydrol. Process.* **2002**, *16*, 3627–3649. [CrossRef]
58. Cosgrove, B.A.; Lohmann, D.; Mitchell, K.E.; Houser, P.R.; Wood, E.F.; Schaake, J.C.; Robock, A.; Marshall, C.; Sheffield, J.; Duan, Q. Real-time and Retrospective Forcing in the North American Land Data Assimilation System (NLDAS) Project. *J. Geophys. Res. Atmos.* **2003**, *108*, D22. [CrossRef]
59. Erickson, T.A.; Williams, M.W.; Winstral, A. Persistence of Topographic Controls on the Spatial Distribution of Snow in Rugged Mountain Terrain, Colorado, United States. *Water Resour. Res.* **2005**, *41*. [CrossRef]
60. Molotch, N.P.; Colee, M.T.; Bales, R.C.; Dozier, J. Estimating the Spatial Distribution of Snow Water Equivalent in an Alpine Basin Using Binary Regression Tree Models: The Impact of Digital Elevation Data and Independent Variable Selection. *Hydrol. Process. Int. J.* **2005**, *19*, 1459–1479. [CrossRef]
61. Liston, G.E.; Elder, K. A Meteorological Distribution System for High-Resolution Terrestrial Modeling (MicroMet). *J. Hydrometeorol.* **2006**, *7*, 217–234. [CrossRef]
62. Liston, G.E.; Elder, K. A Distributed Snow-Evolution Modeling System (SnowModel). *J. Hydrometeorol.* **2006**, *7*, 1259–1276. [CrossRef]
63. Winstral, A.; Marks, D. Long-term Snow Distribution Observations in a Mountain Catchment: Assessing Variability, Time Stability, and the Representativeness of an Index Site. *Water Resour. Res.* **2014**, *50*, 293–305. [CrossRef]
64. Fassnacht, S.R.; Brown, K.S.J.; Blumberg, E.J.; Moreno, J.L.; Covino, T.P.; Kappas, M.; Huang, Y.; Leone, V.; Kashipazha, A.H. Distribution of Snow Depth Variability. *Front. Earth Sci.* **2018**, *12*, 683–692. [CrossRef]
65. Jeong, H.-G.; Ahn, J.-B.; Lee, J.; Shim, K.-M.; Jung, M.-P. Improvement of Daily Precipitation Estimations Using PRISM with Inverse-Distance Weighting. *Theor. Appl. Climatol.* **2020**, *139*, 923–934. [CrossRef]

66. Acar, R.; Sengul, S. The Estimation of Average Areal Snowfall by Conventional Methods and the Percentage Weighting Polygon Method in the Northeast Anatolia Region, Turkey. *Energy Educ. Sci. Technol. Part A-Energy Sci. Res.* **2012**, *29*, 11–22.
67. Martinec, J.; Rango, A.; Roberts, R. *Snowmelt Runoff Model User's Manual*; USDA Jornada Experimental Range, New Mexico State University: Las Cruces, NM, USA, 2008.
68. USACE. *Snow Hydrology*; North Pacific Division; US Army Corps of Engineers: Portland, OR, USA, 1956.
69. Martinec, J. The Degree-Day Factor for Snowmelt Runoff Forecasting. *IUGG Gen. Assem. Hels. IAHS Comm. Surf. Waters* **1960**, *51*, 468–477.
70. Hock, R. Temperature Index Melt Modelling in Mountain Areas. *J. Hydrol.* **2003**, *282*, 104–115. [CrossRef]
71. Davtalab, R.; Mirchi, A.; Khatami, S.; Gyawali, R.; Massah, A.; Farajzadeh, M.; Madani, K. Improving Continuous Hydrologic Modeling of Data-Poor River Basins Using Hydrologic Engineering Center's Hydrologic Modeling System: Case Study of Karkheh River Basin. *J. Hydrol. Eng.* **2017**, *22*, 05017011. [CrossRef]
72. Momblanch, A.; Holman, I.P.; Jain, S.K. Current Practice and Recommendations for Modelling Global Change Impacts on Water Resource in the Himalayas. *Water* **2019**, *11*, 1303. [CrossRef]
73. Parajuli, A.; Nadeau, D.F.; Anctil, F.; Schilling, O.S.; Jutras, S. Does Data Availability Constrain Temperature-Index Snow Models? A Case Study in a Humid Boreal Forest. *Water* **2020**, *12*, 2284. [CrossRef]
74. Kumar, M.; Marks, D.; Dozier, J.; Reba, M.; Winstral, A. Evaluation of Distributed Hydrologic Impacts of Temperature-Index and Energy-Based Snow Models. *Adv. Water Resour.* **2013**, *56*, 77–89. [CrossRef]
75. He, Z.H.; Parajka, J.; Tian, F.Q.; Blöschl, G. Estimating Degree-Day Factors from MODIS for Snowmelt Runoff Modeling. *Hydrol. Earth Syst. Sci.* **2014**, *18*, 4773–4789. [CrossRef]
76. Bobál, P.; Podhorányi, M.; Mudroň, I.; Holubec, M. Mathematical Modelling of the Dynamics of Mountain Basin Snow Cover in Moravian-Silesian Beskydy for Operational Purposes. *Water Resour.* **2015**, *42*, 302–312. [CrossRef]
77. Bhuiyan, H.A.; McNairn, H.; Powers, J.; Merzouki, A. Application of HEC-HMS in a Cold Region Watershed and Use of RADARSAT-2 Soil Moisture in Initializing the Model. *Hydrology* **2017**, *4*, 9. [CrossRef]
78. USACE. SSARR Model, Streamflow Synthesis and Reservoir Regulation. In *User Manual*; US Army Corps of Engineers: Portland, OR, USA, 1991.
79. Bui, C. Application of HEC-HMS 3.4 in Estimating Streamflow of the Rio Grande under Impacts of Climate Change. Master's Thesis, The University of New Mexico, Albuquerque, NM, USA, 2011.
80. McEnroe, B.M. Guidelines for Continuous Simulation of Streamflow in Johnson County, Kansas, with HEC-HMS. Ph.D. Thesis, Department of Civil, Environmental and Architectural Engineering, University of Kansas, Lawrence, KS, USA, 2010.
81. Bicknell, B.R.; Imhoff, J.C.; Kittle, J.L., Jr.; Jobes, T.H.; Donigian, A.S., Jr. *Hydrological Simulation Program-Fortran (HSPF). User's Manual for Release 12. US EPA National Exposure Research Laboratory, Athens, GA, in Cooperation with US Geological Survey*; Water Resources Division: Reston, VA, USA, 2001.
82. Knoben, W.; Freer, J.; Woods, R. Technical Note: Inherent Benchmark or Not? Comparing Nash-Sutcliffe and Kling-Gupta Efficiency Scores. *Hydrol. Earth Syst. Sci. Discuss.* **2019**, *23*, 4323–4331. [CrossRef]
83. Krysanova, V.; Donnelly, C.; Gelfan, A.; Gerten, D.; Arheimer, B.; Hattermann, F.; Kundzewicz, Z.W. How the Performance of Hydrological Models Relates to Credibility of Projections under Climate Change. *Hydrol. Sci. J.* **2018**, *63*, 696–720. [CrossRef]
84. Demirel, M.C.; Mai, J.; Mendiguren, G.; Koch, J.; Samaniego, L.; Stisen, S. Combining Satellite Data and Appropriate Objective Functions for Improved Spatial Pattern Performance of a Distributed Hydrologic Model. *Hydrol. Earth Syst. Sci.* **2018**, *22*, 1299–1315. [CrossRef]

 *water*

Article

Base Flow Variation and Attribution Analysis Based on the Budyko Theory in the Weihe River Basin

Zheng Mu [1,†], Guanpeng Liu [1,†], Shuai Lin [1], Jingjing Fan [1,2,*], Tianling Qin [2,*], Yunyun Li [3], Yao Cheng [1] and Bin Zhou [1]

1 Hebei Collaborative Innovation Center for the Regulation and Comprehensive Management of Water Resources and Water Environment, Hebei University of Engineering, Handan 056002, China; muzheng1981@hebeu.edu.cn (Z.M.); liugp1998@163.com (G.L.); as18833022875@163.com (S.L.); chengyao@hebeu.edu.cn (Y.C.); zhoubin@meichao.com (B.Z.)

2 State Key Laboratory of Simulation and Regulation of Water Cycle in River Basin, China Institute of Water Resources and Hydropower Research, Beijing 100038, China

3 College of Resources and Environmental Engineering, Mianyang Teachers' College, Mianyang 621000, China; liyunyun19900627@163.com

* Correspondence: fanjingjing@hebeu.edu.cn (J.F.); qintl@iwhr.com (T.Q.)

† These authors contributed equally to this work.

Abstract: The composition and change of runoff are closely related to climate change and human activities. To design effective watershed water resources management measures, there is a need for a clear understanding of the impact of climate change and human activities on baseflow and surface runoff. The purpose of this essay is to quantify their impact on the annual total stream flow, surface runoff, and base flow in the Weihe River Basin (WRB) using a two-stage annual precipitation partitioning method, wherein the surface runoff and base flow are separated from the measured total flow by using a one-parameter digital filter method for which the common filter parameter value is 0.925. The stream flow records were split into two periods: 1960–1970 (pre-change period) and 1971–2005 (post-change period) based on the hydrological breakpoints detected. We found that climate change and human activities have different impacts on base flow and surface runoff. We attributed the decrease in surface runoff due to climate change accounting for 76–78%, while we determined that human activities were responsible to the decrease in base flow accounting for 59–73% of the total observed change. We concluded that both climate change and human beings contributed to the hydrologic change through different hydrological processes: climate change dominated the surface runoff change, while human influences controlled the base flow change. To achieve the expected goals of ecological restoration, appropriate measures must be taken by watershed management in the WRB to mitigate the likely impacts of climate change on water hydrology.

Keywords: climate change; human activities; base flow; surface runoff; Weihe River Basin

Citation: Mu, Z.; Liu, G.; Lin, S.; Fan, J.; Qin, T.; Li, Y.; Cheng, Y.; Zhou, B. Base Flow Variation and Attribution Analysis Based on the Budyko Theory in the Weihe River Basin. *Water* **2022**, *14*, 334. https://doi.org/10.3390/w14030334

Academic Editors: Dengfeng Liu, Hui Liu and Xianmeng Meng

Received: 23 December 2021
Accepted: 20 January 2022
Published: 24 January 2022

1. Introduction

The global hydrologic cycle and distribution of water resources are changing on various scales due to climate change and human interference during the past decades [1–3]. The changes in precipitation, temperature, wind speed, humidity, and solar radiation have caused decreases in streamflow in some regions [4,5]. Land use changes to the vegetation distribution structure lead to changes in evapotranspiration and, thus, lead to changes in runoff [6,7]. Dam construction will affect the flow of the river [8,9]; dam and reservoir construction will reduce runoff [10,11]. Deforestation will lead to a decrease in interception capacity and an increase in runoff, while afforestation will lead to an increase in runoff infiltration, a decrease in runoff, and an increase in underground runoff [12,13]; Agricultural water management leads to reduced runoff or groundwater [14]. Therefore, under the influence of the above factors, there is a variation in runoff in different parts of the world. Quantifying the individual impacts of climate change and humans is important for mitigating the negative effects and adapting to novel environments in the future.

Numerous studies have been conducted to quantify the individual effects of climate change and human activities on water resources in different parts of the world [15–19] including the Tarim River [20], Yangtze River [21], and Shiyang River in China [22], the Nzoia River in Kenya [23], and the Athabasca River in Canada [24]. The study in the Haihe River Basin showed that the runoff in the Taolinkou, Zhangjiafen, and Guantai basins decreased by 41.5%, 59.9%, and 73.9%, respectively, mainly affected by human beings [25]. Ma et al. [26] and Zheng et al. [27] reported that climate change contributed over 50% of the decrease in the inflow to the Miyun reservoir during the past decade. Using data from 413 watersheds in the USA, Wang and Hejazi [28] showed that the impacts of climate change outweighed the effects of human activities. Analysis in the Cimarron Skeleton watershed (the South-central Great Plains) in the USA indicated that changes in water use, land use, and land cover accounted for approximately 50% of changes in water flows. [29]. More recently, the study in the Luan River Basin in China illustrated that the impact of climate change contributed 44% in the wet season, but human activities contributed 93% in the dry season to the flow change [30].

The Weihe River Basin (WRB), an important river in Shaanxi Province, known to the Shaanxi people as their mother river, is the largest tributary of the Yellow River in northern China and is a main river for Ningxia and Gansu Provinces in northwestern China that provides the surface water for irrigation and water supplies for several major cities in the arid region. The stream flow in the WRB has significantly declined in the 20th century [31,32]. Based on 13 stream flow gauges in the WRB, the monthly stream flow depth has declined 0.1–2.1 mm during the period from 1960 to 2009, and the main driving factors for the declined stream flow are believed to be related to reservoir operation, vegetation change, surface water consumption, and water and soil conservation [33]. Using an improved climate elasticity method, Zhan [34] found that human activities contributed to 71–78% of the stream flow decline. More recently, Jiang [35] used a framework to identify the two effects on stream flow across the WRB and found that human activities had a significant impact on decreasing runoff.

Most previous studies have focused on the total stream flow, and little attention has been given to how climate or humans alter the flow pathways, i.e., surface flow and groundwater. The arid regions are dominated by rainfall-excess hydrology, and the surface flow generation has rather different mechanisms from the base flow, which is derived from groundwater. We presume that both climate change and human activities may contribute differently to the changes in surface runoff and base flow. Quantifying the effects of climate change and human activities on water resources enables addressing and better management of the water crisis faced by arid regions.

The objective of this study was to quantify the contributions of climate change and human activities to the interpretation of surface runoff and baseflow observations at three hydrologic gauging stations in the WRB. The base flow and surface runoff components were estimated using a two-stage annual precipitation partitioning method. The spatial variation patterns of climate change and human activity with respect to their influences on the base flow and surface flow were also discussed. The results provide a comprehensive analysis climate and human effects on base flow and surface runoff in the WRB.

2. Materials and Methods

2.1. Study Area and Hydrometeorological Data

The WRB (104°00′ E~110°20′ E, 33°50′ N~37°18′ N), located in the semiarid region of China, has a watershed area of 135,000 km^2 and a main river length of 818 km (Figure 1). The WRB starts in the Niaoshu Mountain, flows through Gansu, Ningxia, Shanxi, and eventually joins the Yellow River at Tongguan. The average annual natural runoff is 10.4 billion m^3. The average annual precipitation is about 610 mm, the average annual temperature ranges from 7.8 to 13.5 °C, and the average annual potential evapotranspiration (PET) varies from 800 to 1200 mm [36].

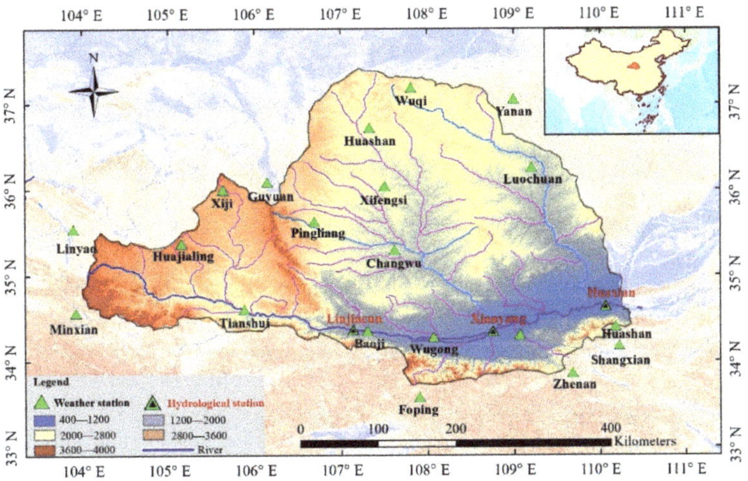

Figure 1. The spatial distribution of 3 hydrologic stations and 22 rain gauges in the Weihe River Basin.

In this study, 22 meteorological stations and three hydrological stations were used to derive daily precipitation, temperature, and stream flow data in the WRB. These stations were located throughout the WRB, representing the characteristics of climate and hydrology for each catchment. Daily data on precipitation and other meteorological variables for the period from 1960 to 2005 were provided by the China Meteorological Administration. The monthly PET series were calculated using the Penman-Monteith method [37] for every meteorological station. The areawide precipitation and PET for each hydrological station was generated by averaging the readings across meteorological stations in and around the controlled watershed. Shaanxi hydrology and Water Resources Bureau provided daily flow data of major rivers in PNG from 1960 to 2005.

The one-parameter digital filter decomposed the daily stream flow into surface runoff and baseflow components with the filter parameter value of 0.925 [38]. In this study, the daily PET, rainfall, surface runoff, and base flow were summarized as annual scale values. The actual annual evaporation values were computed as residuals of the water balance, $E = P - Q$, where E is evapotranspiration, P is precipitation, and Q is streamflow.

2.2. Digital Filtering

The digital filtering method was used to segment the daily scale runoff data of Weihe River Basin. Digital filtering is derived from signal processing technology, and its main principle is the combination of runoff division and digital signal analysis. The fast response and slow response signals in the process of precipitation and runoff are decomposed into high frequency signals and low frequency signals by a digital filter, which represent surface runoff and underground runoff, respectively [39].

The Lyne–Hollick filtering method was first proposed by Lyne and Hollick in 1979. Nathan and Mcmahon introduced hydrologic calculations into the method in 1990; its segmentation equation is:

$$Q_{dt} = f_1 Q_{d\,(t-1)} + \frac{1+f_1}{2}\left[Q_t - Q_{(t-1)}\right], \tag{1}$$

$$Q_{bt} = Q_t - Q_{dt}, \tag{2}$$

where Q_{dt} and Q_{dt-1} are the surface runoff at time t and time $t-1$, respectively; Q_t and $Q_{(t-1)}$ are the runoff at time t and time $t-1$, respectively; Q_{bt} is the base flow at time t; and f_1 is the filter parameter value range 0.90 to 0.95; Nathan selected the filter parameter 0.925 by comparison [38].

2.3. Mann-Kendall Inspection

The Mann-Kendall trend test is a nonparametric statistical test method, also known as the no distribution test [40,41]. For the time series $X = (X_1, X_2, X_3, \ldots, X_n)$, ($n$ is the number of variables), the Mann-Kendall trend test statistic is S.

$$S = \sum_{j=1}^{n-1} \sum_{i=j+1}^{n} \text{sgn}\,(X_i - X_j), \tag{3}$$

$$\text{sgn}(X_i - X_j) = \begin{cases} 1, & X_i - X_j > 0 \\ 0, & X_i - X_j = 0 \\ -1, & X_i - X_j < 0 \end{cases} \tag{4}$$

$$\begin{cases} Z = \frac{S+1}{\sqrt{\frac{n(n-1)(2n+5)}{18}}}; & S > 0 \\ Z = 0 & ;\ S = 0\ , \\ Z = \frac{S+1}{\sqrt{\frac{n(n-1)(2n+5)}{18}}}; & S < 0 \end{cases} \tag{5}$$

In type: S is the normal distribution, Var(S) is variance, and its formula is:

$$\text{Var}(s) = \left[n(n-1)(2n+5) - \sum_{1}^{m} t_k(t_k-1)(2t_k+5)\right]/18 \tag{6}$$

In type: m is the number of groups, and t_k is the number of groups with the same data. When n is greater than 10, the formula of the standardized statistics is:

$$Z = \begin{cases} \frac{s-1}{\sqrt{\text{Var}(s)}}; & if\ S > 0 \\ 0 & ;\ if\ S = 0 \\ \frac{s+1}{\sqrt{\text{Var}(s)}}; & if\ S < 0 \end{cases} \tag{7}$$

During the test, if Z is greater than 0, it indicates a significant upward trend in the series; otherwise, it indicates a significant downward trend in the series. [42]. If we set the significance levels $\alpha = 0.05$, if $|Z| \geq Z_{1-\alpha/2} = 1.96$, then the null hypothesis should be rejected, indicating that the trend of sequence change is significant; otherwise, the null hypothesis should be accepted, indicating that the trend of sequence change is not significant [43].

2.4. Two-Stage Annual Precipitation Partitioning

Assuming that changes in soil water storage can be ignored on an annual scale, precipitation can be divided into two components, namely runoff and evapotranspiration. The water balance can be expressed as:

$$P = Q + E, \tag{8}$$

where P is precipitation, Q is runoff, and E is evapotranspiration.

Following the decomposition method given by L'vovich [44], long-term sedimentation can be broken down into two stages. In the first stage, precipitation is suspended as surface runoff (Q_s) and soil wetting (W), which can be expressed as:

$$P = Q_s + W, \tag{9}$$

In the second stage, the soil wetting is divided into baseflow (Q_b) and evapotranspiration (ET), which can be expressed as:

$$W = Q_b + ET, \tag{10}$$

where the sum of Q_s and Q_b is the total runoff (Q).

Based on the two-stage precipitation partitioning method and the SCS curve number method [45], Ponce and Shetty [46] proposed a method to estimate Q_s and Q_b at the annual scale, and Chen and Wang [47] presented an estimation method on a seasonal scale. In this paper, we focused on the annual scale, and the model can be estimated as:

$$Q_s = \frac{(P - \lambda_s W_p)^2}{P + (1 - 2\lambda_s)W_p} \; (while\ P > \lambda_s W_p), \tag{11}$$

$$Q_b = \frac{(W - \lambda_b V_p)^2}{W + (1 - 2\lambda_b)V_p} (while\ W > \lambda_b V_p), \tag{12}$$

where W_p is the maximum of the total soil wetting, the initial wetting can be explained as a percentage (λ_s) of the soil wetting capacity (W_p), and $\lambda_s W_p$ can be defined as the minimum threshold for precipitation required for surface runoff. Similarly, the wetting threshold for the base flow can be defined as $\lambda_b V_p$.

2.5. The Nash-Sutcliffe Efficiency Coefficient

We used the Nash-Sutcliffe Efficiency coefficient (NSE) function [48] to evaluate model performance:

$$NSE = 1 - \frac{\sum_{i=1}^{n} \left(Q_{obs}^i - Q_{est}^i\right)^2}{\sum_{i=1}^{n} \left(Q_{obs}^i - Q_{ave}\right)^2}, \tag{13}$$

where Q_{obs} is the measured runoff, Q_{est} is the estimated runoff, and Q_{ave} is the mean value of the measured runoff. NSE values can range from $-\infty$ to 1. If NSE = 1, it indicates that the model can perfectly model the measured data, while at a NSE higher than 0.5 the model performance is satisfactory [49].

2.6. Determining the Contribution Rate of Climate Change and Human Activities

As discussed, the datasets were divided into two periods: the pre-change period and post-change period. The pre-change period was the period prior to the point of variation, when there was no significant trend of increase or decrease in total streamflow/surface runoff/baseflow. The post-change period was the period after the point of variation, where there was a clear trend of increase or decrease in total streamflow/surface runoff/baseflow compared to the pre-change period. The proposed quantifying method assumed no human activities for the watershed in the pre-change period. In other words, during the pre-change period, if the total stream flow/surface runoff/base flow changed, it would only be attributed to climate change. During the post-change period, if the total stream flow/surface runoff/base flow changed, it could be caused by climate change or human activities, or both. Under this assumption, the model parameters were based on the observations in the pre-change period. If there were no human activities, the parameters in the post-change period were assumed to be the same as those in the pre-change period. Using the pre-change model parameters and post-change meteorological data, the total stream flow/surface runoff/base flow in the post-change period would be reconstructed, which would not be affected by human factors. The gap between the reconstructed total stream flow/surface runoff/base flow series in the post-change period and the observed ones in the pre-change period could indicate the total stream flow/surface runoff/base flow change caused by climate change. In addition, the gap between the reconstructed total stream flow/surface runoff/base flow series and the observed ones in the post-change period could indicate the total stream flow/surface runoff/base flow change caused by human activities. The quantifying method can be computed as [50,51]:

$$\Delta Q_t = \overline{Q_o} - Q_o, \tag{14}$$

$$\Delta Q_c = \overline{Q_o} - Q_m \tag{15}$$

$$\Delta Q_h = Q_m - Q_o \tag{16}$$

$$\eta_h = \Delta Q_h / \Delta Q_t \times 100\% \tag{17}$$

$$\eta_c = \Delta Q_c / \Delta Q_t \times 100\% \tag{18}$$

where ΔQ_t represents the total change in the total stream flow/surface runoff/base flow, ΔQ_c and ΔQ_h indicate the changes in total stream flow/surface runoff/base flow caused by climate change and human activities, respectively, $\overline{Q_o}$ and Q_o are the observed total stream flow/surface runoff/base flow values in the pre-change period and the post-change period, respectively, Q_m is the modeled total stream flow/surface runoff/base flow values in the post-change period, and η_c and η_h are the climate change and impacts of human activities on total stream flow/surface runoff/base flow changes, respectively.

The advantages of the quantifying method are simple, practical, and can obtain accurate results according to the simulated values, but there is no breakdown of the impact of human activity, such as river damming, irrigation, and urban water diversion.

3. Results

This section provides a concise and accurate description of the experimental results, their interpretation, and some experimental conclusions that can be drawn.

3.1. Base Flow and Surface Runoff

The daily observed total streamflow was decomposed into surface runoff and base flow using the one-parameter digital filter method with the filter parameter value of 0.925 [37], which is widely used in the study area [40]. The observed total annual flow, surface runoff, and base flow at Linjiacun, Xianyang, and Huaxian stations across the WRB are presented in Figure 2. For the three stations, the total stream flow, surface runoff, and base flow at the annual scale tended to decline. The total stream flow decreased faster than either the surface runoff or the base flow, with the surface runoff decreasing the slowest. Both the total flow and base flow were the highest at the Xianyang station. However, its base flow index was almost the same as that at the Huaxian station, with mean annual values of 0.67 at the Xianyang station and 0.66 at the Huaxian station. The evaluation of the base flow index was similar at the Xianyang and Huaxian stations. A significant change for the base flow index at the Linjiacun station occurred in 1970. The mean annual base flow index at Huaxian and Xianyang was 0.67 with a standard deviation of 0.05, while the mean annual base flow index at Linjiacun was 0.63 with a standard deviation of 0.13 (Figure 3). The base flow index was substantially influenced by human activities, climate change, precipitation, and evaporation [40]. Therefore, it is necessary to determine the variation points of total flow, surface runoff, and base flow series for further contribution analysis.

3.2. Breakpoint Analysis

Breakpoints of annual total stream flow, surface runoff, and base flow were detected by the Mann-Kendall test for all three stations from upstream to downstream across the WRB, with significance levels set at $\alpha = 0.05$ ($Z_{1-\alpha/2} = 1.96$), and the results are shown in Table 1. Breakpoints of the total stream flow at the Linjiacun, Xianyang, and Huaxian stations occurred in 1990, 1986, and 1987, respectively. Breakpoints of the surface runoff and base flow at those stations were in 1986, 1989, and 1991, and 1977, 1986, and 1974, respectively. According to the results, the significant change for base flow was earlier than total stream flow, indicating that the base flow was more sensitive to the effects of climate change and human activities than the total stream flow. The breakpoint in the Huaxian station was the earliest, in 1974, among all the watersheds. This is consistent with the base flow index changes in 1970. Therefore, we selected the 1970 points as the variation point to quantify the influence of climate change and human activities on total stream flow, surface runoff, and base flow [52].

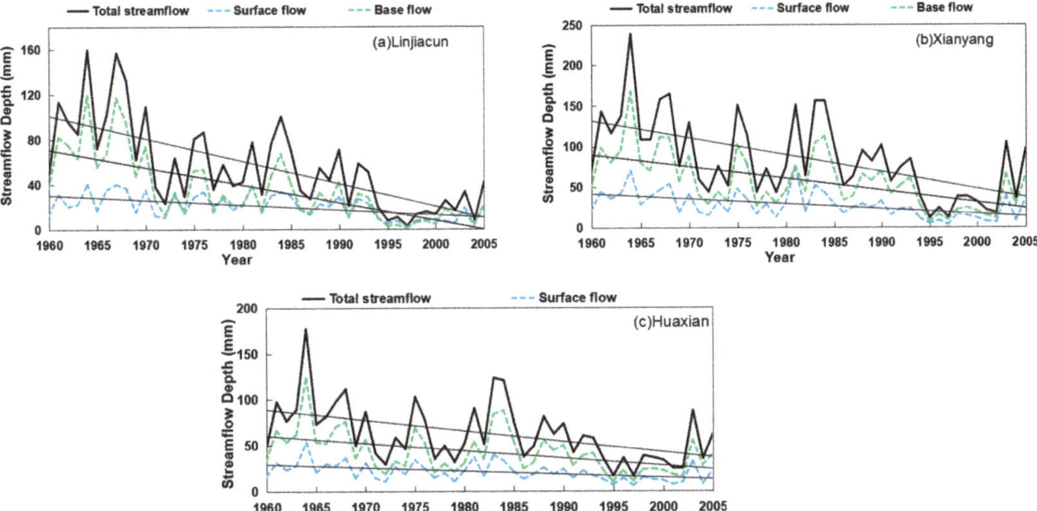

Figure 2. The annual total streamflow, direct runoff, and base flow in (**a**–**c**).

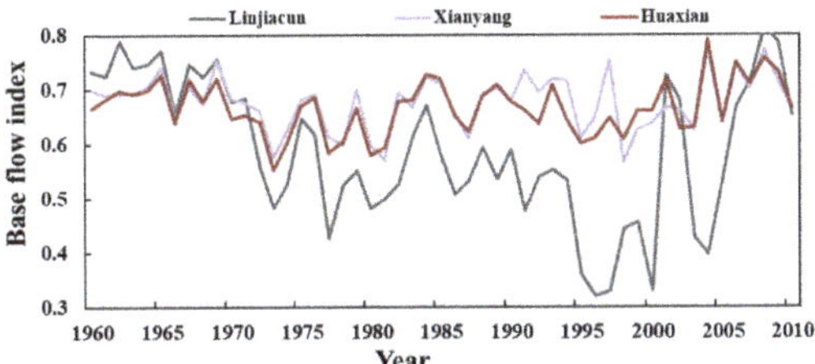

Figure 3. Baseflow Index Change chart of Linjiacun Station, Xianyang Station, and Huaxian Station.

Table 1. The breakpoint detected results of the Mann-Kendall test method.

Station	α	Z	Breakpoint		
			Total Streamflow	Base Flow	Surface Runoff
Linjiacun	0.05	±1.96	1990	1977	1986
Xianyang	0.05	±1.96	1986	1986	1989
Huaxian	0.05	±1.96	1987	1974	1991

3.3. Model Calibration and Stream Flow Reconstruction

The data were divided into two periods, 1960–1970 (pre-change period) and 1971–2005 (post-change period), based on the results of the breakpoint tests to quantify the impact of climate change and humans on surface runoff and baseflow. A two-stage annual precipitation partitioning model was used to estimate the surface runoff and base flow for the post-change period. The model has four parameters: λ_s, Wp, λ_b, and Vp, which need to be estimated. We chose to estimate the parameters in an 11-year period (the whole pre-change period: 1960–1970) instead of in a six-year calibration period (1960–1965) to minimize the uncertainty in the limited data. The values of surface runoff and base flow in the natural period from 1960 to 1970 were used to estimate the model parameters: λ_s, Wp, λ_b, and Vp. The parameters for the three stations in the WRB are listed in Table 2. As shown in Table 2, the parameters λ_s and λ_b do not indicate a significant pattern.

Table 2. Parameter evaluation of the two-stage annual precipitation partitioning model for 3 hydrological stations across the Weihe River.

Site	Wp	Vp	λ_b	λ_s
Linjiacun	8645	3000	0.01	0.01
Xianyang	8645	2616	0.01	0.02
Huaxian	4543	3983	0.06	0.02

The values of NSE and the coefficient of determination (R^2) were used to evaluate the model performance of the base flow and surface runoff. NSE and R^2 were computed in the validation period (1960–1970) for the three sites and are summarized in Table 3. The R^2 values ranged from 0.9 to 0.98, and the NSE values ranged from 0.89 to 0.98. To illustrate the model performance, we used the Linjiacun station as an example (Figure 4). The results showed that the model performed well, although the model led to an underestimation of the results observed during the peak period. Overall, the accuracies of the model calibration and validation were acceptable for the annual surface runoff and base flow estimation in the three stations.

Table 3. The R^2 and Nash-Sutcliffe Efficiency coefficient (NSE) for two-stage annual precipitation partitioning model at the 3 hydrological stations.

Site	Base Flow (mm)		Surface Runoff (mm)	
	R^2	NSE	R^2	NSE
Linjiacun	0.90	0.89	0.96	0.92
Xianyang	0.96	0.96	0.97	0.90
Huaxian	0.98	0.98	0.91	0.90

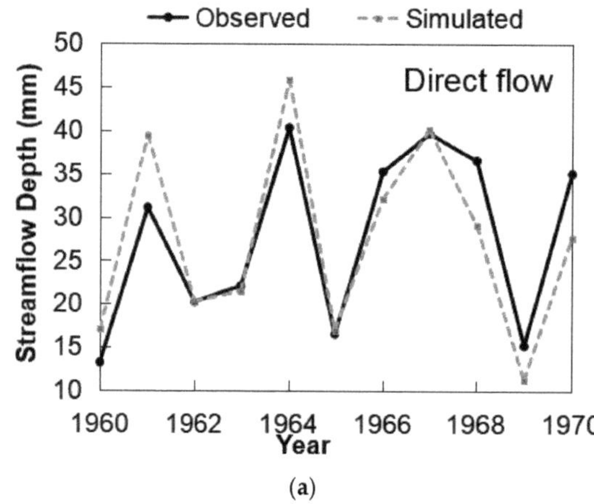

(a)

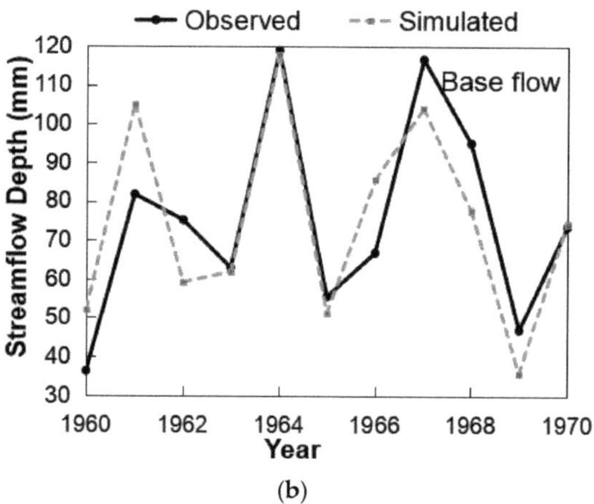

(b)

Figure 4. The comparison of modeled and observed surface runoff (**a**) and base flow (**b**) from 1960 to 1970 at Linjiacun.

Based on the high accuracy of the model estimates, the surface runoff and base flow in the post-change period were reconstructed using pre-change parameter estimates and post-change meteorological data from Linjiacun, Xianyang, and Huaxian stations. Figure 5 shows a comparison of the modeled and observed base flow and surface runoff during the entire period in the WRB. During the pre-change period, data clustered around the 1:1 lines, indicating a good agreement between the modeled and observed data. However, deviations from the 1:1 line were found during the impacted period indicating human activities on base flow. Moreover, the figure shows that human activities affected the base flow more strongly than the surface runoff. The annual precipitation and observed and modeled surface runoff and base flow are shown in Figure 6. The differences between the observed and modeled base flow became larger over time, while the differences between those of the surface runoff remained small.

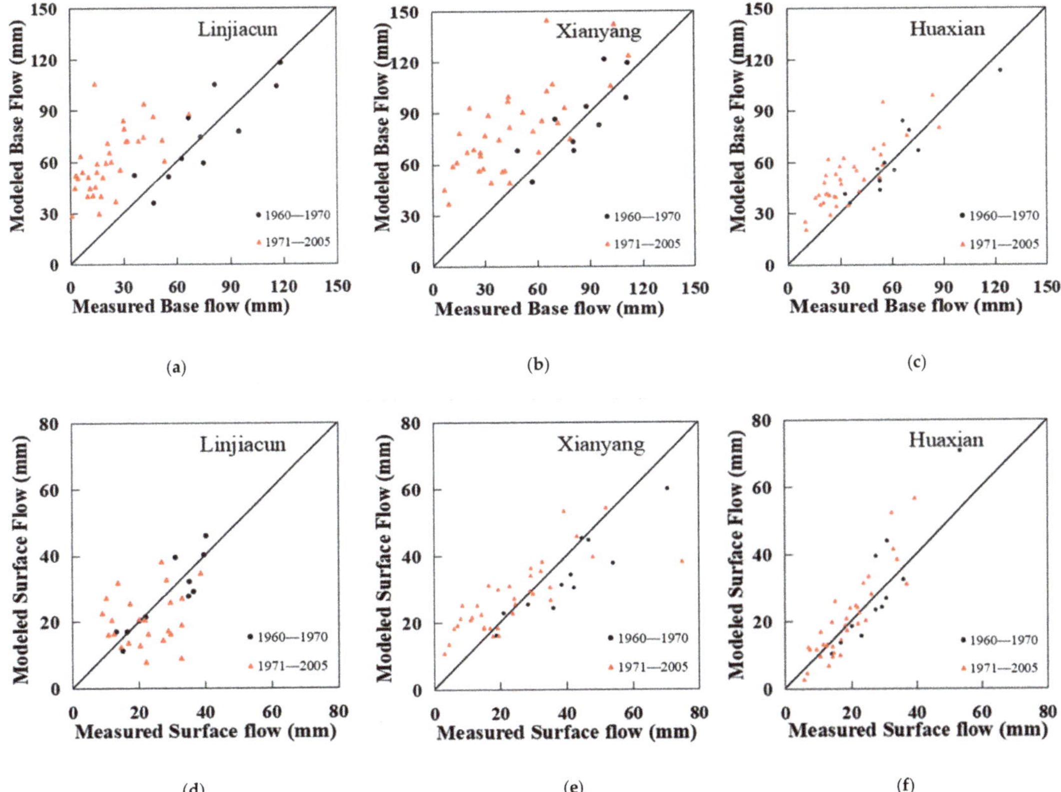

Figure 5. The comparison of the surface runoff and base flow were reconstructed using the parameter estimation from the pre-change period and the meteorological data for the post-change period at Linjiacun, Xianyang, and Huaxian stations across the WRB. The black lines are a 1:1 line. (**a**). measured and modeled baseflow at Linjiacun station; (**b**). measured and modeled baseflow at Xianyang station; (**c**). measured and modeled baseflow at Huaxian station; (**d**). measured and modeled surface flow at Linjiacun station; (**e**). measured and modeled surface flow at Xianyang station; (**f**). measured and modeled surface flow at Huaxian station.

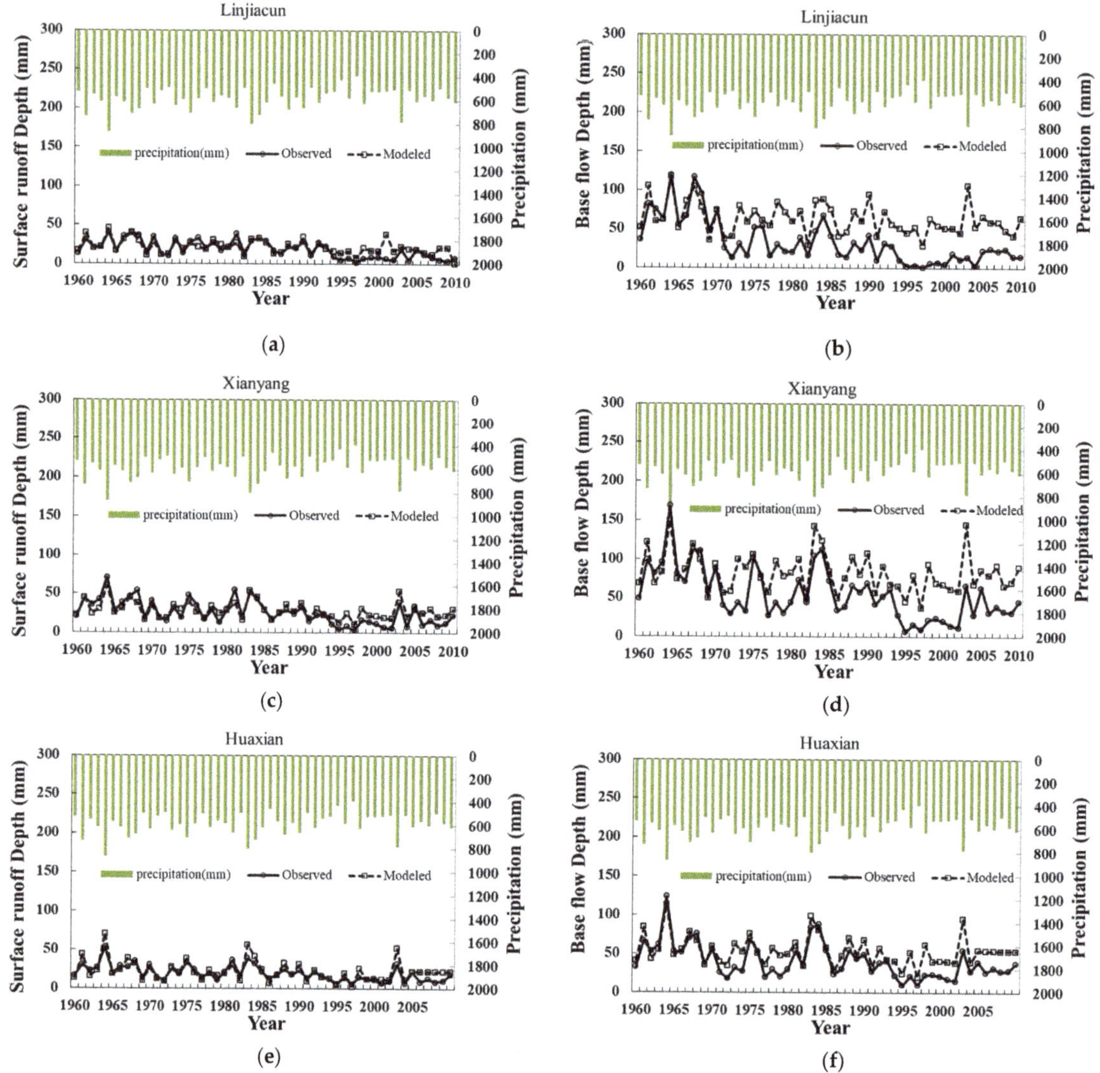

Figure 6. The observed and modeled annual surface runoff and base flow at Linjiacun, Xianyang, and Huaxian stations across the WRB. (**a**). observed and modeled surface flow at Linjiacun station; (**b**). observed and modeled baseflow at Linjiacun station; (**c**). observed and modeled baseflow at xianyang station; (**d**). observed and modeled baseflow at Xianyang station; (**e**). observed and modeled surface flow at Huaxian station; (**f**). observed and modeled baseflow at Huaxian station.

3.4. Attribution Analysis

Based on the reconstructed surface runoff and base flow data (Figure 6), the reconstructed total stream flow was generated by summing the reconstructed surface runoff and the base flow. The gap between the reconstructed total stream flow/surface runoff/base flow series in the post-change period and the observed ones in the pre-change period could indicate the change caused by climate change. The gap between the reconstructed total stream flow/surface runoff/base flow series and the observed ones in the post-change period could indicate the change caused by human activities. Climate change and human activities have different impacts on total runoff, surface runoff, and baseflow changes.

Table 4 shows the effects of climate change and human activities on the total flow changes at the three WRB sites. As the Table 4 shown, during the post-change period, under the two effects, the total stream flow of the Linjiacun, Xianyang, and Huaxian stations decreased significantly, with the values of 61.55 mm, 62.50 mm and 35.92 mm, respectively. The two effects on total stream flow were mainly attributed to human activities at Linjiacuan and Xianyang stations, accounting for 64% and 60%, respectively, while the effect at Huaxian station was 49%. The effects of climate change and human activities on stream flow in WRB have been computed in several studies, and the results of this study are consistent with those of Zuo [32], Zhan [33], and Jiang [34]. Therefore, in the change process of total stream flow, except for the variation at Huaxian Station that was greatly affected by climate change, the other two stations were heavily influenced by human activities.

Table 4. Effects of climate change and human impacts on the annual total streamflow at the Linjiacun, Xianyang, and Huaxian stations.

Site	Period	Total (mm)	Reconstructed (mm)	Total Change (mm)	Human		Climate	
					Values	%	Values	%
Linjiacun	1960–1970	103.17	102.38		39.63	64	21.92	36
	1971–2005	41.63	81.26	61.55				
Xianyang	1960–1970	132.06	125.54		37.59	60	24.91	40
	1971–2005	69.55	107.15	62.50				
Huaxian	1960–1970	90.07	90.83		17.73	49	18.20	51
	1971–2005	54.14	71.87	35.92				

The effects of climate change and human activities on surface runoff and base flow of the three stations are summarized in Tables 5 and 6. Compared to the values of the Linjiacun, Xianyang, and Huaxian stations in the pre-change period, during the post-change period, the surface runoff decreased by 9.1 mm, 16.5 mm, and 9.6 mm, respectively, while the base flow decreased by 52.4 mm, 46.1 mm, and 26.3 mm, respectively. The surface runoff changes of the three stations are two or more times higher than the base flow change. The decrease in surface runoff in Linjiacun, Xianyang, and Huaxian was mainly affected by climate change, accounting for 76%, 77%, and 78%, respectively, while the contribution rate of human activities to base flow was 71%, 73%, and 59%, at Linjiacun, Xianyang, and Huaxian hydrological stations, respectively; so, human activities were the main driving factor of the base flow variation. Climate change controls the surface runoff change and human activities control the base flow change, and the results of this study are consistent with those of Zhang [53]. In summary, climate change greatly affects the change of surface runoff, while human activities greatly influence the change of baseflow.

Table 5. Effects of climate change and human impacts on the annual surface runoff at the Linjiacun, Xianyang, and Huaxian stations.

Site	Period	Surfacel (mm)	Reconstructed (mm)	Surface Change (mm)	Human		Climate	
					Values	%	Values	%
Linjiacun	1960–1970	27.75	27.40					
	1971–2005	18.62	20.81	9.13	2.19	24	6.94	76
Xianyang	1960–1970	40.17	33.79					
	1971–2005	23.72	27.49	16.45	3.77	23	12.68	77
Huaxian	1960–1970	28.13	28.95					
	1971–2005	18.54	20.64	9.59	2.10	22	7.49	78

Table 6. Effects of climate change and human impacts on the annual base flow at the Linjiacun, Xianyang, and Huaxian stations.

Site	Period	Base (mm)	Reconstructed (mm)	Base Change (mm)	Human Values	Human %	Climate Values	Climate %
Linjiacun	1960–1970	75.42	74.98					
	1971–2005	23.02	60.22	52.40	37.20	71	15.20	29
Xianyang	1960–1970	91.89	91.75					
	1971–2005	45.79	79.44	46.10	33.65	73	12.45	27
Huaxian	1960–1970	61.94	28.95					
	1971–2005	35.64	51.16	26.30	15.52	59	10.78	41

4. Discussion

The comparison of changes in total stream flow, surface flow, base flow, and precipitation at the Linjiacun, Xianyang, and Huaxian stations in the pre-change period and the post-change period are shown in Table 7. As seen in Table 7, compared to the pre-change period, the variables in the post-change period decreased. Although the values of total stream flow, surface runoff, base flow, and precipitation decreased, the total stream flow and base flow indicated significant change, with the percentage ranging from 40% to 61% and from 43% to 70%, respectively, while the surface runoff change was slight, with the percentage ranging from 34% to 41%. The decreased values and percentage of the total stream flow and the base flow along the main stream of the WRB displayed a decreasing change from upstream to downstream, with the highest value at Linjiacun station and the lowest one at Huaxian station. Compared to the flow change, the precipitation change was a small range, from 9% to 11%. The surface runoff change was mainly controlled by climate change.

Table 7. The comparison of changes for total stream flow, surface flow, base flow, and precipitation at the Linjiacun, Xianyang, and Huaxian stations in the pre-change period and post-change period.

Index	Linjiacun		Xianyang		Huaxian	
	Change (mm)	Percentage (%)	Change (mm)	Percentage (%)	Change (mm)	Percentage (%)
Total streamflow	63.73	61	62.50	47	35.92	40
Surface flow	9.85	35	16.45	41	9.59	34
Base flow	52.88	70	46.05	50	26.34	43
Precipitation	63.91	11	55.98	9	58.94	9

In addition, a large number of human activities have been implemented since 1970. Soil conservation practices in 1970 covered 2500 km^2; however, the area in 2006 was 33,344 km^2, increased by 30,844 km^2 [34]. It was almost 13 times as large as that in 1970. From 1970 to 2005, just in the Shaanxi province, seven hyper-irrigation areas began irrigation, specifically Baojixia and Jinghuiqu in 1972, Fengjiashan in 1974, Shibaochuan in 1975, Yangmaowan in 1978, Taoqupo in 1980, and Shitouhe in 1981 with a total area of approximately 5900 km^2. Since 1994, the population of the Shaanxi Province has increased by 3.02 million from 20.24 million to 23.26 million, almost 13%. Because of the shortage of water resources, the government has invested human and financial resources in the South-North Water Transfer Project, or WRB, from the Han River to Shaanxi Province [54]. In the Gansu province, rainwater-harvesting agriculture has made great progress, and a number of water harvesting engineering projects started in the 1980s [55]. Su et al. [56] reported rainwater collection, water and soil conservation, and reservoir management are the main factors effecting the stream flow change in the WRB. All the mentioned human activities can cause the stream flow to change directly or indirectly through affecting land use and land cover.

5. Conclusions

In this study, we quantified the effects of climate change and human activities on the total streamflow, surface runoff, and base flow from 1960 to 2005 measured at three hydrological stations: Linjiacun, Xianyang, and Huaxian station in the WRB. Using the breakpoint, the annual data were divided into two periods: the pre-change period and the post-change period. The two-stage model was calibrated and verified by using the pre-change period data, and watershed hydrology was reconstructed for the post-change period. The differences in our study indicated that the main driving factor for the decrease in surface runoff was climate change, contributing over 76%. The contribution rate of human activities to base flow was 71%, 73%, and 59%, respectively; so, human activities were the main driving factor of base flow variation. Overall, climate change and human impact have different effects on different components of the stream flow. For surface flow, climate change has greater impacts than humans; however, for base flow, human activities have greater impacts than climate change. Watershed management in the WRB must incorporate appropriate measures to mitigate the likely impacts of climate change on watershed hydrology and achieve the desired goals of ecological restoration.

This study focuses on the change of historical flow of the Weihe River Basin but does not carry out corresponding research into and prediction of future flow changes. Therefore, future research will focus on the prediction of the future flow of the river basin.

Author Contributions: Conceptualization, Z.M. and B.Z.; methodology, J.F.; software, Y.L.; validation, T.Q. and Y.C.; formal analysis, S.L.; investigation, resources, J.F.; data curation, G.L.; supervision, Y.C.; project administration, T.Q. All authors have read and agreed to the published version of the manuscript.

Funding: This research was supported by the National Key Research and Development Project (Grant No. 2017YFA0605004) , the Innovative Research Group of Heibei Natural Science Foundation, (Grant No. E2019402432, E2019402102), National Science Fund for Young Scholars (Grant No. 52009053), the National Natural Science Foundation of China (Grant No. U1802241), the open project of State Key Laboratory Base of Eco-Hydraulic Engineering in Arid Area, (Grant No. 2019KFKT-4), and Hebei Key Laboratory of Intelligent Water Conservancy in Hebei University of Engineering in Handan.

Conflicts of Interest: The authors declare no conflict of interest.

References

1. Lipczynska-Kochany, E. Effect of climate change on humic substances and associated impacts on the quality of surface water and groundwater: A review. *Total Environ.* **2018**, *640*, 1548–1565. [CrossRef] [PubMed]
2. Nistor, M.M. Climate change effect on groundwater resources in South East Europe during 21st century. *Quat. Int.* **2019**, *504*, 171–180. [CrossRef]
3. Jin, X.; Jin, Y.; Mao, X.; Zhai, J.; Fu, D. Modelling the Impact of Vegetation Change on Hydrological Processes in Bayin River Basin, Northwest China. *Water* **2021**, *13*, 2787. [CrossRef]
4. Wang, D.D.; Yu, X.X.; Jia, G.D.; Wang, H.N. Sensitivity analysis of runoff to climate, variability and land-use changes in the Haihe Basin mountainous area of north China. *Agric. Ecosyst. Environ.* **2019**, *269*, 193–203. [CrossRef]
5. Wang, X.B.; He, K.N.; Dong, Z. Effects of climate change and human activities on runoff in the Beichuan River Basin in the northeastern Tibetan Plateau, China. *Catena* **2019**, *176*, 81–93. [CrossRef]
6. Zheng, Y.T.; Huang, Y.F.; Zhou, S.; Wang, K.Y. Effect partition of climate and catchment changes on runoff variation at the headwater region of the Yellow River based on the Budyko complementary relationship. *Total Environ.* **2018**, *643*, 1166–1177. [CrossRef]
7. Fang, Q.Q.; Wang, G.Q.; Liu, T.X.; Xue, B.L.; Yinglan, A. Controls of carbon flux in a semiarid grassland ecosystem experiencing wetland loss: Vegetation patterns and environmental variables. *Agric. For. Meteorol.* **2018**, *259*, 196–210. [CrossRef]
8. Fang, Q.Q.; Wang, G.Q.; Xue, B.L.; Liu, T.X.; Kiem, A. How and to what extent doesprecipitation on multi-temporal scales and soil moisture at different depths determine carbon flux responses in a water-limited grassland ecosystem? *Sci. Total Environ.* **2018**, *635*, 1255–1266. [CrossRef]
9. Shang, X.X.; Jiang, X.H.; Jia, R.N.; Wei, C. Land use and climate change effects on surface runoff variations in the upper Heihe River basin. *Water* **2019**, *11*, 344. [CrossRef]
10. Vercruysse, K.; Grabowski, R.C. Human impact on river planform within the context of multi-timescale river channel dynamics in a Himalayan river system. *Geomorphology* **2021**, *381*. [CrossRef]

11. Su, C.J.; Chen, X.H. Assessing the effects of reservoirs on extreme flows using nonstationary flood frequency models with the modified reservoir index as a covariate. *Adv. Water Resour.* **2019**, *124*, 29–40. [CrossRef]
12. Wiekenkamp, I.; Huisman, J.A.; Bogena, H.R.; Vereecken, H. Effects of Deforestation on Water Flow in the Vadose Zone. *Water* **2020**, *12*, 35. [CrossRef]
13. Cecilio, R.A.; Pimentel, S.M.; Zanetti, S.S. Modeling the influence of forest cover on streamflows by different approaches. *Catena* **2019**, *178*, 49–58. [CrossRef]
14. Hardie, S.A.; Bobbi, C.J. Compounding Effects of Agricultural Land Use and Water Use in Free-Flowing Rivers: Confounding Issues for Environmental Flows. *Environ. Manag.* **2018**, *61*, 421–431. [CrossRef]
15. Ye, X.; Zhang, Q.; Liu, J.; Li, X.; Xu, C. Distinguishing the relative impacts of climate change and human activities on variation of streamflow in the Poyang Lake catchment, China. *J. Hydrol.* **2013**, *494*, 83–95. [CrossRef]
16. Zolfagharpour, F.; Saghafian, B.; Delavar, M. The impacts of climate variability and human activities on streamflow change at basin scale. *Water Supply* **2020**, *20*, 889–899. [CrossRef]
17. Kim, J.; Choi, J.; Choi, C.; Park, S. Impacts of changes in climate and land use/land cover under IPCC RCP scenarios on streamflow in the Hoeya River Basin, Korea. *J. Hydrol.* **2013**, *452*, 181–195. [CrossRef]
18. Lee, C.H.; Yeh, H.F. Impact of Climate Change and Human Activities on Streamflow Variations Based on the Budyko Framework. *Water* **2019**, *11*, 2001. [CrossRef]
19. Chien, H.; Yeh, J.P.; Knouft, H.J. Modeling the potential impacts of climate change on streamflow in agricultural watersheds of the Midwestern United States. *J. Hydrol.* **2013**, *491*, 73–88. [CrossRef]
20. Deng, X.Y.; Long, A.H.; Ling, H.B.; Deng, M.J.; Zhang, S.P. Effects of Climate Change and Human Activities on Runoff in the Headstream Areas of Tarim River Basin. *J. Environ.* **2015**, *3*, 31–45. [CrossRef]
21. Cheng, J.X.; Xu, L.G.; Fan, H.X.; Jiang, J.H. Changes in the flow regimes associated with climate change and human activities in the Yangtze River. *River Res. Appl.* **2019**, *35*, 1415–1427. [CrossRef]
22. Ma, J.Z.; Chen, L.H.; Zhang, Y.R.; Li, X.H.; Edmunds, W.M. Trends and periodicities in observed temperature, precipitation and runoff in a desert catchment: Case study for the Shiyang River Basin in Northwestern China. *Water Environ. J.* **2013**, *27*, 86–98. [CrossRef]
23. Githui, F.; Gitau, W.; Mutuab, F.; Bauwensa, W. Climate change impact on SWAT simulated streamflow in western Kenya. *Int. J. Climatol.* **2009**, *29*, 1823–1834. [CrossRef]
24. Ghaderpour, E.; Vujadinovic, T.; Hassan, Q.K. Application of the Least-Squares Wavelet software in hydrology: Athabasca River Basin. *J. Hydrol. Reg. Stud.* **2021**, *36*, 100847. [CrossRef]
25. Bao, Z.; Zhang, J.; Wang, G.; Fu, G.; He, R.; Yan, X.; Jin, J.; Liu, Y.; Zhang, A. Attribution for decreasing streamflow of the Haihe River basin, northern China: Climate variability or human activities? *J. Hydrol.* **2012**, *460*, 117–129. [CrossRef]
26. Ma, H.; Yang, D.; Tan, S.K.; Gao, B.; Hua, Q. Impact of climate variability and human activity on streamflow decrease in the Miyun Reservoir catchment. *J. Hydrol.* **2010**, *389*, 317–324. [CrossRef]
27. Zheng, J.K.; Sun, G.; Li, W.H.; Yu, X.X.; Zhang, C.; Gong, Y.B.; Tu, L.H. Impacts of land use change and climate variations on annual inflow into Miyun Reservoir, Beijing, China. *Hydrol. Earth Syst. Sci. Discuss.* **2016**, *20*, 1561–1572. [CrossRef]
28. Wang, D.; Hejazi, M. Quantifying the relative contribution of the climate and direct human impacts on mean annual streamflow in the contiguous United States. *Water Resour. Res.* **2011**, *47*, 16. [CrossRef]
29. Dalea, J.; Zou, C.B.; Andrews, W.J.; Long, J.M.; Liang, Y.; Qiao, L. Climate, water use, and land surface transformation in an irrigation intensive watershed—Streamflow responses from 1950 through 2010. *Agric. Water Manag.* **2015**, *160*, 144–152. [CrossRef]
30. Wang, H.; Chen, L.; Yu, X. Distinguishing human and climate influences on streamflow changes in Luan River basin in China. *Catena* **2016**, *136*, 182–188. [CrossRef]
31. Yang, S.; Kang, T.T.; Bu, J.Y.; Chen, J.H.; Wang, Z.P.; Gao, Y.C. Detection and Attribution of Runoff Reduction of Weihe River over Different Periods during 1961–2016. *Water* **2020**, *12*, 1416. [CrossRef]
32. Li, S.Y.; Yang, G.Y.; Wang, H. The Runoff Evolution and the Differences Analysis of the Causes of Runoff Change in Different Regions: A Case of the Weihe River Basin, Northern China. *Sustainability* **2019**, *11*, 5295. [CrossRef]
33. Zuo, D.; Xu, Z.; Wu, W.; Zhao, J.; Zhao, F. Identification of Streamflow Response to Climate Change and Human Activities in the Wei River Basin, China. *Water Resour. Manag.* **2014**, *28*, 833–851. [CrossRef]
34. Zhan, C.; Jiang, S.; Sun, F.; Jia, Y.; Niu, C.; Yue, W. Quantitative contribution of climate change and human activities to runoff changes in the Wei River basin, China. *Hydrol. Earth Syst. Sci.* **2014**, *18*, 3069–3077. [CrossRef]
35. Jiang, C.; Xiong, L.; Wang, D.; Liu, P.; Guo, S.; Xu, C. Separating the impacts of climate change and human activities on runoff using the Budyko-type equations with time-varying parameters. *J. Hydrol.* **2015**, *522*, 326–338. [CrossRef]
36. Chang, J.X.; Wang, Y.M.; Istanbulluoglu, E.; Bai, T.; Huang, Q.; Yang, D.W.; Huang, S.Z. Impact of climate change and human activities on runoff in the Weihe River basin, China. *Quat. Int.* **2015**, *380*, 169–179. [CrossRef]
37. Monteith, J.L. Solar radiation and productivity in tropical ecosystem. *J. Appl. Ecol.* **1972**, *9*, 747–766. [CrossRef]
38. Nathan, R.J.; McMahon, T.A. Evaluation of automated techniques for base flow and recession analyses. *Water Resour.* **1990**, *26*, 1465–1473. [CrossRef]
39. Zhao, W.; Li, Z.L.; Wang, Y.H. Base flow segmentation and its variation in the upper reaches of Heihe River Basin. *South-to-North Water Transf. Water Sci. Technol.* **2016**, *14*, 26–31. (In Chinese)

40. Cui, Y.H. Study on the Response of Mine Water Inflow to Precipitation in Pingdingshan. No.7 mine. Master's Thesis, North China University of Science and Technology, Qinhuangdao, China, 2018; p. 78. (In Chinese)
41. Dong, X. Trend change analysis of precipitation and runoff in Hejiang River Basin based on Mann-Kendall test. *Guangdong Water Resour. Hydropower* **2021**, *4*, 73–78. (In Chinese)
42. Chen, Z.R. Application Research of Mann-Kendall Test Method in the Analysis of Eutrophication Trend of Xinxihe Reservoir, Anhui Agricultural Science Bulletin. *Anhui Agric. Sci. Bull.* **2019**, *25*, 99–100+145. (In Chinese)
43. Mann, B. Non-parametric test against trend. *Econometricn* **1945**, *13*, 245–259. [CrossRef]
44. L'vovich, M.I.; Nace, R.L., Translators; *World Water Resources and Their Future*; Wiley: Hoboken, NJ, USA, 1979; p. 415.
45. US Department of Agriculture Soil Conservation Service (USDA SCS). *National Engineering Handbook*; Section 4: Hydrology; SCS: Washington, DC, USA, 1985.
46. Ponce, V.M.; Shetty, A.V. A conceptual model of catchment water balance. 1. Formulation and calibration. *J. Hydrol.* **1995**, *173*, 27–40. [CrossRef]
47. Chen, X.; Wang, D. Modeling seasonal surface runoff and base flow based on the generalized proportionality hypothesis. *J. Hydrol.* **2015**, *527*, 367–379. [CrossRef]
48. Nash, J.E.; Sutcliffe, J.V. River forcasting using conceptual models. 1: A discussion of principles. *J. Hydrol.* **1970**, *10*, 280–290. [CrossRef]
49. Moriasi, D.N.; Arnold, J.G.; Liew, M.V.; Bingner, R.L.; Harmel, R.D.; Veith, T.L. Model evaluation guidelines for systematic quantification of accuracy in watershed simulations. *Trans. ASABE* **2007**, *50*, 885–900. [CrossRef]
50. Wang, G.; Zhang, J.; He, R. Impacts of environmental change on runoff in Fenhe river basin of the middle Yellow River. *Adv. Water Sci.* **2006**, *17*, 853–858. (In Chinese)
51. Zhang, A.; Zhang, C.; Fu, G.; Wang, B.; Bao, Z.; Zheng, H. Assessments of impacts of climate change and human activities on runoff with SWAT for the Huifa River Basin, Northeast China. *Water Resour. Manag.* **2012**, *26*, 2199–2217. [CrossRef]
52. Rawshan, A.; Alban, K.; Shadan, A.; Shadan, A.; Ozgur, K. Long-Term Trends and Seasonality Detection of the Observed Flow in Yangtze River Using Mann-Kendall and Sen's Innovative Trend Method. *Water* **2019**, *9*, 1855. [CrossRef]
53. Zhang, Y. *Study on the Impact of Baseflow on Watershed Water Balance*; Northwest A&F University: Yangling, China, 2014. (In Chinese)
54. Chang, J.X.; Jiang, J. Water dispatch of the south to north water transfer project in Shaanxi province. *J. Nat. Resour.* **2011**, *26.1*, 110–118. (In Chinese)
55. Xiao, G.; Wang, J. Research on progress of rainwater harvesting agriculture on the Loess Plateau of China. *Acta Ecol. Sin.* **2003**, *23*, 1003–1011. (In Chinese)
56. Su, X.L.; Kang, S.Z.; Wei, X.M.; Xing, D.W.; Cao, H.X. Impact of climate change and human activity on the runoff of Wei River basin to the Yellow River. *J. Northwest Sci.-Tech. Univ. Agric. For.* **2007**, *35*, 153–159. (In Chinese)

 water MDPI

Article

Hydrological Modeling in the Chaohu Lake Basin of China—Driven by Open-Access Gridded Meteorological and Remote Sensing Precipitation Products

Junli Liu [1], Yun Zhang [2], Lei Yang [3] and Yuying Li [2,*]

1 Hangzhou Institute of Technology, Xidian University, Hangzhou 311200, China; liujunli@xidian.edu.cn
2 International Joint Laboratory for Watershed Ecological Security and Collaborative Innovation Center of Water Security for Water Source Region of Middle Route Project of South-North Water Diversion in Henan Province, College of Water Resource and Environment Engineering, Nanyang Normal University, Nanyang 473061, China; zy174812@163.com
3 Provincial Geomatics Centre of Jiangsu, Nanjing 210013, China; yangleimyself@163.com
* Correspondence: lyying200508@163.com

Abstract: This study assessed the performance of two well-known gridded meteorological datasets, CFSR (Climate Forecast System Reanalysis) and CMADS (China Meteorological Assimilation Driving Datasets), and three satellite-based precipitation datasets, TRMM (Tropical Rainfall Measuring Mission), CMORPH (Climate Prediction Center morphing technique), and CHIRPS (Climate Hazards Group InfraRed Precipitation with Station data), in driving the SWAT (Soil and Water Assessment Tool) model for streamflow simulation in the Fengle watershed in the middle–lower Yangtze Plain, China. Eighteen model scenarios were generated by forcing the SWAT model with different combinations of three meteorological datasets and six precipitation datasets. Our results showed that (1) the three satellite-based precipitation datasets (i.e., TRMM, CMORPH, and CHIRPS) generally provided more accurate precipitation estimates than CFSR and CMADS. CFSR and CMADS agreed fairly well with the gauged measurements in maximum temperature, minimum temperature, and relative humidity, but large discrepancies existed for the solar radiation and wind speed. (2) The impact of precipitation data on simulated streamflow was much larger than that of other meteorological variables. Satisfactory simulations were achieved using the CMORPH precipitation data for daily streamflow simulation and the TRMM and CHIRPS precipitation data for monthly streamflow simulation. This suggests that different precipitation datasets can be used for optimal simulations at different temporal scales.

Keywords: hydrological modelling; evaluation; satellite rainfall; climatic variables; simulation

Citation: Liu, J.; Zhang, Y.; Yang, L.; Li, Y. Hydrological Modeling in the Chaohu Lake Basin of China—Driven by Open-Access Gridded Meteorological and Remote Sensing Precipitation Products. *Water* **2022**, *14*, 1406. https://doi.org/10.3390/w14091406

Academic Editor: Renato Morbidelli

Received: 26 March 2022
Accepted: 26 April 2022
Published: 28 April 2022

1. Introduction

Meteorological data such as air temperature and precipitation are important inputs to hydrological models. With a common-knowledge "Garbage in, garbage out" approach, meteorological data of good quality are prerequisites to achieve good simulation results using hydrological models and thus further to achieve reasonable decision support based on model outputs [1,2]. The traditional and common sources of meteorological data are ground measurements from gauge stations; such point-based measurements are considered as the most accurate data over the limited representative areas. The modelers need measurements from a dense network of gauge stations to adequately characterize the spatial and temporal variability of meteorological variables at the basin scale [3]. The in situ data collection and maintenances are usually time-consuming and labor-/resources-intensive, the gauge stations are unevenly distributed, and overall the number of stations is declining at the global scale [1]. As a result, modelers often encounter the challenge to obtain sufficient in situ measurements, as they expect. In developing countries and remote areas, gauge stations are very sparse and in situ measurements are not even available over some regions. Many scientific communities have been stressing the need for more in situ measurements; one of the feasible ways to meet this need is to promote innovations and multidisciplinary cooperation in designing low-cost monitoring devices and in developing or combining monitoring techniques. In this regard, some concrete efforts are underway, for example, the working group Measurements and Observations in the XXI century (MOXXI) was established in 2013 with the specific aims of targeting innovation in all realms of hydrological measurements from ground-based to remote sensing [4,5].

In recent years, various freely available gridded meteorological datasets at different spatial and temporal resolutions over the global or quasi-global scales have been developed and released to the public [6–11]. For example, the Climate Forecast System Reanalysis (CFSR) dataset is such a global meteorological dataset covering the 39-year period from 1979 to 2017. The CFSR data were produced by a global, high resolution, coupled atmosphere–ocean–land surface–sea ice system [10]. The meteorological variables include precipitation, air temperature, wind speed, relative humidity, and solar radiation. The China Meteorological Assimilation Driving Dataset (CMADS) is a country-scale gridded meteorological dataset containing the same types of data as CFSR, which used more measurement data in China and integrated the Climate Prediction Center (CPC) morphing technique (CMORPH) satellite-based rainfall product [12]. Several agencies have preprocessed CFSR and CMADS products to generate the datasets in the desired input format of the widely used hydrological model, i.e., the Soil and Water Assessment Tool (SWAT) model [13]. This makes these data sets very convenient to use for the modelling community [14–17]. There are also many gridded precipitation datasets such as the TRMM (Tropical Rainfall Measuring Mission) multi-satellite precipitation analysis (TMPA) product [18] and CHIRPS (Climate Hazards Group InfraRed Precipitation with Station data) precipitation dataset [19]. In recent years, with the rapid development of machine learning and especially deep learning techniques [20], the accuracy of gridded meteorological and precipitation datasets is expected to be improved dramatically in the near future [21].

The increasing availability of gridded meteorological datasets has attracted attention to use them in driving hydrological models for streamflow simulation [22–25]. As the accuracy of these gridded meteorological datasets varies among regions [2,26–30], it is necessary to evaluate these datasets before their application in specific areas. In this regard, many evaluation studies have been conducted to assess the performance of open-access weather data in hydrological simulations by using the best available gauge data as a reference [31,32]. For CFSR, it was found to be able to drive the hydrological model to yield satisfactory streamflow simulation in Lake Tana Basin (the upper part of the Upper Blue Nile basin) [16], the Bahe River Basin of the Qinling Mountains, China [33], four small watersheds in the USA, and the Gumera watershed in Ethiopia [19]. However, unsatisfactory results of streamflow simulation using CFSR as forcing data were also reported in the upstream watersheds of Three Gorges Reservoir in China [34] and two watersheds in the USA [28]. For CMADS, most evaluation studies using it as forcing data showed satisfactory results, such as those conducted in the Yellow River Source Basin located in the Qinghai–Tibet Plateau [17], the Lijiang watershed in South China [14], and the Jing and Bortala River Basin in Northwest China [35]. It is well recognized that a certain product's performance would vary from region to region, and evaluation of a certain product in various environments is essential for understanding its global performance [7].

This study focuses on the Chaohu Lake basin in the middle–lower Yangtze Plain, China. This region has been facing serious water pollution problems due to non-point-source pollution caused by intense agricultural activities (e.g., pesticide and fertilizer use). [36]. Watershed simulation and scenario analysis are expected to provide valuable instructions for water quality control and water resources management. As water is an important medium for mass transport, adequate modeling of hydrological processes is a prerequisite to characterizing the nutrient migration processes at the watershed scale [37]. Reliable meteorological input data are the premise of the hydrological model setup. Considering that measurements from meteorological and rainfall stations are usually hard to retrieve for many reasons (e.g., data-sharing policy), it is necessary to find out whether open-access gridded meteorological data can meet the requirements of hydrological modelling. In this study, a subbasin of the Chaohu Lake basin, where measurements from meteorological and rainfall stations are relatively complete, was selected to evaluate the performance of two mainstream, open-access, gridded meteorological datasets (i.e., CFSR and CMADS) and three popular, satellite-based, gridded precipitation datasets (i.e., TRMM, CMORPH, and CHIRPS) in driving SWAT in streamflow simulation in this region.

2. Study Area

The Chaohu Lake, located in Anhui Province, China, is the fifth largest freshwater lake in China and it is of great importance in terms of water resources and agriculture [38]. The Fengle river is a main tributary of the Chaohu Lake Basin (Figure 1). The drainage area of the Fengle watershed is 1500 km^2 with elevations ranging from 7 to 455 m above mean sea level, and the main stream length is about 50 km. The land use types include agricultural lands (about 45%), forests (39%), built-up lands (10%), and water areas (i.e., ponds and rivers, 6%). There are no large cities or industry factories in this river basin. Based on the available gauge precipitation data during 2008–2014, the mean annual precipitation is 1096 mm/year. The inter-annual distribution of precipitation is uneven, with the most precipitation occurring in spring and summer. Based on gauged data between 2008 and 2014, the average daily maximum and minimum air temperature are 21.1 °C and 12.3 °C, respectively, and the daily mean air temperature is 16.7 °C.

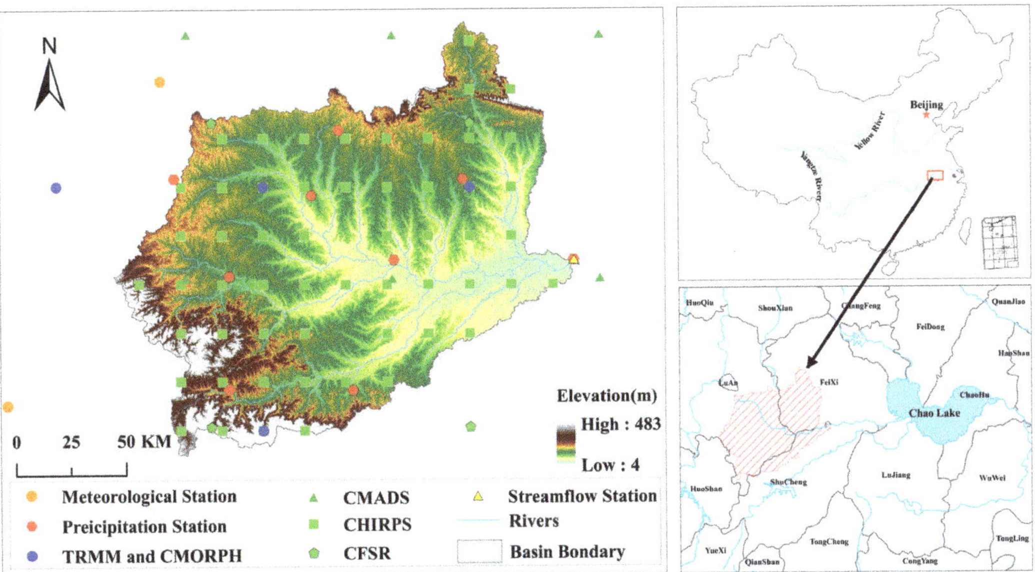

Figure 1. Locations of the Fengle river basin, gauge stations, and the center points of grid cells of the gridded meteorological and precipitation datasets.

3. Datasets and Methods

3.1. In Situ Meteorological Measurements

In situ measurements of meteorological data from three observation stations were obtained from the China Meteorological Administration. Ground measurements of precipitation from nine rain gauge stations and the measured daily streamflow from the hydrological station at the outlet of the Fengle watershed were obtained from the Hefei Bureau of hydrology and water resources (Figure 1). In terms of gridded meteorological/precipitation data, the numbers of grid cells are 8, 4, 9, 9, and 55 for CMADS, CFSR, TRMM, CMORPH, and CHIRPS, respectively. After a rigorous analysis of available data, the simulation period was set as 2008–2014.

3.2. The CFSR and CMADS Meteorological Data

CFSR is the third-generation reanalysis product of the National Center for Environmental Prediction (NCEP) and was derived from a global coupled atmosphere–ocean–land surface–sea ice system [10]. The system provides a range of atmospheric, oceanic, and land surface output products around the world at hourly time resolution. The spatial resolution of CFSR global atmospheric products is ~38 km, with 64 levels extending from the surface to 0.26 hPa. The CFSR data covers the period from 1979 to the present with continuous updates. It is popularized by the SWAT official website that provides ready-to-use weather data in the desired format at the data portal http://globalweather.tamu.edu (accessed on 16 May 2020).

The CMADS meteorological dataset was constructed based on nearly 40,000 regional automatic stations and the CMORPH global precipitation products [39]. These solid data sources make CMADS have wide applicability in China. A variety of methods, such as loop nesting of data, projection of resampling models, and bilinear interpolation, were used. The CMADS provides daily data for a 9-year period from 2008 to 2016 with a spatial resolution of 0.25° for version 1.1, which can be freely downloaded at http://westdc.westgis.ac.cn (accessed on 16 May 2020). CMADS version 1.1 was used in this study. The locations of the center points of CFSR and CMADS grid cells are shown in Figure 1.

3.3. The TRMM, CMORPH, and CHIRPS Precipitation Datasets

TRMM, CMORPH, and CHIRPS are three quasi-global gridded precipitation datasets. The TRMM TMPA products provide precipitation for the spatial coverage of 50° N–S from 1998 to the present with a spatial resolution of 0.25° × 0.25° [18]. In this study, the TRMM 3B42 product was used. The original temporal resolution of this dataset is 3 h, and the daily aggregated TRMM 3B42 product can be obtained from the NASA Goddard Earth Sciences (GES) Data and Information Services Center (DISC) (https://disc.gsfc.nasa.gov, accessed on 20 May 2020). The CMORPH products provide precipitation for the spatial coverage of 60° N–S from 1998 to the present [40]. The lasted version 1.0 includes three different products, including raw (satellite-only precipitation estimates), bias-corrected (CRT), and gauge-satellite blended (BLD). In this study, the CMORPH BLD product with a spatial resolution of 0.25° × 0.25° was used, which can be downloaded from https://ftp.cpc.ncep.noaa.gov/precip/CMORPH_V1.0/BLD (accessed on 20 May 2020). The CHIRPS product provides precipitation for the spatial coverage of 50° N–S with a high spatial resolution (i.e., 0.05° × 0.05°) from 1981 to the present [19]. The CHIRPS version 2.0 data were downloaded from https://data.chc.ucsb.edu/products/CHIRPS-2.0/global_daily (accessed on 20 May 2020) and used in this study.

3.4. SWAT Modelling Procedures

Developed by the Agricultural Research Service of the United States Department of Agriculture, Agricultural Research Service [41], the Soil and Water Assessment Tool (SWAT) is a semi-distributed, process-based, and time-continuous watershed model. SWAT is capable of modeling hydrological processes, soil erosion, and water quality at basin scales [42]. In SWAT, the river basin is first divided into subbasins, then further to the hydrologic response units (HRUs) that represent a unique combination of soil type, land use, and slope. More details on the description of SWAT can be found in many sources [25,42,43], and thus this description was not repeated here for conciseness. As an easy-to-use toolbar in the QGIS interface, QSWAT (Version 2017.02_1.4, Texas A&M University, College Station, TX, USA) was used to set up the SWAT model in this study.

To set up the SWAT model (Figure 2), the following data sources were used: (1) the elevation data was obtained from the Digital Elevation Model (DEM) product of ALOS

World 3D–30 m (AW3D30) which was released by the Japan Aerospace Exploration Agency (JAXA) with a horizontal spatial resolution of approximately 30 m. These data were downloaded from http://www.eorc.jaxa.jp/ALOS/en/aw3d30/ (accessed on 6 September 2019). (2) The land use map in 2010, with a scale of 1:100,000, was obtained from the National Earth System Science Data Sharing Platform of China (http://www.geodata.cn, accessed on 6 September 2019). (3) The soil map, with a scale of 1:500,000, was obtained from China Soil Database (CSDB, http://vdb3.soil.csdb.cn/, accessed on 6 September 2019). This study used eighteen scenarios that are generated through the combination of three meteorological (excluding precipitation) datasets (Gauge, CFSR, and CMADS) and six precipitation datasets (Gauge, CFSR, CMADS, TRMM, CMORPH, and CHIRPS), to study the impact of different input data on the streamflow simulation.

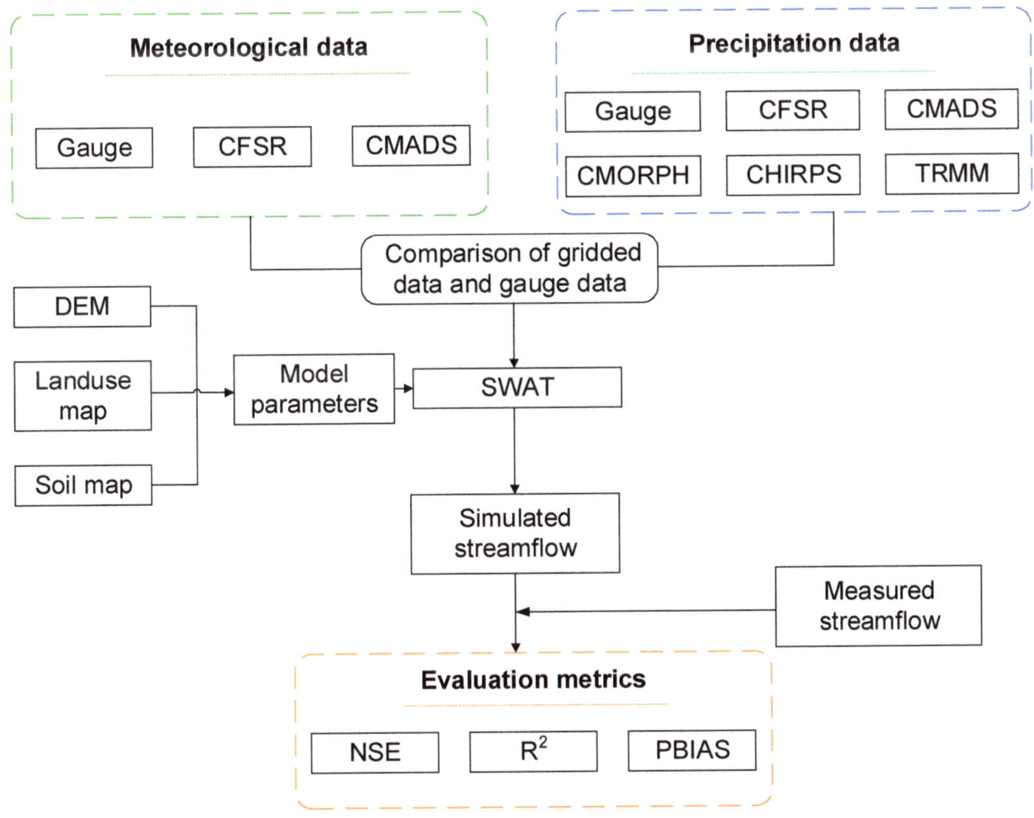

Figure 2. Flowchart of streamflow simulation driven by open-access gridded meteorological and remote sensing precipitation products.

The SWAT model was run at the daily time scales for the period 1 January 2008–31 July 2014. The first year (2008) was used as a warm-up period to alleviate the impact of initial hydrological simulation conditions. From 1 January 2009 to 31 December 2011, as a calibration period, sensitivity parameters were identified and further calibrated to make the simulated streamflow fit the observation as close as possible. The period 1 January 2012–31 July 2014 was used as the independent validation period to test the validity of calibrated parameters. In this study, the sequential uncertainty fitting algorithm version 2 (SUFI-2) in the SWAT-CUP tool [44,45] was adopted to perform sensitivity analysis and automatic calibration. Finally, 12 highly sensitive parameters for model calibration were selected, as presented in Table 1. It should be noted that antecedent moisture conditions have considerable impacts on runoff generation, and more runoff will be generated under wetter conditions [46]. Even for forest ecosystems, finite amounts of precipitation can be retained during extreme rainfall events [47–49]. The SCS curve number (CN) in Table 1 is for average soil moisture conditions (CN2), the SWAT model updates CN values on each day according to the current soil moisture levels, and whether the ground is frozen. The CN value is the highest (referred to as CN3) when the soil is at field capacity, and the lowest (referred to as CN1) when the soil is at wilting point. Both CN1 and CN3 are functions of CN2, and the details can be found in the theoretical documentation of SWAT [42].

Table 1. List of parameters used for calibration and their default values and ranges for calibration ("a__", "v__", and "r__" mean an absolute increase, a replacement, and a relative change from the initial parameter values, respectively).

Parameters	Description	Default	Range
v_ALPHA_BF.gw	Baseflow alpha factor (1/days)	0.048	0–1
v_GW_DELAY.gw	Groundwater delay [days]	31	0–500
v_GW_REVAP.gw	Groundwater "revap" coefficient	0.02	0.02–0.2
v_ALPHA_BNK.rte	Baseflow alpha factor for bank storage (days)	0	0–1
v_CH_K2.rte	Effective hydraulic conductivity [mm/hr]	0	5–130
v_CH_N2.rte	Manning's "n" value for the main channel	0.014	0–0.3
r_SOL_AWC.sol	Available water capacity of the soil layer [mm H_2O/mm soil]	Soil layer specific	±60%
r_SOL_BD.sol	Moist bulk density (Mg/m^3 or g/cm^3)	Soil layer specific	±60%
r_SOL_K.sol	Saturated hydraulic conductivity (mm/hr)	Soil layer specific	±60%
r_CN2.mgt	SCS runoff curve number	HRU specific	−30–10%
v_SFTMP.bsn	Snowfall temperature (°C)	1	−5–5
r_SLSUBBSN.hru	Average slope length (m)	HRU specific	0–20%

Two iterations were carried out in the process of calibration, with 1000 simulations in each iteration (a total of 2000 simulations were carried out in the calibration period) by using the objective function of the Nash–Sutcliffe efficiency coefficient (NSE) [50]. After each iteration, the parameter ranges were updated based on the new parameter ranges recommended by SWAT-CUP and the physical boundaries of the parameters. The best model result among the 2000 simulations was selected as the final model output to evaluate and compare different models' performance. Three indicators, including NSE, the coefficient of determination (R^2), and the percent bias (PBIAS), were used to evaluate and compare the model outputs. This study referred to the widely used guideline [51] to classify the model performance in terms of NSE; models would be classified as unsatisfactory (NSE $\leq$ 0.50), satisfactory (0.50 < NSE = 0.65), good (0.65 < NSE = 0.75), and very good (NSE > 0.75). PBIAS measures the average tendency of the simulated data to be larger or smaller than their observed counterparts. PBIAS is ideally zero, with positive values indicating model overestimation and negative values indicate model underestimation. The low-magnitude PBIAS values indicate accurate model simulations [52].

4. Results

4.1. Comparison of Different Meteorological Inputs

Figure 3 illustrates the cumulative fraction of daily precipitation at the watershed scale from six sources (i.e., Gauge, CFSR, CMADS, TRMM, CMORPH, and CHIRPS) during the period 2009–2014. The probability of dry day for Gauge was 54%, and the values were much lower (i.e., 32%) for CMADS and much higher for CHIRPS (i.e., 72%). The other three datasets had similar values to that of Gauge (i.e., 45% for CFSR, 51% for TRMM, and 48% for CMORPH). For precipitation intensity within 1–5 mm/day, CFSR had obviously lower frequencies than Gauge, while CHIRPS had obviously higher frequencies. CMADS had higher frequencies of precipitation within 5–20 mm/day. The average annual precipitation for Gauge, CFSR, CMADS, TRMM, CMORPH, and CHIRPS were 1178, 1230, 966, 1260, 1208, and 1238 mm/year during the entire period 2009–2014.

Figure 4 shows the monthly precipitation of six datasets averaged over the watershed from 2009 to 2014. All six datasets displayed similar rainy seasons centered from June to September. CMADS had generally less monthly precipitation than Gauge, and there were considerably higher peaks in several months in the CFSR precipitation, for instance, August in 2009 and 2012. The three satellite-based precipitation datasets (i.e., TRMM, CMORPH, and CHIRPS) were generally close to gauge measurements, except for some lower peaks, such as in July and September of 2010.

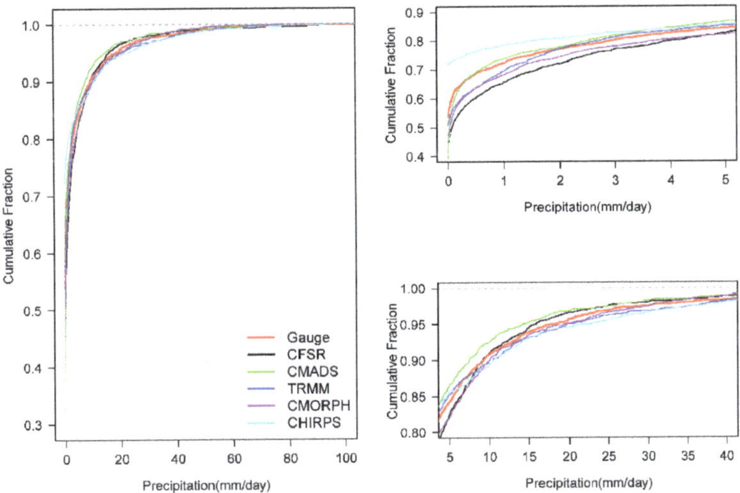

Figure 3. The cumulative fraction of daily precipitation from six datasets (Gauge, CFSR, CMADS, TRMM, CMORPH, and CHIRPS) at the watershed scale during 2009–2014. The left subfigure shows the cumulative fraction for daily precipitation ranging from 0 to 100 mm/day; in order to show more details, the two subfigures in the right show the cumulative fraction for daily precipitation ranging from 0 to 5 mm/day and 5 to 40 mm/day, respectively.

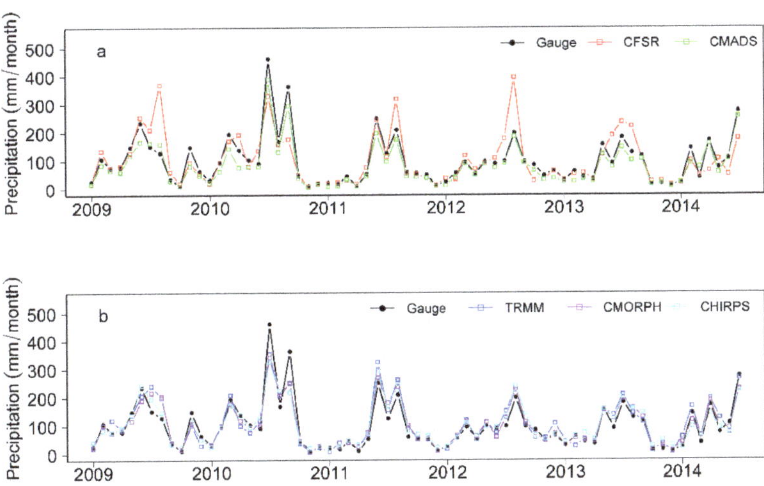

Figure 4. Comparison of monthly precipitation totals from six datasets at the watershed scale during 2009–2014, including (**a**) Gauge CFSR, and CMADS, (**b**) Gauge, TRMM, CMORPH, and CHIRPS.

Figure 5 shows the cumulative fraction of daily maximum and minimum air temperature, solar radiation, wind speed, relative humidity, and potential evapotranspiration (PET) over the basin from three datasets between 2009 and 2014. Figure 6 displays the monthly mean of daily maximum and minimum air temperature, solar radiation, wind speed, relative humidity, and the sum of PET over the entire period. Taking an overall look at Figures 5 and 6, despite some fluctuations, three datasets had a fairly good agreement in the maximum temperature, minimum temperature, and relative humidity. However, large discrepancies among the three datasets can be clearly seen for the solar radiation and wind speed. For solar radiation, the two gridded products had larger fluctuations than Gauge. For wind speed, CMADS was considerably lower than the Gauge and CFSR datasets which display very good agreement with each other. To compute the PET in SWAT, the Penman–Monteith method was used, which requires air temperature, solar radiation, relative humidity, and wind speed as inputs. Despite these large discrepancies in daily solar radiation and wind speed, overall monthly PET totals from the three datasets were in good agreement, as shown in Figures 5 and 6. This suggests the discrepancies in individual weather variables (wind speed and solar radiation in particular in this case) cancelled each other to a certain degree when integrated into a further output, the PET in this case. Since PET would affect the computation of water balance in SWAT, considering good agreement in monthly PET totals from all the three datasets, it was expected that solar radiation and wind speed inputs from three datasets had little influence on the SWAT modelling results in this studied basin.

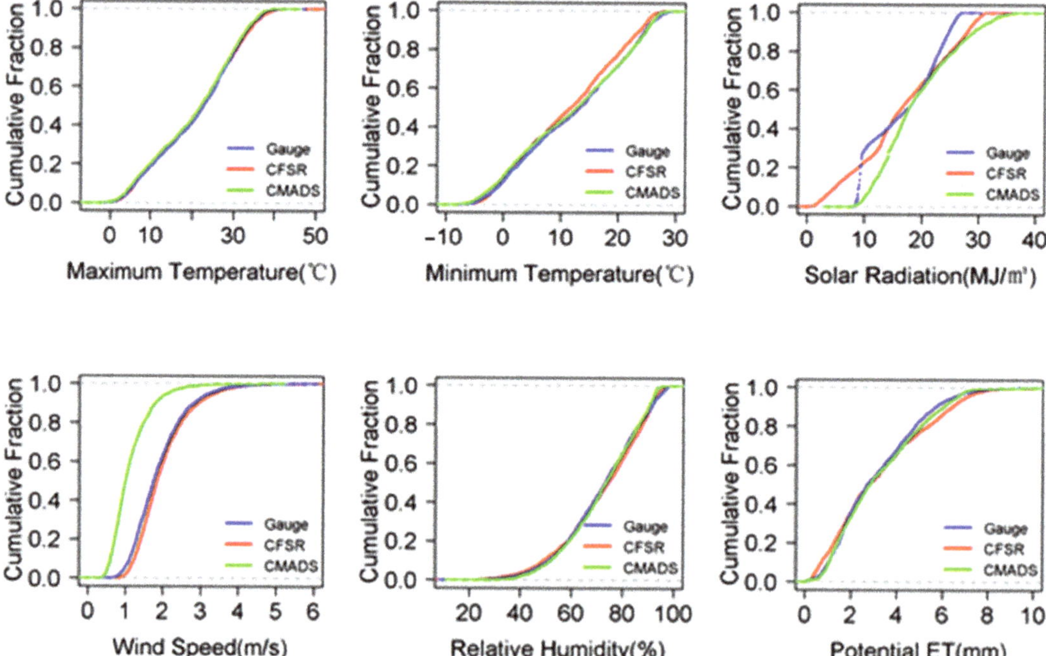

Figure 5. The cumulative fraction of daily maximum and minimum air temperature, solar radiation, wind speed, relative humidity, and potential ET from three different datasets (Gauge, CFSR, and CMADS) at the watershed scale during 2009–2014.

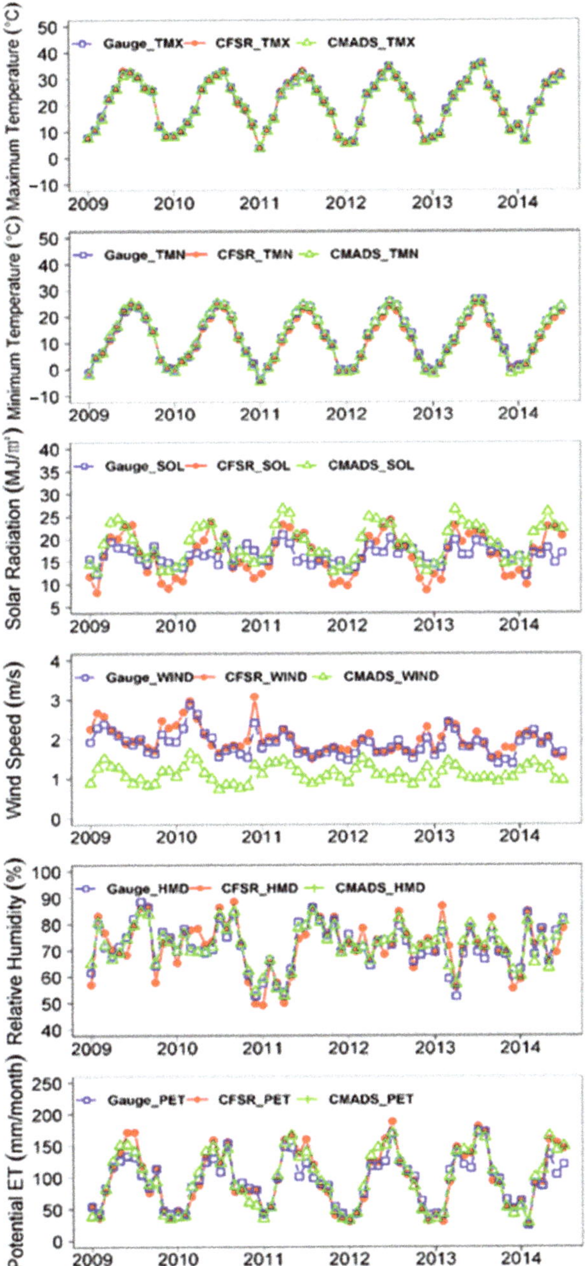

Figure 6. Comparison of the monthly mean of daily maximum (TMX) and minimum (TMN) air temperature, solar radiation (SOL), wind speed (WIND), and relative humidity (HMD) from the Gauge, CFSR, and CMADS datasets, and their resulting potential evapotranspiration (PET) estimates at the watershed scale during 2009–2014.

4.2. Simulation Results Using Different Meteorological Inputs

Tables 2 and 3 summarize the model performance for the eighteen scenarios at daily and monthly timescale, respectively. It can be found that using the same precipitation data but different meteorological data, the streamflow simulation results are similar. Using different precipitation data and the same meteorological data, the results tend to be quite different. This finding shows the dominant role of precipitation data in streamflow simulation. The scenarios using gauged precipitation data had the best performance for both daily and monthly streamflow simulation. During the calibration and validation periods, almost all the simulated results using gauged precipitation data reached the level of very good performance (NSE > 0.75) at daily and monthly timescales (Tables 2 and 3). It is worth noting that among the three scenarios using gauged precipitation data, the scenario with gauged precipitation data together with gauged meteorological data yielded the best performance with NSE higher than 0.87 for the calibration period and with NSE of 0.82 for the validation period.

Table 2. Evaluation statistics for the performance of eighteen scenarios in daily streamflow simulation.

Precipitation Data	Meteorological Data (Excluding Precipitation)	2009–2011 (Calibration)			2012–2014 (Validation)		
		NSE	R^2	PBIAS (%)	NSE	R^2	PBIAS (%)
Gauge	Gauge	0.87	0.88	−23.0	0.82	0.83	−24.1
	CFSR	0.86	0.88	−28.2	0.81	0.83	−32.6
	CMADS	0.85	0.88	−29.4	0.79	0.83	−38.1
CFSR	Gauge	0.32	0.34	−26.0	0.15	0.45	2.6
	CFSR	0.32	0.34	−34.2	0.15	0.44	−10.9
	CMADS	0.30	0.32	−27.6	0.19	0.44	−3.0
CMADS	Gauge	0.64	0.79	−63.8	0.65	0.72	−55.8
	CFSR	0.64	0.79	−64.3	0.61	0.70	−58.4
	CMADS	0.64	0.79	−63.5	0.64	0.72	−58.0
TRMM	Gauge	0.50	0.53	−25.3	0.38	0.44	−13.5
	CFSR	0.54	0.59	−39.1	0.33	0.43	−22.1
	CMADS	0.45	0.48	−30.6	0.37	0.44	−23.0
CMORPH	Gauge	0.56	0.69	−49.2	0.65	0.68	−37.8
	CFSR	0.56	0.72	−52.0	0.65	0.69	−39.7
	CMADS	0.55	0.70	−53.2	0.64	0.68	−43.5
CHIRPS	Gauge	0.40	0.42	−34.7	0.21	0.30	−22.7
	CFSR	0.44	0.46	−35.7	0.22	0.32	−20.2
	CMADS	0.38	0.40	−37.0	0.21	0.29	−27.8

The models using CMADS and CMORPH precipitation data as inputs performed satisfactorily in simulating daily streamflow with NSE larger than 0.55 for both the calibration and validation periods. The models using CMADS precipitation data had larger bias values than those using CMORPH precipitation data, so CMORPH is preferred for daily streamflow simulation. For monthly streamflow simulation, the models using TRMM and CHIRPS precipitation data as inputs performed satisfactorily with all except one NSE value larger than 0.5. These results suggested that different gridded precipitation datasets should be used to achieve optimal results for daily and monthly streamflow simulations. Models using CFSR precipitation data as inputs performed the worst among all considered scenarios. All NSE values were less than 0.50 with even negative values for monthly simulation during the validation period (Table 3), suggesting the unsatisfactory model performance.

Table 3. Evaluation statistics for the performance of eighteen scenarios in monthly streamflow simulation.

Precipitation Data	Meteorological Data (Excluding Precipitation)	2009–2011 (Calibration)			2012–2014 (Validation)		
		NSE	R^2	PBIAS (%)	NSE	R^2	PBIAS (%)
Gauge	Gauge	0.89	0.95	−23.0	0.82	0.89	−24.2
	CFSR	0.87	0.95	−28.2	0.73	0.86	−32.6
	CMADS	0.86	0.95	−29.4	0.67	0.85	−38.2
CFSR	Gauge	0.40	0.44	−26.2	−0.40	0.42	2.2
	CFSR	0.37	0.44	−34.4	−0.34	0.38	−11.4
	CMADS	0.36	0.41	−27.7	−0.38	0.39	−3.4
CMADS	Gauge	0.52	0.91	-63.8	0.45	0.85	−56.0
	CFSR	0.50	0.89	−64.3	0.37	0.85	−58.6
	CMADS	0.51	0.90	−63.6	0.41	0.87	−58.2
TRMM	Gauge	0.61	0.73	−25.4	0.58	0.61	−13.1
	CFSR	0.64	0.83	−39.2	0.51	0.59	−21.7
	CMADS	0.59	0.72	−30.8	0.54	0.60	−22.7
CMORPH	Gauge	0.49	0.80	−49.3	0.67	0.85	−37.9
	CFSR	0.48	0.84	−52.1	0.62	0.83	−39.7
	CMADS	0.45	0.81	−53.3	0.57	0.82	−43.6
CHIRPS	Gauge	0.58	0.69	−34.7	0.59	0.66	−22.4
	CFSR	0.58	0.71	−35.7	0.56	0.61	−19.8
	CMADS	0.55	0.68	−36.9	0.47	0.57	−27.4

Although the models using different meteorological datasets had comparable performance, Gauge performed the best and CFSR usually performed better than CMADS, especially on the monthly scale. This may be related to the large bias of CMADS in wind speed (Figure 4). These results suggest that the CFSR meteorological data should be used in this region when gauge measurements are not available.

Figures 7 and 8 compare the measured and simulated streamflow obtained from the CFSR meteorological (excluding precipitation) data and different precipitation data at the daily and monthly timescales. These diagrams give information about the temporal consistency between the measured streamflow and simulated streamflow. In general, most of the models can well reproduce the seasonal changes in streamflow. The consistencies of simulated streamflow and measured streamflow using all the open-access gridded precipitation datasets were worse than those using gauged precipitation data. Underestimation persists in the simulation of the peak flow, especially for the extreme precipitation events such as those in July and September of 2010. This is due to the underestimation of the precipitation amount of these extreme events. CMADS had the best performance among open-access gridded precipitation datasets in capturing the precipitation amounts and streamflow during extreme events, indicating CMADS should be used for the simulation of extreme events if gauge measurements are not available.

It should be noted that the PBIAS values in this study were large with values beyond the generally acceptable thresholds (i.e., −15% < PBIAS < +15%) in most cases (Tables 2 and 3). There may be two reasons for the large biases: firstly, the Nash–Sutcliffe efficiency coefficient (NSE) was adopted as the target of parameter calibration, and PBIAS was not set as an optimization goal. Secondly, paddy fields and ditches are widely distributed in the study area, and the related hydrological processes are not well represented in the SWAT model.

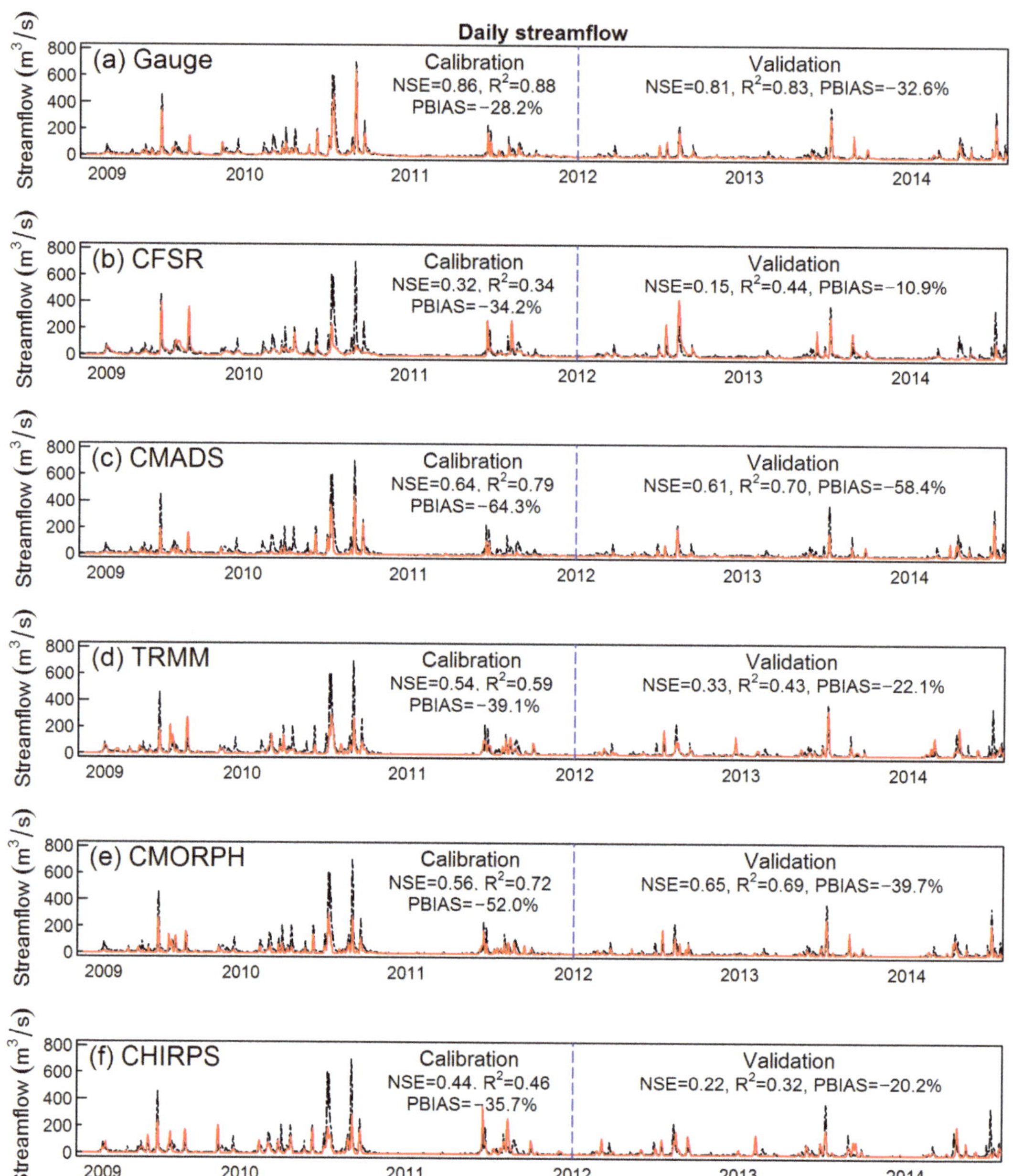

Figure 7. Comparison of daily measured (black line) and simulated (red line) streamflow from models using the CFSR meteorological (excluding precipitation) datasets and six different precipitation datasets for the period 2009–2014, including (**a**) Gauge, (**b**) CFSR, (**c**) CMADS, (**d**) TRMM, (**e**) CMORPH, and (**f**) CHIRPS.

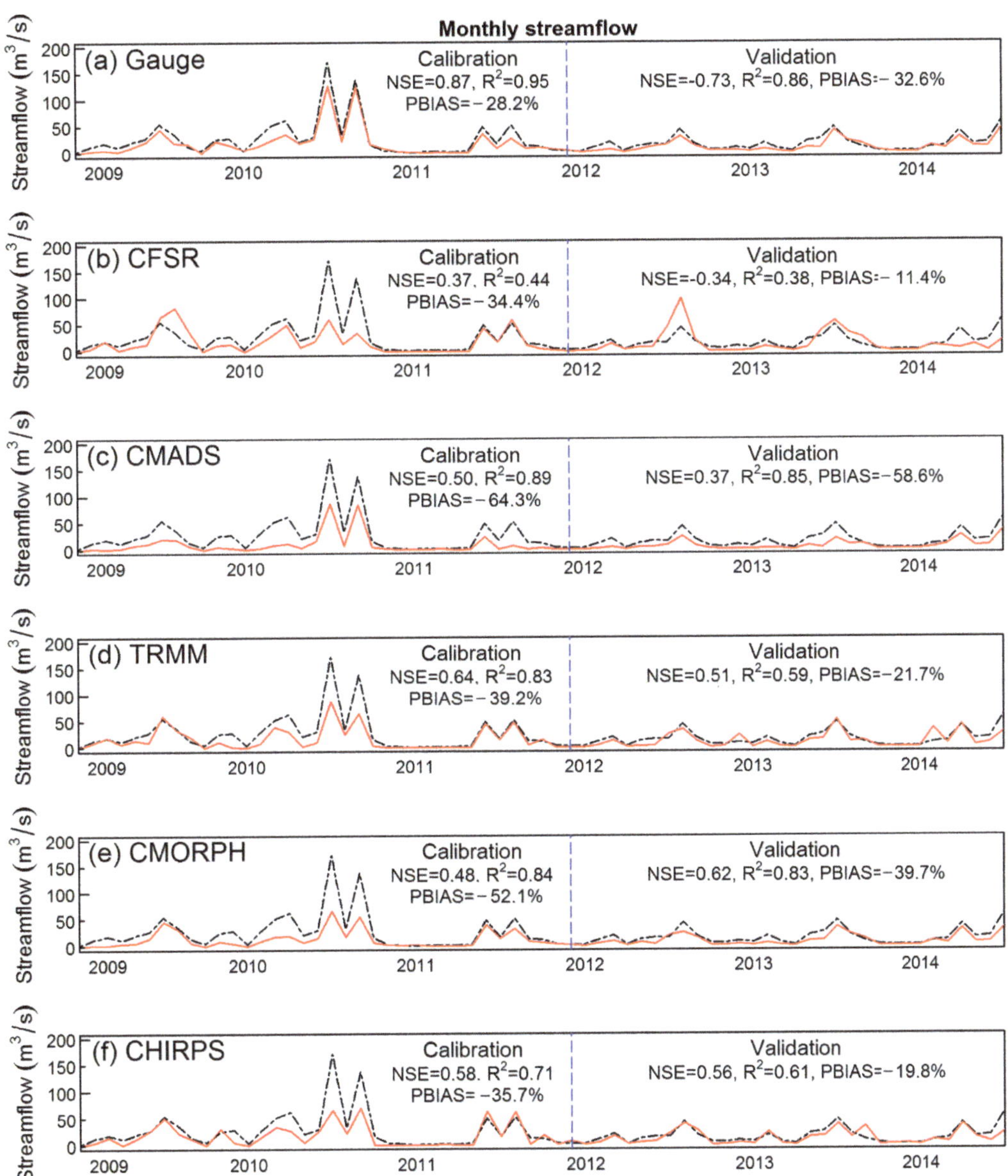

Figure 8. Comparison of monthly measured (black line) and simulated (red line) streamflow from models using the CFSR meteorological (excluding precipitation) datasets and six different precipitation datasets for the period 2009–2014, including (**a**) Gauge, (**b**) CFSR, (**c**) CMADS, (**d**) TRMM, (**e**) CMORPH, and (**f**) CHIRPS.

4.3. Comparison of Calibrated Parameters and Water Balance Components

Tables 4 and 5 show the optimal parameters for the eighteen scenarios calibrated by SWAT-CUP. The scenarios using the same precipitation data (e.g., Gauge, CFSR, CMADS, and CMORPH) and different meteorological (excluding precipitation) data often had almost the same optimal parameters, which showed the dominant role of precipitation to drive the model. The scenarios using the CFSR meteorological (excluding precipitation) data and two different precipitation datasets (i.e., TRMM and CHIRPS) also had the same parameters, which suggested that the influence of meteorological data also should not be overlooked. The parameters related to channel routing (e.g., ALPHA_BNK, CH_K2, and CH_N2) were relatively close to each other among different scenarios, while there were large differences for parameters related to soil properties (e.g., SOL_BD, SOL_K, and SOL_AWC) and groundwater (e.g., GW_DELAY). These results suggested that the parameters related to subsurface water movement have large uncertainty because they are difficult to measure in a spatially explicit way.

Table 4. Optimal parameters calibrated for the scenarios using the Gauge, CFSR, and CMADS precipitation data. The scenarios using the same precipitation data and different meteorological (excluding precipitation) data often had similar optimal parameters (with the different ones displayed in bold). "Gauge_P" represents using gauged precipitation data, "Gauge_M" represents using gauged meteorological data, and so on.

Parameters	Gauge_P and Gauge_M	Gauge_P and CFSR_M	Gauge_P and CMADS_M	CFSR_P and Gauge_M	CFSR_P and CFSR_M	CFSR_P and CMADS_M	CMADS_P and Gauge_M	CMADS_P and CFSR_M	CMADS_P and CMADS_M
r__CN2.mgt	0.0181	0.0181	0.0181	−0.1025	−0.1025	−0.1025	0.0518	0.0518	0.0518
v__ALPHA_BF.gw	0.5350	0.5350	0.5350	0.5207	0.5207	0.5207	0.4736	0.4736	0.4736
v__GW_DELAY.gw	65.5835	65.5835	65.5835	8.6958	8.6958	8.6958	208.5404	208.5404	208.5404
v__GW_REVAP.gw	0.0976	0.0976	0.0976	0.0550	0.0550	0.0550	0.0594	0.0594	0.0594
v__ALPHA_BNK.rte	0.2412	0.2412	0.2412	0.3264	0.3264	0.3264	0.3553	0.3553	0.3553
v__CH_K2.rte	6.3179	6.3179	6.3179	6.7284	6.7284	6.7284	6.8055	6.8055	6.80545
v__CH_N2.rte	0.0912	0.0912	0.0912	0.0762	0.0762	0.0762	0.0766	0.0766	0.0766
r__SOL_AWC().sol	−0.4144	−0.4144	−0.4144	−0.7113	−0.3789	−0.7113	−0.1629	−0.2983	−0.2983
r__SOL_BD().sol	0.0204	0.0204	0.0204	0.5426	0.5426	0.5426	−0.5752	−0.5752	−0.5752
r__SOL_K().sol	−0.0827	−0.0827	−0.0827	0.5185	0.5185	0.5185	0.2755	0.2755	0.2755
v__SFTMP.bsn	2.975	3.8442	3.8442	0.1092	0.1092	0.1092	4.2758	4.2758	4.2758
r__SLSUBBSN.hru	0.0761	0.0761	0.0761	0.1164	0.1164	0.1164	0.0361	0.0361	0.0361

Table 5. Optimal parameters calibrated for the scenarios using the TRMM, CMORPH, and CHIRPS precipitation data. The scenarios using the same precipitation data and different meteorological (excluding precipitation) data often had similar optimal parameters (with the different ones displayed in bold). "Gauge_P" represents using gauged precipitation data, "Gauge_M" represents using gauged meteorological data, and so on.

Parameters	TRMM_P and Gauge_M	TRMM_P and CFSR_M	TRMM_P and CMADS_M	CMORPH_P and Gauge_M	CMORPH_P and CFSR_M	CMORPH_P and CMADS_M	CHIRPS_P and Gauge_M	CHIRPS_P and CFSR_M	CHIRPS_P and CMADS_M
r__CN2.mgt	0.0580	0.0031	0.0181	0.0676	0.0676	0.0676	0.0006	0.0031	0.0006
v__ALPHA_BF.gw	0.8654	0.6998	0.5350	0.8782	0.8782	0.8782	0.3026	0.6998	0.3026
v__GW_DELAY.gw	154.7278	2.7892	65.5835	318.0529	318.0529	318.0529	2.2707	2.7892	2.2707
v__GW_REVAP.gw	0.0441	0.0694	0.0976	0.0250	0.0250	0.0250	0.0760	0.0694	0.0760
v__ALPHA_BNK.rte	0.3662	0.3804	0.2412	0.6512	0.6512	0.6512	0.3525	0.3804	0.3525
v__CH_K2.rte	5.3577	8.5511	6.3179	7.5657	7.5657	7.5657	8.6323	8.5511	8.6323
v__CH_N2.rte	0.0897	0.0930	0.0916	0.0840	0.0840	0.0840	0.1252	0.0930	0.1252
r__SOL_AWC().sol	0.0008	0.0062	−0.4144	0.1718	0.1718	0.1718	0.1800	0.0062	0.1800
r__SOL_BD().sol	0.4754	0.5861	0.0204	0.0106	0.0106	0.0106	0.0098	0.5861	0.0098
r__SOL_K().sol	0.4548	0.5931	−0.0827	0.5160	0.5160	0.5160	0.4455	0.5931	0.4455
v__SFTMP.bsn	0.7148	−3.4315	3.8442	3.4774	3.4774	3.4774	0.5679	−3.4315	0.5679
r__SLSUBBSN.hru	0.1733	0.1205	0.0760	0.0901	0.0901	0.0901	0.0464	0.1205	0.0464

Figure 9 shows the comparison of different water balance components of the eighteen modelling scenarios. There are eight subgraphs, each of which represents the values of a water balance component of eighteen modelling scenarios. As we can see, the scenarios using the same precipitation data and different meteorological (excluding precipitation) data usually yielded similar water balance components. The scenarios using the same meteorological data but different precipitation data had an obvious difference in WYLD (water yield, which is the net amount of water that leaves the subbasin and contributes to streamflow in the reach), SUR_Q (surface runoff contribution to streamflow), GW_Q (groundwater contribution to streamflow), LAT_Q (lateral flow contribution to streamflow), and PERC (water percolating past the root zone) components, slight differences in the precipitation and AET (actual evapotranspiration) components, and almost no difference in the PET components. The scenarios using the TRMM and CHIRPS precipitation data had similar values to those using gauged precipitation data in WYLD, GW_Q, and PERC, while the scenarios using the CMADS precipitation data had large differences from those using gauged precipitation data in most balance components. The scenarios using the TRMM and CHIRPS precipitation data also had similar spatial distributions in precipitation to those using gauged precipitation data (Figure 10). The scenarios using all the gridded precipitation data can well simulate the pattern that the southwest part of the watershed had the largest WYLD values (Figure S1) because the steep terrain in that region is conducive to runoff generation.

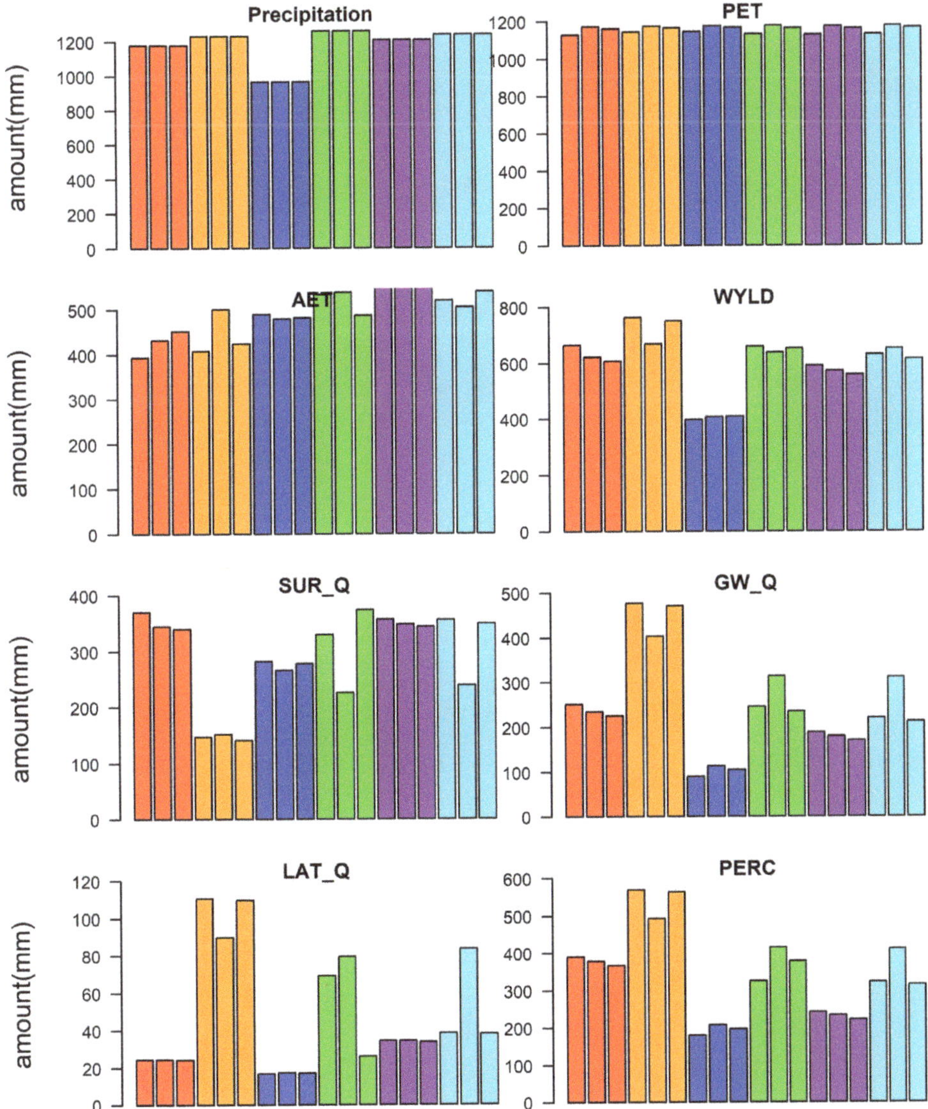

Figure 9. Annual mean water balance components for simulations using eighteen scenarios during 2009–2014. The different colors represent using different precipitation datasets (Gauge, CFSR, CMADS,

TRMM, CMORPH, CHIRPS), and the three bars displayed in the same color used the Gauge, CFSR, and CMADS meteorological data from left to right, respectively. AET—actual evapotranspiration; WYLD—water yield, the net amount of water that leaves the subbasin and contributes to streamflow in the reach; SUR_Q—surface runoff contribution to streamflow; GW_Q—groundwater contribution to streamflow; LAT_Q—lateral flow contribution to streamflow; PERC—water percolating past the root zone.

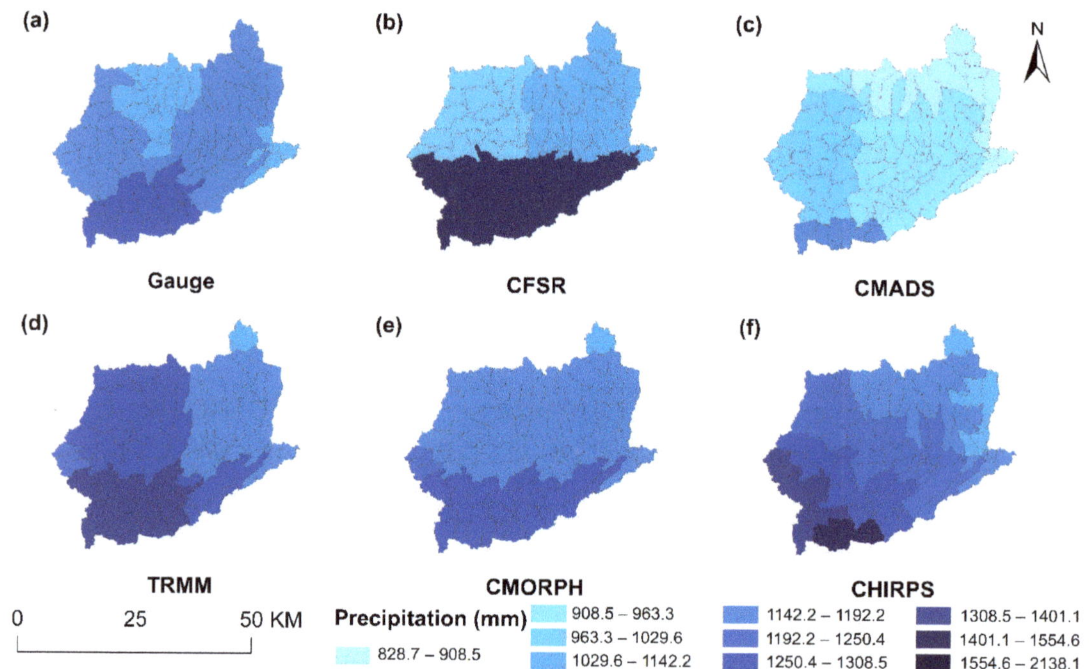

Figure 10. Spatial distributions of annual mean precipitation using different precipitation data during 2009–2014, including (**a**) Gauge, (**b**) CFSR, (**c**) CMADS, (**d**) TRMM, (**e**) CMORPH, and (**f**) CHIRPS.

It should be noted that, although the scenarios using the CFSR precipitation data yielded similar WYLD to the scenarios using gauged precipitation data, there are great differences among the water balance components (Figure 9). This shows the limitation of model calibration only based on measured streamflow at the outlet, which is very common in state-of-the-art hydrological modelling, as streamflow is usually the only measured water balance component available in many watersheds. Therefore, the simulation results of individual water balance components calibrated with only outlet streamflow may contain large uncertainty [1,53]. This also highlights that there is a clear need for measurements of other water balance components with the community's effort for innovation in all realms from ground-based to remote sensing [5]. Once such data are more available, multivariable and multisite calibration is a promising way to reduce the aforementioned uncertainty.

5. Discussion

5.1. Comparison with Existing Studies

The performance of gridded meteorological and precipitation datasets often varied with location. It was reported that CFSR typically performs well in the United States and South America, but performs poorly for Asia and Africa compared with other products, such as CMADS [54]. For example, Fuka et al. (2014) reported that the stream discharge simulations forced by CFSR precipitation and temperature data performed as good as or better than those forced by data from traditional weather gauging stations in four small watersheds in the USA [19]. Liu et al. (2018) and Zhang et al. (2020) found CMADS had better performance than CFSR for streamflow simulation in the Yellow River Source Basin, China, and the Muda River Basin, Malaysia [17,55]. While this study found that CFSR meteorological data usually performed better than CMADS, especially at the monthly scale caused by the large bias of CMADS in wind speed, but the CFSR precipitation data performed the worst among all the precipitation datasets evaluated in this study. Gao et al. (2018) also found the performance of CFSR precipitation data was not satisfactory in the Xiang River Basin of China [56], and similar results were also reported in two watersheds in the USA [24,26] and the upstream watersheds of Three Gorges Reservoir in China [30,32]. The model driven by gauged precipitation data together with gauged meteorological data yielded the best performance, which further confirms previous findings that ground measurements are the most reliable input data compared to open-access gridded products [1,16,21].

An important finding of this study is that the CMORPH precipitation data had overall the best performance for daily streamflow simulation among gridded precipitation datasets; TRMM had overall the best performance for monthly streamflow simulation; and CMADS performed the best in capturing the precipitation amounts and streamflow during extreme events. Therefore, different gridded precipitation datasets should be used for different aims. The good performance of CMORPH on the daily scale and that of TRMM on the monthly scale were also reported in the Meichuanjiang watershed of the Ganjiang Basin, China [2], in the Adige Basin, Italy [7], and in Australia [57]. It was also reported that the performances of CMORPH and TRMM were similar in some regions (e.g., Central Thailand) [58]. The capability of CMADS precipitation data in capturing extreme events can be attributed to the incorporation of in situ precipitation measurements from dense weather stations [36].

5.2. Uncertainty of the Evaluation

It has been widely acknowledged that considerable uncertainties exist in the results of hydrological modeling, which come from model structure, parameters, and inputs [59–61]. To quantify these uncertainties, various approaches, such as GLUE (generalized likelihood uncertainty estimation [62]) and MCMC (Markov chain Monte Carlo [63]), have been proposed. Using these methods, a confidence band instead of a single curve can be obtained to represent the uncertainty of a hydrological variable, such as streamflow. Nevertheless, searching for a single optimal parameter set through model calibration is still the commonly used approach for the inter-comparison of gridded meteorological and precipitation datasets [1,16,28]. In this study, a single-model structure was used following previous researchers, as this approach is relatively simple and at the same time can reasonably reflect the accuracy of different meteorological/precipitation products. In order to obtain more comprehensive and reliable evaluation results, uncertainty analysis should be conducted in the future. In addition, the usage of the differential split-sample test (the DSS-test, e.g., using periods with apparent different climate conditions, e.g., dry/wet or cold/warm, where calibration is performed on one period and validation is performed on another period) will also improve the reliability of evaluation [64].

6. Conclusions

This study evaluated the performance of two mainstream open-access gridded meteorological datasets (i.e., CFSR and CMADS) and three satellite-based precipitation datasets in driving the SWAT model in streamflow simulation at daily and monthly timescales in the Fengle watershed, Anhui Province, China. Eighteen modelling scenarios, which are generated through the combination of six precipitation datasets (i.e., Gauge, CFSR, CMADS, TRMM, CMORPH, and CHIRPS) and three meteorological (excluding precipitation) datasets (i.e., Gauge, CFSR, and CMADS), were used to study the impact of different input data on simulation results. After comprehensive comparison of these scenarios, the following conclusions can be drawn:

(1) In this study area, the three satellite-based precipitation datasets (i.e., TRMM, CMORPH, and CHIRPS) were generally close to gauged data except for some lower peaks, while CMADS had overall lower precipitation than gauged data and CFSR had poor temporal consistency with gauged data. For the other meteorological variables, excluding precipitation, CFSR and CMADS had fairly good agreement with gauged data in the maximum temperature, minimum temperature, and relative humidity, but there are large discrepancies among them for the solar radiation and wind speed. In particular, for solar radiation, gauged data had smaller fluctuations than the other two datasets; for wind speed, CMADS was considerably lower than the gauged and CFSR datasets. However, despite these discrepancies, overall monthly PET totals from the three datasets were in good agreement, suggesting that the discrepancies in individual weather variables cancelled each other to a certain degree.

(2) The impact of precipitation data on simulated streamflow is much larger than that of other meteorological data. In this study, the best simulation results were obtained using gauged data for both precipitation and other meteorological variables. At the same time, this study also got satisfactory daily simulation results using the CMORPH precipitation data and monthly simulation results using the TRMM and CHIRPS precipitation data. These results suggested that different gridded precipitation datasets should be used to obtain optimal results for daily and monthly streamflow simulations. Although the models using different meteorological datasets had comparable performance, CFSR usually performed better than CMADS especially at the monthly scale in this area.

(3) There were considerable differences in the calibrated optimal parameters and water balance components among the eighteen scenarios even for the scenarios with similar water yield to streamflow (e.g., the scenarios using gauged precipitation data and those using CFSR precipitation data). This highlights the inherent limitations of model calibration only based on measured streamflow at the outlet, which should be reduced through multivariable and multisite calibration once data allows.

Currently, the development of data fusion and machine learning techniques provides unprecedented opportunities to design novel methods to reduce the uncertainties of the precipitation dataset. While the focus of this study was to evaluate the performance of existing datasets on hydrological modeling, in the future, novel methods should be developed to construct more accurate precipitation datasets. In addition, it is important to obtain the right simulation results for the right reasons, especially for policy makers. Although the simulation results in this study were generally good based on indicators such as NSE and R^2, the SWAT model did not adequately represent the impacts of widely distributed paddy fields and ditches on hydrological processes in this study area. In the future, the model should be improved to better characterize these agriculture-related processes and make the simulation results more reliable for decision making.

Supplementary Materials: The following supporting information can be downloaded at: https://www.mdpi.com/article/10.3390/w14091406/s1, Figure S1: Annual mean WYLD (water yield) for simulations using CFSR meteorological data and different precipitation data during 2009–2014, including (a) Gauge, (b) CFSR, (c) CMADS, (d) TRMM, (e) CMORPH, and (f) CHIRPS.

Author Contributions: Conceptualization, J.L. and Y.L.; methodology, J.L., L.Y. and Y.L.; software, J.L. and L.Y.; validation, L.Y.; formal analysis, J.L. and L.Y.; investigation, Y.Z. and L.Y.; resources, L.Y.; data curation, J.L. and L.Y.; writing—original draft preparation, J.L. and L.Y.; writing—review and editing, J.L., L.Y. and Y.L.; visualization, Y.Z. and L.Y.; supervision, Y.L.; project administration, Y.L.; funding acquisition, Y.L. All authors have read and agreed to the published version of the manuscript.

Funding: This research was funded by the National Natural Science Foundation of China (Project No.: 51879130 and 42077449), Discipline Innovation and Talent Introduction Base Project of Colleges and Universities in Henan Province of China (Project No.: CXJD2019001).

Institutional Review Board Statement: Not applicable.

Informed Consent Statement: Informed consent was obtained from all subjects involved in the study.

Data Availability Statement: The data used to support the findings of this study are included within the article.

Acknowledgments: We are thankful to the anonymous reviewers for their valuable comments.

Conflicts of Interest: The authors declare no conflict of interest.

References

1. Duan, Z.; Chen, Q.; Chen, C.; Liu, J.; Gao, H.; Song, X.; Wei, M. Spatiotemporal analysis of nonlinear trends in precipitation over Germany during 1951–2013 from multiple observation-based gridded products. *Int. J. Climatol.* **2019**, *39*, 2120–2135. [CrossRef]
2. Liu, J.; Duan, Z.; Jiang, J.; Zhu, A. Evaluation of three satellite precipitation products TRMM 3B42, CMORPH, and PERSIANN over a subtropical watershed in China. *Adv. Meteorol.* **2015**, *2015*, 151239. [CrossRef]
3. Dai, Q.; Yang, Q.; Zhang, J.; Zhang, S. Impact of gauge representative error on a radar rainfall uncertainty model. *J. Appl. Meteorol. Climatol.* **2018**, *57*, 2769–2787. [CrossRef]
4. Duan, Z.; Gao, H.; Ke, C. Estimation of lake outflow from the poorly gauged Lake Tana (Ethiopia) using satellite remote sensing data. *Remote Sens.* **2018**, *10*, 1060. [CrossRef]
5. Tauro, F.; Selker, J.; Van De Giesen, N.; Abrate, T.; Uijlenhoet, R.; Porfiri, M.; Manfreda, S.; Caylor, K.; Moramarco, T.; Benveniste, J. Measurements and observations in the XXI century (MOXXI): Innovation and multi-disciplinarity to sense the hydrological cycle. *Hydrol. Sci. J.* **2018**, *63*, 169–196. [CrossRef]
6. Cornes, R.C.; van der Schrier, G.; van den Besselaar, E.J.; Jones, P.D. An ensemble version of the E-OBS temperature and precipitation data sets. *J. Geophys. Res. Atmos.* **2018**, *123*, 9391–9409. [CrossRef]
7. Duan, Z.; Liu, J.; Tuo, Y.; Chiogna, G.; Disse, M. Evaluation of eight high spatial resolution gridded precipitation products in Adige Basin (Italy) at multiple temporal and spatial scales. *Sci. Total Environ.* **2016**, *573*, 1536–1553. [CrossRef]
8. Duan, Z.; Tuo, Y.; Liu, J.; Gao, H.; Song, X.; Zhang, Z.; Yang, L.; Mekonnen, D.F. Hydrological evaluation of open-access precipitation and air temperature datasets using SWAT in a poorly gauged basin in Ethiopia. *J. Hydrol.* **2019**, *569*, 612–626. [CrossRef]
9. Harris, I.; Jones, P.D.; Osborn, T.J.; Lister, D.H. Updated high-resolution grids of monthly climatic observations–the CRU TS3. 10 Dataset. *Int. J. Climatol.* **2014**, *34*, 623–642. [CrossRef]
10. Saha, S.; Moorthi, S.; Wu, X.; Wang, J.; Nadiga, S.; Tripp, P.; Behringer, D.; Hou, Y.-T.; Chuang, H.-Y.; Iredell, M. The NCEP climate forecast system version 2. *J. Clim.* **2014**, *27*, 2185–2208. [CrossRef]
11. Tapiador, F.J.; Turk, F.J.; Petersen, W.; Hou, A.Y.; García-Ortega, E.; Machado, L.A.; Angelis, C.F.; Salio, P.; Kidd, C.; Huffman, G.J. Global precipitation measurement: Methods, datasets and applications. *Atmos. Res.* **2012**, *104*, 70–97. [CrossRef]
12. Meng, X.; Wang, H.; Shi, C.; Wu, Y.; Ji, X. Establishment and evaluation of the China meteorological assimilation driving datasets for the SWAT model (CMADS). *Water* **2018**, *10*, 1555. [CrossRef]
13. Arnold, J.G.; Moriasi, D.N.; Gassman, P.W.; Abbaspour, K.C.; White, M.J.; Srinivasan, R.; Santhi, C.; Harmel, R.; Van Griensven, A.; Van Liew, M.W. SWAT: Model use, calibration, and validation. *Trans. ASABE* **2012**, *55*, 1491–1508. [CrossRef]
14. Cao, Y.; Zhang, J.; Yang, M.; Lei, X.; Guo, B.; Yang, L.; Zeng, Z.; Qu, J. Application of SWAT model with CMADS data to estimate hydrological elements and parameter uncertainty based on SUFI-2 algorithm in the Lijiang River Basin, China. *Water* **2018**, *10*, 742. [CrossRef]
15. Chen, M.; Yue, S.; Lü, G.; Lin, H.; Yang, C.; Wen, Y.; Hou, T.; Xiao, D.; Jiang, H. Teamwork-oriented integrated modeling method for geo-problem solving. *Environ. Model. Softw.* **2019**, *119*, 111–123. [CrossRef]
16. Dile, Y.T.; Srinivasan, R. Evaluation of CFSR climate data for hydrologic prediction in data-scarce watersheds: An application in the Blue Nile River Basin. *JAWRA J. Am. Water Resour. Assoc.* **2014**, *50*, 1226–1241. [CrossRef]
17. Liu, J.; Shanguan, D.; Liu, S.; Ding, Y. Evaluation and hydrological simulation of CMADS and CFSR reanalysis datasets in the Qinghai-Tibet Plateau. *Water* **2018**, *10*, 513. [CrossRef]
18. Huffman, G.J.; Bolvin, D.T.; Nelkin, E.J.; Wolff, D.B.; Adler, R.F.; Gu, G.; Hong, Y.; Bowman, K.P.; Stocker, E.F. The TRMM multisatellite precipitation analysis (TMPA): Quasi-global, multiyear, combined-sensor precipitation estimates at fine scales. *J. Hydrometeorol.* **2007**, *8*, 38–55. [CrossRef]
19. Fuka, D.R.; Walter, M.T.; MacAlister, C.; Degaetano, A.T.; Steenhuis, T.S.; Easton, Z.M. Using the climate forecast system reanalysis as weather input data for watershed models. *Hydrol. Process.* **2014**, *28*, 5613–5623. [CrossRef]
20. Reichstein, M.; Camps-Valls, G.; Stevens, B.; Jung, M.; Denzler, J.; Carvalhais, N. Prabhat. Deep learning and process understanding for data-driven earth system science. *Nature* **2019**, *566*, 195–204. [CrossRef]
21. Moraux, A.; Dewitte, S.; Cornelis, B.; Munteanu, A. A deep learning multimodal method for precipitation estimation. *Remote Sens.* **2021**, *13*, 3278. [CrossRef]
22. Gilewski, P.; Nawalany, M. Inter-comparison of rain-gauge, radar, and satellite (IMERG GPM) precipitation estimates performance for rainfall-runoff modeling in a mountainous catchment in Poland. *Water* **2018**, *10*, 1665. [CrossRef]

23. Sapountzis, M.; Kastridis, A.; Kazamias, A.; Karagiannidis, A.; Nikopoulos, P.; Lagouvardos, K. Utilization and uncertainties of satellite precipitation data in flash flood hydrological analysis in ungauged watersheds. *Glob. Nest J.* **2021**, *23*, 388–399.
24. Bitew, M.M.; Gebremichael, M.; Ghebremichael, L.T.; Bayissa, Y.A. Evaluation of high-resolution satellite rainfall products through streamflow simulation in a hydrological modeling of a small mountainous watershed in Ethiopia. *J. Hydrometeorol.* **2012**, *13*, 338–350. [CrossRef]
25. Tuo, Y.; Duan, Z.; Disse, M.; Chiogna, G. Evaluation of precipitation input for SWAT modeling in Alpine catchment: A case study in the Adige river basin (Italy). *Sci. Total Environ.* **2016**, *573*, 66–82. [CrossRef]
26. Gao, F.; Zhang, Y.; Chen, Q.; Wang, P.; Yang, H.; Yao, Y.; Cai, W. Comparison of two long-term and high-resolution satellite precipitation datasets in Xinjiang, China. *Atmos. Res.* **2018**, *212*, 150–157. [CrossRef]
27. Mahmoud, M.T.; Hamouda, M.A.; Mohamed, M.M. Spatiotemporal evaluation of the GPM satellite precipitation products over the United Arab Emirates. *Atmos. Res.* **2019**, *219*, 200–212. [CrossRef]
28. Radcliffe, D.; Mukundan, R. PRISM vs. CFSR precipitation data effects on calibration and validation of SWAT models. *JAWRA J. Am. Water Resour. Assoc.* **2017**, *53*, 89–100. [CrossRef]
29. Zhang, L.; Meng, X.; Wang, H.; Yang, M. Simulated runoff and sediment yield responses to land-use change using the SWAT model in northeast China. *Water* **2019**, *11*, 915. [CrossRef]
30. Zhu, Q.; Hsu, K.L.; Xu, Y.P.; Yang, T. Evaluation of a new satellite-based precipitation data set for climate studies in the Xiang River basin, southern China. *Int. J. Climatol.* **2017**, *37*, 4561–4575. [CrossRef]
31. Jiang, S.; Liu, R.; Ren, L.; Wang, M.; Shi, J.; Zhong, F.; Duan, Z. Evaluation and hydrological application of CMADS reanalysis precipitation data against four satellite precipitation products in the Upper Huaihe River Basin, China. *J. Meteorol. Res.* **2020**, *34*, 1096–1113. [CrossRef]
32. Yong, B.; Hong, Y.; Ren, L.L.; Gourley, J.J.; Huffman, G.J.; Chen, X.; Wang, W.; Khan, S.I. Assessment of evolving TRMM-based multisatellite real-time precipitation estimation methods and their impacts on hydrologic prediction in a high latitude basin. *J. Geophys. Res. Atmos.* **2012**, *117*, D09108. [CrossRef]
33. Hu, S.; Qiu, H.; Yang, D.; Cao, M.; Song, J.; Wu, J.; Huang, C.; Gao, Y. Evaluation of the applicability of climate forecast system reanalysis weather data for hydrologic simulation: A case study in the Bahe River Basin of the Qinling Mountains, China. *J. Geogr. Sci.* **2017**, *27*, 546–564. [CrossRef]
34. Yang, Y.; Wang, G.; Wang, L.; Yu, J.; Xu, Z. Evaluation of gridded precipitation data for driving SWAT model in area upstream of three gorges reservoir. *PLoS ONE* **2014**, *9*, e112725. [CrossRef]
35. Li, Y.; Wang, Y.; Zheng, J.; Yang, M. Investigating spatial and temporal variation of hydrological processes in western China driven by CMADS. *Water* **2019**, *11*, 435. [CrossRef]
36. Zhang, Z.; Gao, J.; Gao, Y. The influences of land use changes on the value of ecosystem services in Chaohu Lake Basin, China. *Environ. Earth Sci.* **2015**, *74*, 385–395. [CrossRef]
37. Liu, J.; Liu, Z.; Yin, Y.; Croke, B.F.W.; Chen, M.; Qin, C.-Z.; Tang, G.; Zhu, A.X. A hybrid vector-raster approach to drainage network construction in agricultural watersheds with rice terraces and ponds. *J. Hydrol.* **2021**, *601*, 126585. [CrossRef]
38. Kong, X.; He, Q.; Yang, B.; He, W.; Xu, F.; Janssen, A.B.; Kuiper, J.J.; van Gerven, L.P.; Qin, N.; Jiang, Y. Hydrological regulation drives regime shifts: Evidence from paleolimnology and ecosystem modeling of a large shallow Chinese lake. *Glob. Chang. Biol.* **2017**, *23*, 737–754. [CrossRef]
39. Meng, X.; Wang, H. Significance of the China meteorological assimilation driving datasets for the SWAT model (CMADS) of East Asia. *Water* **2017**, *9*, 765. [CrossRef]
40. Xie, P.; Joyce, R.; Wu, S.; Yoo, S.-H.; Yarosh, Y.; Sun, F.; Lin, R. Reprocessed, bias-corrected CMORPH global high-resolution precipitation estimates from 1998. *J. Hydrometeorol.* **2017**, *18*, 1617–1641. [CrossRef]
41. Arnold, J.G.; Srinivasan, R.; Muttiah, R.S.; Williams, J.R. Large area hydrologic modeling and assessment part I: Model development 1. *JAWRA J. Am. Water Resour. Assoc.* **1998**, *34*, 73–89. [CrossRef]
42. Neitsch, S.L.; Arnold, J.G.; Kiniry, J.R.; Williams, J.R. *Soil and Water Assessment Tool Theoretical Documentation Version 2009*; Texas Water Resources Institute: College Station, TX, USA, 2011.
43. Song, X.; Duan, Z.; Kono, Y.; Wang, M. Integration of remotely sensed C factor into SWAT for modelling sediment yield. *Hydrol. Process.* **2011**, *25*, 3387–3398. [CrossRef]
44. Abbaspour, K.C.; Johnson, C.; Van Genuchten, M.T. Estimating uncertain flow and transport parameters using a sequential uncertainty fitting procedure. *Vadose Zone J.* **2004**, *3*, 1340–1352. [CrossRef]
45. Abbaspour, K.C.; Yang, J.; Maximov, I.; Siber, R.; Bogner, K.; Mieleitner, J.; Zobrist, J.; Srinivasan, R. Modelling hydrology and water quality in the pre-alpine/alpine Thur watershed using SWAT. *J. Hydrol.* **2007**, *333*, 413–430. [CrossRef]
46. Chow, V.T.; Maidment, D.R.; Mays, L.W. *Applied Hydrology*; McGraw-Hill Book Company: New York, NY, USA, 1988.
47. Bredemeier, M.; Cohen, S.; Godbold, D.L.; Lode, E.; Pichler, V.; Schleppi, P. *Forest Management and the Water Cycle: An Ecosystem-Based Approach*; Springer Science & Business Media: Berlin/Heidelberg, Germany, 2010; Volume 212.
48. Kastridis, A.; Theodosiou, G.; Fotiadis, G. Investigation of flood management and mitigation measures in ungauged NATURA protected watersheds. *Hydrology* **2021**, *8*, 170. [CrossRef]
49. Schleppi, P. Forested water catchments in a changing environment. In *Forest Management and the Water Cycle*; Springer: Berlin/Heidelberg, Germany, 2010; pp. 89–110.

50. Nash, J.E.; Sutcliffe, J.V. River flow forecasting through conceptual models part I—A discussion of principles. *J. Hydrol.* **1970**, *10*, 282–290. [CrossRef]
51. Moriasi, D.N.; Arnold, J.G.; Van Liew, M.W.; Bingner, R.L.; Harmel, R.D.; Veith, T.L. Model evaluation guidelines for systematic quantification of accuracy in watershed simulations. *Trans. ASABE* **2007**, *50*, 885–900. [CrossRef]
52. Peng, J.; Liu, T.; Huang, Y.; Ling, Y.; Li, Z.; Bao, A.; Chen, X.; Kurban, A.; De Maeyer, P. Satellite-based precipitation datasets evaluation using gauge observation and hydrological modeling in a typical arid land watershed of Central Asia. *Remote Sens.* **2021**, *13*, 221. [CrossRef]
53. Bitew, M.M.; Gebremichael, M. Evaluation of satellite rainfall products through hydrologic simulation in a fully distributed hydrologic model. *Water Resour. Res.* **2011**, *47*, W06526. [CrossRef]
54. Tan, M.L.; Gassman, P.W.; Liang, J.; Haywood, J.M. A review of alternative climate products for SWAT modelling: Sources, assessment and future directions. *Sci. Total Environ.* **2021**, *795*, 148915. [CrossRef]
55. Zhang, D.; Tan, M.L.; Dawood, S.R.; Samat, N.; Chang, C.K.; Roy, R.; Tew, Y.L.; Mahamud, M.A. Comparison of NCEP-CFSR and CMADS for hydrological modelling using SWAT in the Muda River Basin, Malaysia. *Water* **2020**, *12*, 3288. [CrossRef]
56. Gao, X.; Zhu, Q.; Yang, Z.; Wang, H. Evaluation and hydrological application of CMADS against TRMM 3B42V7, PERSIANN-CDR, NCEP-CFSR, and gauge-based datasets in Xiang River Basin of China. *Water* **2018**, *10*, 1225. [CrossRef]
57. Chua, Z.-W.; Kuleshov, Y.; Watkins, A. Evaluation of satellite precipitation estimates over Australia. *Remote Sens.* **2020**, *12*, 678. [CrossRef]
58. Yang, W.-T.; Fu, S.-M.; Sun, J.-H.; Zheng, F.; Wei, J.; Ma, Z. Comparative evaluation of the performances of TRMM-3B42 and climate prediction centre morphing technique (CMORPH) precipitation estimates over Thailand. *J. Meteorol. Soc. Jpn. Ser. II* **2021**, *99*, 1525–1546. [CrossRef]
59. Darbandsari, P.; Coulibaly, P. Inter-comparison of lumped hydrological models in data-scarce watersheds using different precipitation forcing data sets: Case study of Northern Ontario, Canada. *J. Hydrol. Reg. Stud.* **2020**, *31*, 100730. [CrossRef]
60. Kumari, N.; Srivastava, A.; Sahoo, B.; Raghuwanshi, N.S.; Bretreger, D. Identification of suitable hydrological models for streamflow assessment in the Kangsabati River Basin, India, by using different model selection scores. *Nat. Resour. Res.* **2021**, *30*, 4187–4205. [CrossRef]
61. Pappenberger, F.; Beven, K.J. Ignorance is bliss: Or seven reasons not to use uncertainty analysis. *Water Resour. Res.* **2006**, *42*, W05302. [CrossRef]
62. Beven, K.; Binley, A. The future of distributed models: Model calibration and uncertainty prediction. *Hydrol. Process.* **1992**, *6*, 279–298. [CrossRef]
63. Vrugt, J.A.; ter Braak, C.J.F.; Diks, C.G.H.; Robinson, B.A.; Hyman, J.M.; Higdon, D. Accelerating markov chain monte carlo simulation by differential evolution with self-adaptive randomized subspace sampling. *Int. J. Nonlinear Sci. Numer. Simul.* **2009**, *10*, 273–290. [CrossRef]
64. Refsgaard, J.C.; Madsen, H.; Andréassian, V.; Arnbjerg-Nielsen, K.; Davidson, T.A.; Drews, M.; Hamilton, D.P.; Jeppesen, E.; Kjellström, E.; Olesen, J.E.; et al. A framework for testing the ability of models to project climate change and its impacts. *Clim. Chang.* **2014**, *122*, 271–282. [CrossRef]

 water

Article

Delineating Groundwater Recharge Potential through Remote Sensing and Geographical Information Systems

Ahsen Maqsoom [1,*], Bilal Aslam [2], Nauman Khalid [1], Fahim Ullah [3], Hubert Anysz [4,*], Abdulrazak H. Almaliki [5], Abdulrhman A. Almaliki [6] and Enas E. Hussein [7]

1. Department of Civil Engineering, COMSATS University Islamabad, Wah Cantt 47040, Pakistan; fa17-cve-048@cuiwah.edu.pk
2. School of Informatics, Computing, and Cyber Systems, Northern Arizona University, Flagstaff, AZ 86011, USA; ba924@nau.edu
3. School of Surveying and Built Environment, University of Southern Queensland, Springfield, QLD 4300, Australia; fahim.ullah@usq.edu.au
4. Faculty of Civil Engineering, Warsaw University of Technology, 00-637 Warsaw, Poland
5. Civil Engineering Department, College of Engineering, Taif University, Taif 21944, Saudi Arabia; a.almaliki@tu.edu.sa
6. Independent Researcher in Computer Science, Jeddah 12462, Saudi Arabia; a.almaliki1222@gmail.com
7. National Water Research Center, P.O. Box 74, Shubra El-Kheima 13411, Egypt; enas_el-sayed@nwrc.gov.eg

* Correspondence: ahsen.maqsoom@ciitwah.edu.pk (A.M.); hubert.anysz@pw.edu.pl (H.A.)

Citation: Maqsoom, A.; Aslam, B.; Khalid, N.; Ullah, F.; Anysz, H.; Almaliki, A.H.; Almaliki, A.A.; Hussein, E.E. Delineating Groundwater Recharge Potential through Remote Sensing and Geographical Information Systems. *Water* **2022**, *14*, 1824. https://doi.org/10.3390/w14111824

Academic Editors: Dengfeng Liu, Hui Liu and Xianmeng Meng

Received: 19 April 2022
Accepted: 2 June 2022
Published: 6 June 2022

Abstract: Owing to the extensive global dependency on groundwater and associated increasing water demand, the global groundwater level is declining rapidly. In the case of Islamabad, Pakistan, the groundwater level has lowered five times over the past five years due to extensive pumping by various departments and residents to meet the local water requirements. To address this, water reservoirs and sources need to be delineated, and potential recharge zones are highlighted to assess the recharge potential. Therefore, the current study utilizes an integrated approach based on remote sensing (RS) and GIS using the influence factor (IF) technique to delineate potential groundwater recharge zones in Islamabad, Pakistan. Soil map of Pakistan, Landsat 8TM satellite data, digital elevation model (ASTER DEM), and local geological map were used in the study for the preparation of thematic maps of 15 key contributing factors considered in this study. To generate a combined groundwater recharge map, rate and weightage values were assigned to each factor representing their mutual influence and recharge capabilities. To analyze the final combined recharge map, five different assessment analogies were used in the study: poor, low, medium, high, and best. The final recharge potential map for Islamabad classifies 15% (136.8 km^2) of the region as the "best" zone for extracting groundwater. Furthermore, high, medium, low, and poor ranks were assigned to 21%, 24%, 27%, and 13% of the region with respective areas of 191.52 km^2, 218.88 km^2, 246.24 km^2, and 118.56 km^2. Overall, this research outlines the best to least favorable zones in Islamabad regarding groundwater recharge potentials. This can help the authorities devise mitigation strategies and preserve the natural terrain in the regions with the best groundwater recharge potential. This is aligned with the aims of the interior ministry of Pakistan for constructing small reservoirs and ponds in the existing natural streams and installing recharging wells to maintain the groundwater level in cities. Other countries can expand upon and adapt this study to delineate local groundwater recharge potentials.

Keywords: geographical information systems; groundwater assessment; groundwater recharge; remote sensing; Islamabad

1. Introduction and Background

Groundwater is necessary to sustain various forms of life [1]. It is defined as a form of water occupying all the voids within a geological stratum [2]. It is one of the important water sources for agriculture, industry, and domestic use worldwide [3]. The groundwater level is naturally maintained through precipitation that balances the water cycle, which is crucial for all multicellular life forms. The occurrence of groundwater in a geological formation and the scope for its exploitation primarily depend on the formation porosity [2]. The aquifers rely upon soil and fissured rocks as the medium of pores for the consistent flow between them [4]. In these complex networks of interconnected pores, fractures, cracks, joints, crushed zones (such as faults zones or shear zones), or solution cavities, rainwater can easily percolate through them and maintain groundwater tables [5].

In the past few decades, the greater reliance on groundwater has decreased groundwater table levels. Globally, more than 60% of agricultural practices depend on groundwater as a water source [6]. In developing countries in Asia, groundwater-based irrigation has grown up to 500% [7]. Moreover, due to the rapid increase in population, the demand for groundwater resources increases due to the inadequate availability of useable surface water resources. Furthermore, increased industrial and agricultural activities pollute water resources by directly releasing untreated waste into channels [8]. This eventually results in the unavailability of clean surface water, causing extreme dependency on the groundwater table. Therefore, the recharge of groundwater is of extreme importance to meet the global population's needs.

Groundwater/aquifer recharge is defined as water entry from the unsaturated zone to the saturated zone [9]. The degree of the recharge by natural means primarily depends on the amount of rainfall in a region that is considered a prime element for groundwater recharge [4]. The relationship between rainfall and the natural groundwater recharge is mainly governed by the region's topography, soil moisture content, rock structures, geology, the extent of fractures, elevation, slope, drainage patterns and density, landform, and land-use/land-cover and climatic conditions [3,4,10]. As a result of climate change, the overall global precipitation has decreased, resulting in a decrease in groundwater recharge [11,12]. Furthermore, the rapid worldwide urbanization also results in transforming once natural landscapes into urban water-impervious lands [12]. This limits the availability of freshwater resources but also causes hindrance in the recharge of the available water resources [13]. This puts tremendous pressure on the groundwater table considering the continuous use of groundwater to sustain essential life forms [10].

The aforementioned factors are resulting in water scarcity around the globe and are emerging as a major concern globally [14]. To temporarily maintain the groundwater levels and meet the ever-increasing water demand, artificial methods for recharging the aquifers have been employed. These methods are considered a prerequisite for sustainable groundwater management [3,15]. For this purpose, a new technique called managed aquifer recharge (MAR) has been gaining popularity lately. It is an efficient means of recycling storm water or treated sewage effluent for non-potable and indirect potable reuse in urban and rural areas [16]. Despite these artificial methods, a more sustainable approach must be adopted, and focus must be put on the natural means of groundwater recharge in line with the United Nations Sustainable Development Goals (UNSDGs).

In the case of Pakistan, the agriculture sector is the prime contributor to the country's GDP, with an overall contribution of 21% [17]. The surface water supplies are sufficient to irrigate 27% of the cultivable area, whereas the remaining 73% is directly or indirectly irrigated using groundwater. This is evident since out of Pakistan's total estimated annual groundwater extraction of 60 billion cubic meters [18,19], more than 85% is used for agricultural purposes compared to 40% in the rest of the world [20,21]. This makes Pakistan the third-largest user of groundwater for irrigation in the world [17]. Irrigation and agricultural usage have caused excessive groundwater abstraction in Pakistan, leading to water scarcity [7]. This growing deficiency of groundwater and ever-widening consumption for food production could weaken agriculture-dependent economies such as Pakistan [22,23]. In addition to the great agricultural and industrial demand for water, the increased urbanization [12] and overpopulation in Pakistan have also led to the overexploitation of ground and underground water. This, in turn, affects the water level/table and thus its availability [13].

Furthermore, the reduction of natural water pervious landscapes due to urbanization [13] and the natural reduction of precipitation due to climate change also prevent proper groundwater recharge [12]. Due to these facts, Pakistan is affected by acute groundwater shortages similarly to most developing countries [24,25]. As a result, the local groundwater levels are falling, increasing pumping costs and deteriorating groundwater quality. Thus, it is high time to carry out studies to delineate potential groundwater recharge zones in the country to use the resulting data to devise mitigation strategies [8].

Researchers have used different criteria for delineating potential groundwater zones in previous studies. Examples include the use of lineament and hydro geomorphology [26], geophysical data with geospatial information [27–33], delineation of artificial recharges sites using the use of remote sensing (RS) and geographic information system (GIS) [28,34,35], and the use of RS and GIS for geomorphic features and lineaments [36–42]. These techniques are important tools for enabling the appropriate management of crucial groundwater resources [43]. They are used to integrate various data to delineate potential groundwater zone and solve associated groundwater problems. Furthermore, these technologies are rapid and cost-effective in producing valuable data on geology, geomorphology, lineaments, slope, etc., which are important parameters for groundwater exploration, exploitation, and devising management strategy. Therefore, recent studies have used RS, satellite imagery, and GIS for hydrogeological and hydro-geomorphological investigations.

Several studies have also applied RS and GIS applications to delineate groundwater resources and potential recharge zones [8,34,44–58]. Some specific examples include a study by Saraf et al. [59], which used GIS technology to process and interpret groundwater quality data. In other studies, GIS and RS integrated with multi-criteria decision making (MCDM) have been successfully used to uncover potential recharge zones [60]. Such integration has also been used for district groundwater modeling [61], identification of water zones [62], climatic analysis for groundwater recharge [63], and aquifer analysis for recharge [64]. Selvam et al. [65] used similar techniques to decipher the groundwater recharge potential zones in a coastal area of India, which is geographically closer to our case study area. Other relevant studies using GIS have been described in Table 1 along with their respective limitations.

Table 1. Studies outlining techniques for groundwater recharge.

Technique Used	Usage and Findings	Key Factors/Parameters	Limitations	Ref
GIS and RS with fuzzy analytic hierarchy process (AHP)	Fuzzy AHP was used to delineate groundwater recharge zones. Several parameters were considered, and GIS and RS techniques were applied.	Drainage, Geomorphology, Geology, Land Use/Land Cover (LULC), Lineament, Permeability, Slope, Soil Texture, Soil Depth, Rainfall.	Fuzzy AHP brings more complexity and fuzziness to the decision-making process, thereby affecting outcomes.	[66,67]
GIS and RS with MCDM	MCDM was integrated with RS and GIS to delineate and map potential groundwater zones.	Density, Drainage Geology, Geomorphology, Lineament, LULC, Soil, Slope, Rainfall.	Various MCDM models can provide conflicting rankings of the alternatives for a common set of information.	[66,68]
GIS and RS with frequency ratio (FR)	FR, RS, and GIS were combined to delineate and map the potential groundwater zones.	Drainage Density, Soil Density, Geomorphology, Lineament Lithology, Land-use Pattern, Slope, Soil Texture, Rainfall.	The FR method utilizes past trends to predict the future outcome, making this approach depend on historical data that may not always be available.	[69–71]
Thermal infrared imagery	A thermal infrared multispectral scanner was used to delineate potential groundwater recharge zones.	Hydrogeology, Height, Thermal Parameters	Thermal activities around artificial structures such as power plants and industrial zones, clouds, and other distractions can lead to inaccurate data.	[68,72]

Table 1 shows various factors considered in respective studies for delineating groundwater resources. In this respect, a more accurate predicting model can be devised by increasing the number of influencing factors used and improving the data collection procedures. The current study uses an integrated RS and GIS technologies approach to delineate the potential recharge zones and categorize the study area into regions with high, moderate, low, and very low recharge potential. These techniques were employed in combination with the influencing factor (IF) technique, which has been previously used for studies related to semi-arid areas [10] and coastal areas [65]. However, it has not been employed in a noncoastal terrain such as the study area in the current research.

Moreover, compared to the previous studies, more factors have been introduced to increase the accuracy of the predicted results in the current study. The key assessment factors are overlaid with the spatial analysis tool of ArcGIS 9.3 to produce a combined thematic map uncovering the zones with their potential recharge. To further improve the model efficiency, more data were taken for the factors affected by temporal variations such as rainfall, etc. For other factors, data from a decade were taken and averaged before being used in the model development to nullify the effect of temporal variations. Further, thematic maps of larger spatial scales and the digital elevation model (DEM) data of a smaller resolution were used to study the targeted area comprehensively and accurately.

This study has practical applications for water management in developing and developed countries. For example, the groundwater delineation process paves the way for the relevant authorities to develop infrastructure and devise critical policies and committees to better manage the local groundwater sources. Furthermore, it can help policymakers, town planners, and construction stakeholders to plan future cities with a focus on sustainability and preserving the natural landscape required for proper groundwater recharge. Moreover, artificial structures could also be constructed to meet the associated groundwater demand and enable groundwater flow towards the region of lower concentration systematically. Such planned groundwater management will help meet the ever-increasing and widespread water demand among the country's residential, commercial, and agricultural zones. Moreover, sophisticated systems such as the one proposed in this study have lower costs and can easily interpret data to identify and suggest water contributing zones and factors. Accordingly, the applications in developing countries are numerous, which are usually concerned about the budgets of such projects. This provides incentives for developing countries such as Pakistan to use these sophisticated and integrated systems for groundwater delineation.

Further, this research contributes to the existing literature by providing an efficient integrated approach of RS and GIS coupled with the IF technique to identify the potential groundwater zones in a non-coastal study area. A similar approach was used to identify groundwater recharge zones in the coastal areas [73] and near the watershed [66]. However, such a study has not been conducted in non-coastal areas in a developing country. This presents a research gap that has been targeted in the current study. Moreover, a distinguishing element of this study is the introduction of more factors coupled with the use of more data (of a decade) for the temporal affected factors to nullify the temporal influence and variations. This was reported as a limitation in multiple similar studies. This study considers a larger spatial scale and finer resolution compared to other published works. This study can be extended to other non-coastal cities around the globe.

The main objective of this research is to identify the potential influencing factors that may impact groundwater recharge. Further, the potential groundwater recharge zones are determined by incorporating all influencing factors using the IF weightage technique. This will help the policymakers manage the groundwater resources and help researchers understand the utilization of remote sensing and GIS for groundwater analysis.

2. Study Area

The case study area of this research is Islamabad, the capital city of Pakistan, located at the edge of the Potohar plateau. It is located 14 km northeast of Rawalpindi in the province of Punjab. In terms of map reference, it is located at 33°49′ north and 72°24′ east of Greenwich [74,75]. Islamabad lies at an altitude range of 457–610 m and has 906.50 km^2 [76]. The climate of the area is humid and subtropical. May, June, and July are the warmest months, with average temperatures ranging from 36 °C to 42 °C, with temperatures sometimes as high as 48 °C. In comparison, the coldest months are December and January, with mean minimum temperatures ranging from 3 °C to 5.5 °C [77].

In Islamabad, groundwater is mainly used for drinking and agriculture purposes [78]. Since its announcement as the capital on 14 August 1967, the urbanization in and around Islamabad has been growing rapidly, leading to the development of multiple residential sectors (Sectors D to I) and more new ones being proposed, such as sectors A to C and sub-sectors I-14 to I-16 [74]. This is due to the increased migration of people in hopes of better facilities and high-end, luxurious lifestyles. According to the 2017 census, Islamabad recorded a population growth rate of 4.91 percent, and its population increased from 0.81 million in 1998 to 2.0 million in 2017 [79]. Such a mass-level migration to Islamabad increases the demand and reliance on groundwater to sustain life necessities [80].

Moreover, since Islamabad rests on the Potohar Plateau and consists of a hard rock terrain, its surface does not allow enough permeable surface for groundwater tables to be properly recharged [70]. As a result, the groundwater levels of Islamabad are depleting rapidly on an annual basis, as reported by the metropolitan corporation of Islamabad [80]. The Interior Ministry of Pakistan reported a 6 ft decrease in Islamabad groundwater in 2013, followed by a 10 ft, 16 ft, 23 ft, and 30ft from 2014 to 2017, respectively. It is estimated that groundwater levels in Islamabad have decreased by five times as of 2018 [80]. Therefore, it is imperative that new and reliable water sources must be found. Accordingly, it is necessary to carry out a study to delineate the potential groundwater zones in the city. This can help the policymakers and town planners to preserve such zones with permeable strata in the city to mitigate this groundwater recharge issue or alternatively better plan the construction activities around such areas.

Figure 1 shows the Islamabad map that is divided into five zones: zone 1 to zone 5 [74]. These zones are the administrative boundaries of the study area. They can be used as a reference for policymakers for decision making for each zone with respect to findings of this research. The city infrastructure has been planned in nine sectors in total, and an alphabet from A–I represents each sector. Every sector covers an area of approximately 2 km^2 and is further subdivided into four sub-sectors, each containing a central shopping mall, public park, and other amenities [74,81]. These sectors are the gridded divisions of the city to subdivide the capital into small units. It is similar to municipalities in developed countries and presents a grid division of the city. Out of the 5, zone 4 has the largest area, 282.5 km^2 [82], while zone 1 has the most developed residential area [83]. Zone 2 has an area of 9804 acres. Since CDA apportioned this zone to a private and cooperative housing scheme for improvement, zone 2 has become the city's most alluring space [83]. Zone 3 (203.9 km^2) is one of the most beautiful areas of Islamabad. Vacation spots such as Daman-e-Koh and Peer Sohawa are situated in this zone [84]. Zone 5 (157.9 km^2) is near the old airport and is one of the most populated zones [85].

Islamabad continues to experience expansion to accommodate the increasing population. The territorial limits of Islamabad have expanded by 87.31 km^2 from 1972 to 2009, with a significant reduction in the forest covers and other natural habitats [86]. As a result, Islamabad has registered the highest population growth rate of 4.91 percent, and the population has increased from 0.81 million in 1998 to 2.0 million in 2017 [79]. This rapid urbanization has led to many development projects being initiated within the city, including the extension of transportation systems, revision of the city master plan, and industrial and real estate development [12,75] that provide job opportunities to the residents [87,88].

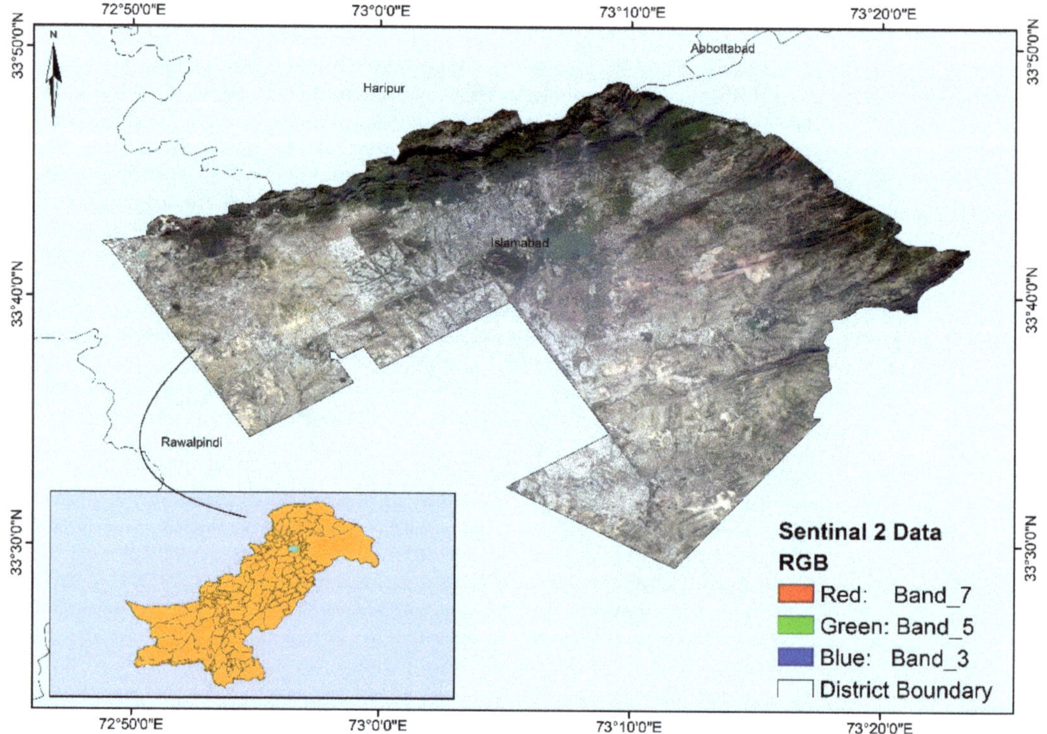

Figure 1. The study area (Islamabad) and its zones.

Due to this rapid increase in population, Islamabad has undergone many predicted and unpredictable changes [74]. One such change is the higher water demand in the region [80]. The main water resources for Islamabad are surface- and groundwater. Simli Dam and Khanpur Dam are major water resources for Islamabad. Along with the surface water, the Capital Development Authority (CDA) supplies groundwater extracted from 180 tube wells to Islamabad. Private and municipal wells are also used to fulfill the local water requirements [79]. Despite the aforementioned resources, the increased population has heightened the reliance on groundwater since it is one of the primary sources for domestic use [89]. The resulting extensive use of groundwater in the region leads to the depletion of natural groundwater resources [80].

Moreover, considering that the study area is situated in the Potohar Plateau, where the terrain is geologically composed of tertiary sandstone, limestone, and alluvial deposits [77], the recharge capacity of the region is not good. Thus, groundwater does not recharge properly, resulting in the depletion and unavailability of clean drinking water. The areas facing severe water shortage include sectors G6, G7, H8, G13, I-10, [90], and I-8/1 [91]. Thus, it is important to manage the regional groundwater resources [78]. For this purpose, the current study delineates Islamabad's potential groundwater recharge zones. The obtained potential recharge map provides the information to help improve the local management of groundwater resources. Such an assessment is important for future planning and development policies in the area and devising strategies for efficiently utilizing natural resources such as groundwater.

3. Factors Affecting Groundwater Recharge Potential

Groundwater is affected by multiple factors such as land use, slope, and lineament [92]. In addition, the study area's rainfall, soil conditions, and soil types also influence the groundwater [93]. In this study, 15 influencing factors (Ifs) were identified and used to develop potential zones to produce an error-free diverse outcome instead of a single influencing factor outcome, which provides a limited outcome in terms of accuracy [65]. Broadly, these factors can be grouped into four key groups: (1) elevation and slope, (2) rainfall and drainage, (3) land-use/land-cover and soil characteristics, and (4) faults, as listed in Table 2. The influencing factors (IFs) are the factors that can affect some features of the target object, system, or phenomenon [94]. IFs can be used as control variables to determine the key influencing factors of an object, system, or phenomenon. These have been used in various studies. In water-related studies, IFs have been used to assess the seasonal changes in water quality [95], water transport through cracks in concrete [96], distribution characteristics of microplastics in urban tap water [97], comprehensive evaluation and urban agglomeration water resources carrying capacity [98], and others. Accordingly, in the current study, IFs are used to delineate potential groundwater recharge zones in Islamabad, Pakistan. These 15 key factors are listed in Table 2 and discussed subsequently.

Table 2. Factors influencing groundwater recharge classified criteria.

Group	Key Factors	Source of Categorization	Selected Ref
Elevation and slope	Elevation	Height value	[99]
	Slope	Slope gradient	[100]
	Slope length	Measurement of slope lengthwise	[100]
	Aspect	Aspects of area	[70]
	Total wetness index	Runoff collection and infiltration	[101]
Rainfall and drainage	Rainfall	Zones with rainfall recement	[93]
	Drainage distance	Distance to drainage networks	[102]
	Drainage density	Density values for drainage	[103]
Land use/land cover and soil characteristics	Land use/land cover	Satellite imageries	[104]
	Soil	Textures	[72]
	Lithology	Rock type details	[105]
	Plan curvature	Detailed area curvature	[70]
	Profile curvature	Flow categorization	[61]
Faults	Distance to faults	Lineament distance	[106]
	Fault density	Density for lineaments	[107]

3.1. Elevation

Surface elevation plays an important part in groundwater recharge. It is the primary source for triggering the water flow under gravity [99]. Elevation studies highlight the regions contributing to the groundwater flow; i.e., higher slopes allow less water infiltration. Islamabad has variable elevation, as it is composed of both mountainous regions and flat surfaces. The mountainous regions have higher slopes that transfer water from higher elevation to lower elevation. A similar study found designated slope as a very important factor in groundwater recharge [108]. Previous research has indicated that gentle slopes and flat surfaces have higher recharge potential compared to inclined surfaces and higher slopes [100]. Therefore, the inclusion of surface elevation signifies the groundwater flow and determines the flow direction as it induces the flow under gravity [108–110]. A major part of the current study area consists of mountainous regions with high surface elevations. Therefore, it is used as a key factor in the current study.

3.2. Slope

Slope defines the extent to which groundwater can be recharged with the precipitated water [100]. The regions with higher slopes experience rapid water running over the surface, hindering the absorption of precipitated water into the groundwater [65]. Conversely, in areas involving lower slopes and vegetation, the water cannot run off the surface rapidly, and thus, more of it is absorbed in between the pores and adds to the groundwater table [100]. In relevant studies, it has been established that the topographical feature of the slope impacts the directional flow of water and indicates its accumulation. Further, the flat surfaces with gentle slopes displayed the highest infiltration capacity [109,111], thus contributing to an increase in the groundwater table. Our study area, Islamabad, comprises high-slope areas, as the northern outskirt is predominant with the mountain region, making the slope one of the important factors for the current study. Accordingly, the slope has been included as one of the key factors in this study.

3.3. *Slope Length*

Slope length indicates the physical characteristic of the slope in terms of its extension and magnitude. It helps determine the flow and highlight possible regions of groundwater retention [100]. Being a primary factor for groundwater contribution, slope length determines runoff strength and the groundwater flow direction. Slope length also indicates the amount of rainfall that would reach the groundwater table through infiltration [100,108,109]. Gentle slopes have greater infiltration capacity, displaying greater groundwater recharge potential and vice versa [100]. Slope lengths help understand the flow of precipitation as the water runs off from higher elevation towards the lower elevation. Considering that our study area is predominantly sloped in the northern parts due to mountain ranges, this is an important factor in this research.

3.4. *Aspect*

The front-facing side of a slope, or generally the face of the slope, is defined as the aspect [109]. When combined with the slope and slope length maps, the aspect can indicate the extension of a particular slope in a specified direction to unveil the potential flow of groundwater [70]. The aspect proceeded by flat surfaces or gentle slopes allows the precipitated water to flow smoothly and streamlined, thereby maximizing the area's infiltration capacity, leading to greater recharge [70,109]. Islamabad is composed of higher elevations at the northern outskirt that stretches predominantly towards the east. The aspect is proceeded by the gentle and flat surfaces containing the residential zone of Islamabad. The aspect can indicate the flow of precipitation and groundwater accumulation towards the inner zones in Islamabad. Therefore, it is used as a key factor in the current study.

3.5. *Topographic Wetness Index (TWI)*

The topographic wetness index (TWI) is a steady-state wetness index used to quantify topographic control on hydrological processes [101]. TWI indicates control over the groundwater processes, such as flow and retention in a specified zone. Several studies have been published explaining the process to calculate the TWI [101,111]. TWI provides detail about the flow of groundwater considering the effect of the slope. TWI can impact groundwater flow and its occurrence in a varied elevation areas such as Islamabad. Numerous studies have linked TWI, slope, and elevation effects to the water recharge potential [65,100,109]. TWI gives an indirect indication of water moisture availability and potential recharge zones. Therefore, this has been used as a key factor in the current study.

3.6. *Rainfall*

Rainfall or precipitation positively affects the groundwater table because of larger water infiltration [93]. Rainfall has always been a reliable source of freshwater [65]. Previous research has linked both the movement and occurrence of ground and surface water to mainly depend upon rainfall [108,111]. Considering that the rainfall quantities of the study area can indicate the movement of groundwater and can depict the flow and accumulation of water bodies, it is important to include this factor while investigating groundwater recharge zones [109]. Therefore, it is considered significant for Islamabad as well and used in the current study as a key factor. Further, since Islamabad is a rainy area, and some mountainous regions in the area receive more rainfall than other parts of Pakistan, rainfall is a key factor dictating the local climate and recharging the water sources.

3.7. Drainage Distance

Drainage distance is crucial for water studies, such as its occurrence and flow assessments. Drainage distance highlights the geological distance between successful drainage zones. The drainage density indicates the drainage condition of the water shed [109]. Groundwater movement beneath the surface can be unfolded by uncovering the drainage networks according to lineaments such as underground fractures and faults. Lineaments impact groundwater movement within the surface [65,102]. For studies relating to groundwater recharge, the inclusion of drainage distance is crucial because of its relationship with permeability which is the property that describes the flow of water bodies beneath the earth's surface [108]. A similar study prioritized areas comprising more considerable drainage distances for the groundwater recharge potential and vice versa [109]. Accordingly, drainage distance has been shortlisted as a key factor in the current study for the study area of Islamabad.

3.8. Drainage Density

Drainage density is the ratio of all the streams over the area to the total area [65]. It indicates the drainage capacity and measures the drainage over a particular watershed [103]. A higher drainage density region indicates a well-distributed water flow area with multiple streams contributing to the flow and recharge and vice versa. A similar study has linked higher drainage density to greater groundwater recharge potential [108]. According to the previous research, the drainage density contributes toward the groundwater recharge as it describes the flow pattern and the occurrence of water beneath the surface [65,109]. As Islamabad receives higher rainfall towards the northern outskirts, and the density of the drainage network would greatly influence the flow and occurrence of groundwater in the region, drainage density is selected as a key factor for this study.

3.9. Land Use/Land Cover

Land use/land cover involves several elements, including soils, human settlements, vegetation cover, waste lands, etc. [112]. The settlement in an area affects the groundwater due to the human-made structures. The land vegetation covering is one of the major groundwater factors used for retaining water [65]. Depending upon the porosity and permeability, the soil conditions of an area also control groundwater seepage through the surface. RS and GIS usage for land mapping has gained popularity recently [6,104]. With the help of land use/land cover, a similar study has linked the best and most abundant agricultural practices with groundwater availability over the study region [109]. For Islamabad, the regions should be studied based on their demand for groundwater, thereby necessitating the inclusion of land use/land cover in this study.

3.10. Soil

Soil is one of the most important factors for groundwater recharge since groundwater movement through the surface is controlled by soil type and properties [65]. Accordingly, parameters such as porosity and permeability are of utmost importance and are crucial to groundwater flow [72]. Moreover, the soil is also responsible for the filtering or buffering activities between the atmosphere and the groundwater in the biosphere [65]. Therefore, it is considered one of the prime influencing factors in groundwater recharge analysis. Considering that soil properties vary in each region, large-scale test data of the soil type might be required. In previous research consisting of a variable soil type for groundwater recharge, higher weightage has been allocated to the soil as a contributing factor. Accordingly, it has been declared as one of the high IF [109,111]. Furthermore, greater variations of the soil types were seen influencing the groundwater recharge potential in relevant studies [108]. In the current study area, the terrain has high soil variation; the northern outskirts are predominant with mountainous soil, and the southern outskirts are predominant with loamy soils. Thus, soil type is selected as a key factor in this study.

3.11. Lithology

Lithology refers to the physical appearance of rocks. Rock characteristics impact the movement of water beneath the surface [105]. In smaller rocks, the water finds more passage for movement and vice versa. If the grains are arranged in a well-graded manner, there is no passageway for water and vice versa [65]. Lithology plays an important part in dictating groundwater flow via channels, permeability, and occurrence [104]. This factor has been considered in a similar groundwater recharge study outlining the influence of rock type, soil type, and the higher permeability on groundwater movement and occurrence [105,109]. Several other factors may influence the lithological characterization and its impact on groundwater recharge. However, this research is limited to lithological information and does not have permeability, porosity, or grain size information. Further, it is based on a literature review for assigning weightages of lithologies. The terrain is composed of various rock types in our study area, including tertiary sandstone, limestone, and alluvial deposits [84]. Lithology contributes to groundwater flow and is included in the current study [105].

3.12. Plan Curvature

Plan curvature explains the geometry of a particular region. It helps understand the way contours intersect the horizontal region and their impact on the slope inclination of a particular zone [70]. It explains the flow of groundwater and helps establish a generalized flow pattern. Plan curvature approximates the inclination of various zones that impacts groundwater recharge through topographical influence [111]. The inclination of the area is marked with a slope that runs from the region of higher inclination towards the lower inclination, thus indicating groundwater flow [100,109]. The region of Islamabad is higher in inclination towards the northern region that goes down towards the southern zones. This is because the northern area is comprised of mountainous regions, and the southern zone consists of high-populous flat regions, establishing a generalized pattern of inclination decrease [111]. The inclination and gentle slopes and the presence of flat surfaces greatly influence groundwater recharge [108]. Therefore, plan curvature has been included as a key factor in this study.

3.13. Profile Curvature

Profile curvatures define the nature of the ground zones under study: linear, concave, and convex. It is defined as the line parallel to the direction of the maximum slope. Patterns might indicate a general linear formation with a defined value approaching zero. A positive value indicates an upward concave profile, while the negative region represents an upward convex profile [70]. The profile curvature helps classify the area into lower or higher water-retention zones depending upon its convexity and concavity. Accordingly, the regions comprising elevated convex profiles within center zones are regarded as less water holding and vice versa [110]. The curvature of the study area is included in this study to assess its effect on the water-retention capability of the zone following related studies [109,110]. Considering the variability of Islamabad's surface in terms of slope and elevation, it is important to consider the influence of profile curvature on groundwater recharge in this region. Therefore, this factor has been used in the current study.

3.14. Distance to Fault

Faults describe the change in geological composition in a particular zone [106]. These indicate the movement and change a particular rock surface has undergone in a specified period. For example, earthquake-induced faults can indicate rapid geological movement beneath the surface. The parameters of faults can have vast ranges. Distance to faults impacts the flow and occurrence of groundwater [108]. It is important, as it indicates groundwater flow and can highlight the zones contributing to underground-water flow [106,109]. In our study area, Islamabad and nearby regions have more faults that influence the groundwater recharge. Thus, this factor is included in the current study.

3.15. Fault Density

The magnitude of faults (density) indicates the potential groundwater regions. In a similar study, lineaments such as faults have been reported to impact the groundwater recharge potential zones and are considered key IF [108]. Fault density helps determine the occurrence and movement of groundwater beneath the surface. Many relevant studies have included fault density as a key factor in assessing groundwater recharge potentials [60,65,66]. As previously discussed, Islamabad has higher faults than the rest of the country. Therefore, fault density is included as a key factor in the study.

4. Methodology

The current study follows a four-step approach. In the first step, the relevant thematic layers are identified. First, the thematic layers used for the study were extracted that act as input data for the eventual delineation of recharge zones. These thematic maps present the geographical map of the study region in accordance with the subject matter. The current study utilizes thematic maps for 15 hydrological factors. These include distance to faults, land use, lithology, drainage density, slope, soil, rainfall, plan curvature, fault density, profile curvature, TWI, elevation, aspect (the front-facing direction of a slope), drainage distance, and slope length. These factors were extracted from previous literature [60,66,87,103,113] considering the geological properties of the study area as listed and are discussed in Section 3 of the study.

The thematic maps used in the research were generated at a 1:200,000 scale considering that this would eventually increase accuracy. In addition, the majority of the data sets were available at this scale. The differing scales were later normalized for the sake of uniformity. The digital elevation model (DEM) data are used on a global scale at 30 m $\times$ 30 m resolution for topographic analysis. This resolution is highly important, as it contributes to how sharply the objects can be seen in an image. It represents the size of the tiniest feature captured by a satellite sensor or portrayed in a satellite photo. It is commonly expressed as a single number representing the length of one of the sides of a square (grid) [12]. In addition to the normalization of the input data, uniformity is ensured in their format for easy integration of these thematic maps into the GIS platform. For this purpose, the acquired maps are converted into raster form before integration with the GIS.

In the second step, the pre-processing of the thematic layers was performed to ensure uniform projection and resolution. This is followed by the assignment of scores and suitable weightage to each factor. During weightage overlay analysis, the ranking was given for each parameter of each thematic map, and weights were assigned according to the influences (following IF technique) of the feature on the hydrogeological environment of the area coupled with that parameter's contribution toward the groundwater recharge as shown in previous researches [65,108,109].

The IF technique was used to assign scores and get a diverse and error-free outcome. A diversely produced thematic map considers the input from multiple hydrological procedures, thus not relying on a single hydrological process where the outcome can be manipulated and is prone to error. Moreover, due to finer resolution, any errors in the weighted overlay analysis within the ArcGIS were eliminated since such resolutions result in finer interpretation.

The third step involves using ArcGIS to deploy the thematic layers to get the processed images containing the potential zones. In this step, all the scored thematic maps along are integrated by employing the "Spatial Analysis tool" in ArcGIS 9.3, whereby rankings are assigned to all the thematic maps. Then, these weighted thematic maps are overlaid using ArcGIS to highlight the potential recharge zones. In the fourth (last) step, the study area was categorized based on the potential groundwater rechargeability into five different classes: poor, low, medium, high, and best in terms of their capability for the groundwater recharge potential.

Figure 2 shows a flowchart summarizing the methodology used in this study. The associated steps include acquiring the data, converting to raster, preprocessing (confirming projection and resolution coupled with assigning scores and weights), integrating GIS for final output, and categorizing the study area based on groundwater recharge capability. Figure 2 also shows the source of the acquired data. Accordingly, the thematic maps are acquired from Landsat-8 TM Satellite, Aster DEM, and soil and geological maps of Pakistan. The following sections explain the IFs used in this study in detail, their sources, and the procedure for assigning weights to each of these factors.

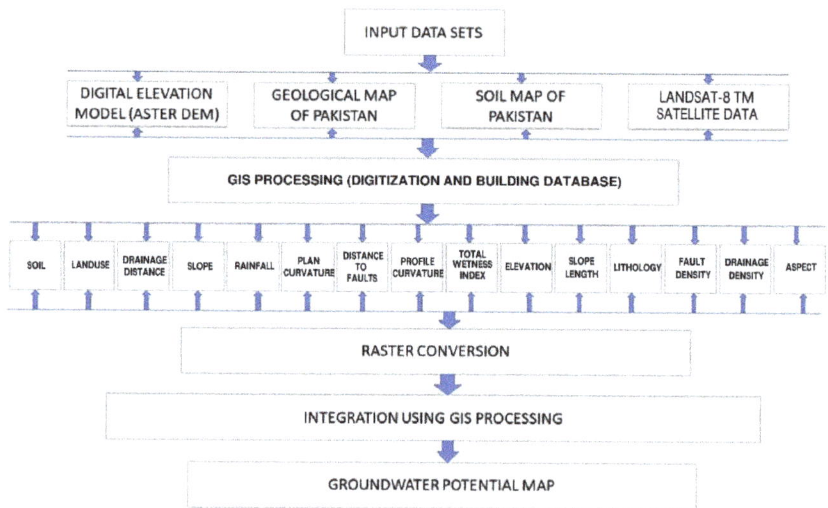

Figure 2. Flowchart for potential groundwater assessment using integrated remote sensing and GIS techniques.

4.1. Acquisition of Thematic Maps for Contributing Factors

Table 3 below enlists the sources for acquiring thematic maps for all the contributing factors. The soil thematic map was generated using the Soil Map of Pakistan [114]. Land use, rainfall, and TWI thematic maps were generated using Landsat 8TM satellite data. Drainage distance, slope, plan curvature, profile curvature, slope length, elevation, drainage density, and aspect thematic maps were generated using ASTER global DEM. Finally, distance to faults, lithology, and fault density thematic maps were generated using data from the geological map of Pakistan on a scale of 1:200,000 [115].

Table 3. Acquisition of Thematic Maps for Contributing Factors.

Factors (units)	Sources of Acquisition for Thematic Maps
Soil	Soil map of Pakistan
Land use, TWI, Rainfall (mm/y)	Landsat-8 TM satellite data
Drainage distance (m), Slope (degree), Plan curvature, Profile curvature, Slope length (m), Elevation (m), Drainage density, Aspect	ASTER GDEM
Distance to faults (m), Lithology, Fault density	Geological map of Pakistan

These thematic map data were cross-checked using ground surveys for cross-validation. The imagery was visually interpreted to delineate rainfall, land use, and other factors with the help of slandered characteristic image-interpretation elements such as tone, texture, shape, size, pattern, and association using the Landsat 8 satellite data products. These data sets are used for assessing groundwater recharge potential [65,109,111].

4.2. Weightage Assignment via IF Technique

Weights and rates were assigned to the factors to obtain a final combined recharge potential map. Using the IF technique, the influence of various factors was taken into account, and the level of impact they have on the hydrological aspect of groundwater flow and its occurrence was assessed. A weightage approach was included as used by [65] to assign weightage to the factors that would ultimately define the control they can assert over the groundwater recharge of the study area. The current study follows a similar approach. In assigning weights to the considered factors, five major descriptive levels were plotted for each factor ranging from very high to very low, including some interrelated levels. These weightage values range from 10 to 1 point, i.e., a very high range is assigned a score of 10, and the minimum level is 1 following relevant groundwater studies [113,116]. These weights for each factor were assigned based on their degree of impact on groundwater recharge as extracted from relevant literature [6,10,63,113].

5. Results and Discussions

This section presents the results and discussions in line with the adopted method.

5.1. Spatial Analysis of Considered Key Factors

Figure 3 represent the resulting thematic maps of the 15 considered factors for the current study area. Figure 3a highlights the wells or water extraction points in the study area. These are primarily located in the residential zones and plain areas of Islamabad. Figure 3b shows the thematic map of rainfall for Islamabad. The resulting map highlights that Islamabad receives ample rainfall. Further, it shows a rhythmic increase in rainfall volume from south to north. The northeast outskirts receive the highest rainfall, consisting of regions from Rawat to Crore Village. Low-rainfall regions are evident in the southwest. Considering the high rainfall in the northeastern regions, there are more chances for more groundwater recharge and high groundwater levels in alluvial plains [64], thus displaying a higher potential for groundwater recharge. Moreover, the map shows that around 44% of the area receives less than 882 mm of rainfall, 16% area receives rainfall between 882–999 mm, 10% area receives rainfall between 999–1116 mm, 9% area receives rainfall between 1116–1233 mm, while 21% of the area receives most rainfall ranging between 1233–1350 mm. This shows that around 40% of Islamabad receives good rainfall. This assessment can help policymakers preserve the natural terrain in the region receiving more rainfall and utilize it for groundwater recharge.

Figure 3c shows Islamabad's thematic layer of plan curvature data. The figure categorizes the regions based on concavity and convexity. The map shows that the northeast region of Islamabad is composed of higher convexity, whereas a systematic decrease in convexity is observed from north to south. This indicates a higher surface and altitude in the north and a gradual decrease towards the south. This heavily contributes to the groundwater flow from north to south, where a gentler slope and plain area can accumulate this water and get recharged. A similar study accounted for alluvial plain and gentle slopes to be more promising for groundwater potential due to large infiltration rates, high porosity, and permeability [116].

Figure 3d shows the thematic layer of soil data for Islamabad, where the region is classified based on soil composition. Soil types impact groundwater flow directly, but they also impact other important phenomena, such as infiltration [117], which ultimately impact groundwater recharge. The soil conditions define permeability, which impacts groundwater infiltration and soil porosity. For example, the calcareous loamy soil is abundant in arid and densely populated areas. Figure 3d shows that the mountainous soil forms the northern edge of Islamabad that receive a decent amount of rainfall. Such soil helps infiltration, enabling the groundwater to flow towards the inner zones. While no definite pattern exists throughout the study area, calcareous soils are mostly reported for various regions.

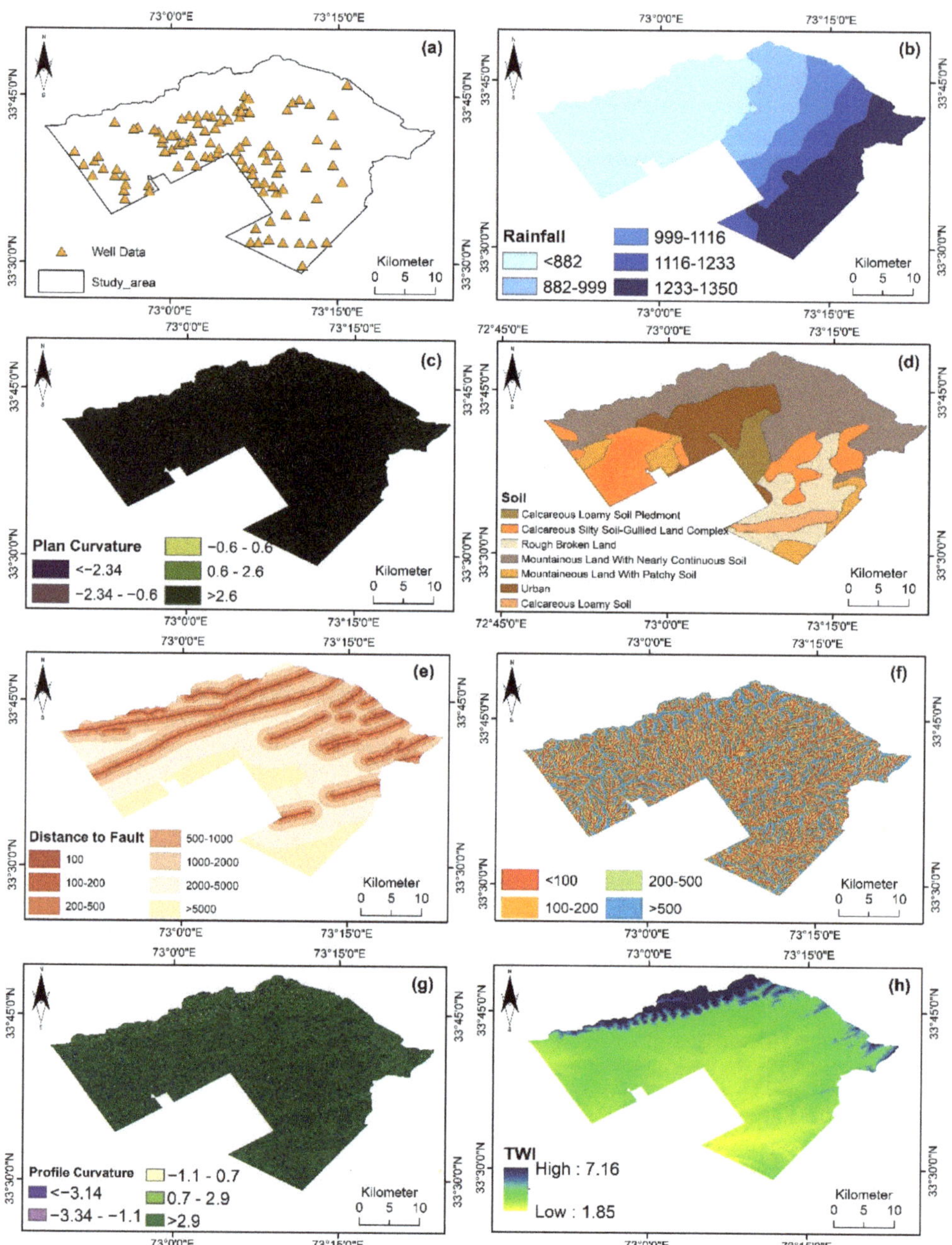

Figure 3. *Cont.*

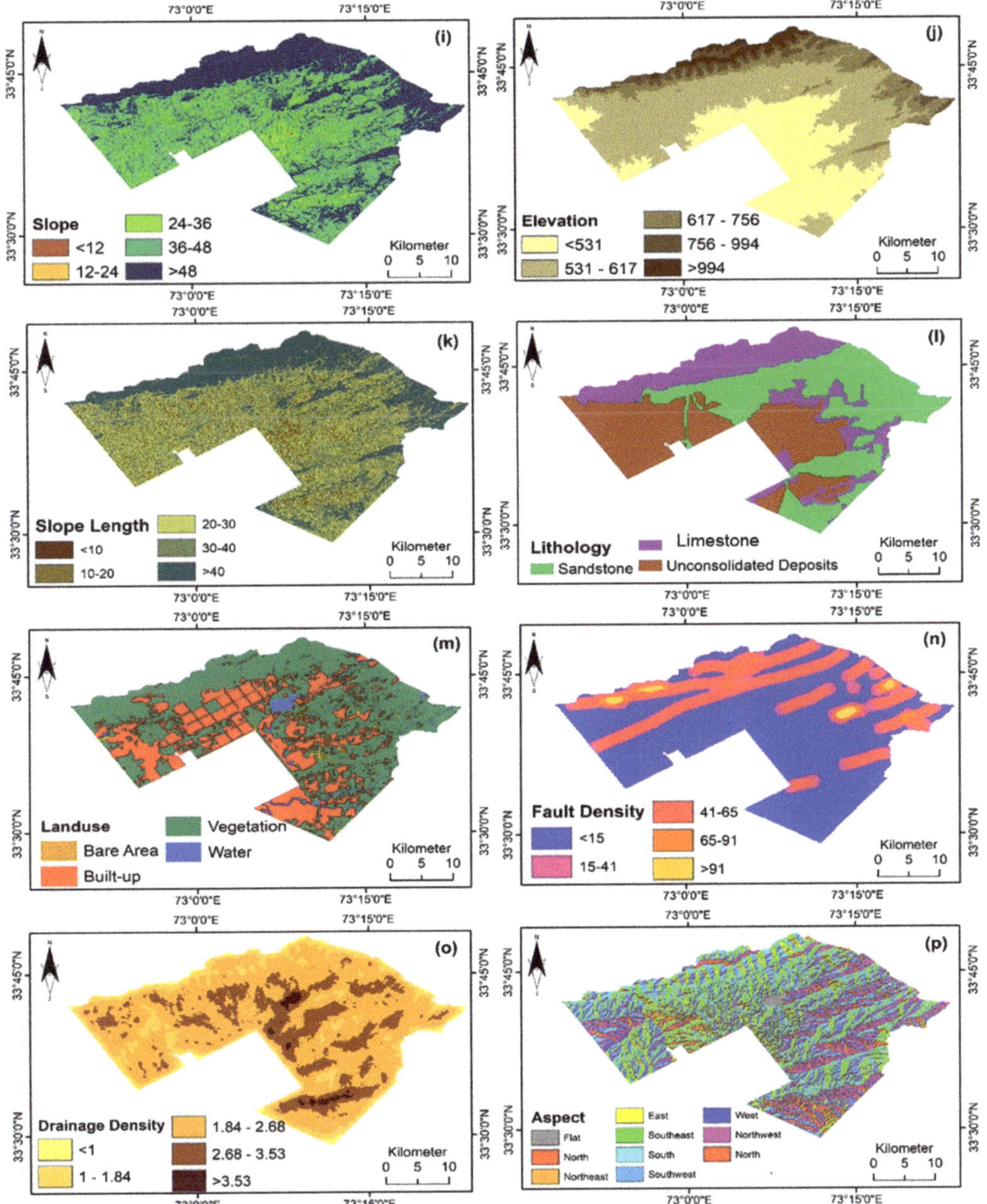

Figure 3. Thematic layers of selected factors for Islamabad's data (part 1), (**a**) well data, (**b**) rainfall data, (**c**) plan curvature, (**d**) soil data, (**e**) distance to fault, (**f**) drainage distance, (**g**) profile curvature, (**h**) TWI, (**i**) slope, (**j**) elevation, (**k**) slope length, (**l**) lithology, (**m**) land use, (**n**) fault density, (**o**) drainage density, (**p**) aspect.

Figure 3e shows the thematic layer map of distance to fault for Islamabad. This map categorizes regions with respect to distances to faults. Considering that the faults act as points with more recharge capability, more distance from faults implies less recharge capability and vice versa. In this respect, Figure 3e shows that the major faults are all located on the outskirts of Islamabad. The zones comprising convex geological features and landscapes have nearby faults, whereas the southern regions comprising more land use and less geological convexity comprise low distances to faults. This aligns with several studies that have established patterns with lineaments and groundwater recharge potential [10,118]. Overall, the southern regions with less distance to faults display more recharge potential in the current study area.

Figure 3f shows the thematic layer of drainage distance for Islamabad. It categorizes the study area based on the distance of various zones from the drainage networks. Figure 3f shows that the study area comprises abundant and closely located drainage networks. However, there is no defined pattern for the drainage distances in the study area. Considering that a lesser distance from the drainage pathway displays higher groundwater recharge potential [119], the drainage distance thematic map suggests that the study area has a larger potential for groundwater recharge. Further, there is a well-distributed groundwater flow throughout the region. Figure 3g shows the thematic layer of profile curvature for Islamabad. It highlights the geological characteristics of Islamabad and depicts the concavity and convexity of the region. It is indicated that the outskirts of the northern region are higher in altitude and contribute to the groundwater flow under gravity. A higher profile value indicates a rising elevation, ensuring a systematic flow towards the inner edges with the highest and densest land use in Islamabad. This is in line with a previous study's findings that suggest a higher potential of gentle slopes for groundwater recharge [116].

Figure 3h shows the thematic layer of TWI for Islamabad. The TWI map shows the impact of geology on the hydrological aspects. The outskirts, shaded in deep blue in Figure 3h, show the zones with geological makeup that impact regional hydrology. Following our thematic maps for the land use, well data, and rainfall, the TWI highlights Islamabad's northern outskirts as the areas directly reaching the groundwater. The inner edges with lower index value contribute little to the groundwater flow, while the geological makeup of the outermost skirts contributes greatly to the groundwater flow towards the center, housing the area with the highest and densest land use. A direct relationship between the higher TWI value was also established by another study [120]. Following our findings, a higher TWI value suggests a better groundwater recharge potential in the Islamabad region.

Figure 3i shows the thematic layer of slope data for Islamabad, showing that the northern outskirts of Islamabad have the highest slope. The slope plays an important part in determining the runoff direction of groundwater. The thematic map indicates that 23% of the region has a slope greater than 48 degrees, 38% has a slope ranging from 36 to 48, 16% has a slope ranging from 24 to 36, 9% has a slope ranging from 12 to 24, and 4% of the region has a slope less than 12 degrees. The figure shows that the outskirts of Islamabad in the northern region comprise the highest slopes due to mountains that promote a rapid runoff towards the south. While some water is lost during the runoff, infiltration takes water to the deep soil layers, contributing to recharging the local groundwater table.

Islamabad's outskirts comprise Attock, Wah Cantt, and Taxila in the west; Murree in the northeast; Haripur in the north; Gujar Khan, Rawat, Mandrah, and Kahuta in the southeast; Rawalpindi to the south and southwest; and other Punjab regions in the east. The greater slope in the northern region ensures a flow of water towards the south with the highest settlement and greatest water recharge potential.

Figure 3j shows the thematic layer of elevation data for Islamabad. Islamabad is high on the northern edge due to the mountains that decrease towards the south. The area with residential zones, i.e., the inner edges, and that towards Rawalpindi has higher population density and low elevation. This systematic decrease of elevation contributes directly to the groundwater flow as the water flows under the action of gravity. The higher elevation area also receives greater rainfall, as shown in our rainfall thematic map, ensuring infiltration and surface runoff towards the inner edge. Thus, the area with higher elevation retains rainwater for a lesser time duration and generates more runoff towards the residential areas in Islamabad, in line with published studies [64].

Figure 3k shows the thematic layer of slope length data for Islamabad that highlights the lengths of slopes in the region. Longer slope lengths are evident on the northern outskirts, while a rhythmic slope length decrease can be observed towards the south. The area with the highest land use comprises regions with lower slope length values. The groundwater flows from the northern sides with the highest slope lengths promoting recharge potentials and infiltration. The gradually decreasing slopes towards the center help with groundwater recharge to meet the requirements of the local population.

Figure 3l shows the thematic layer of lithology data for Islamabad that shows regions with limestone and unconsolidated deposits to be abundant in the area. However, there is no defined pattern, and the data are scattered throughout the region. The concentrated regions are highlighted in red, green, and purple colors in Figure 3l. There is a presence of sandstone in the northeastern region along the dense mountainous regions that continues towards the northwestern region.

Further sandstone and unconsolidated deposits are seen within the areas of highest land use towards the southwest. Past glacial activity has contributed to the unconsolidated deposits in the region due to the weathering of rocks. A previous study also established a pattern between the weathering of rocks towards the increased groundwater recharge potential [121]. An increased recharge was also observed in the area of higher unconsolidated deposits in another study [120]. Accordingly, there is a greater potential for groundwater recharge in the study area.

Figure 3m shows the thematic layer of land-use data for Islamabad, showing areas such as bare land, water bodies, built up, and vegetative regions. Such a map displays the variation of population density and associated water demand throughout the study area [10]. The thematic map for Islamabad indicates that 4% of the region comprises bare land, 36% is built-up region, 51% is vegetative, while 9% of the study area is composed of water bodies. Further, it can be observed that most of the built-up region is around the inner region of Islamabad. This region falls towards the city of Rawalpindi, which has a far greater population density than Islamabad. The runoff from the northern region infiltrates into the groundwater table around these internal regions, where there is a greater need for water.

Figure 3n shows the thematic layer of fault density for Islamabad that highlights geological features induced by the movement of rock bodies. These faults govern groundwater flow following their complex and favorable topography. Accordingly, the fault densities for the area include 43 % area with less than 15 fault density, 6% area ranging from 15 to 41, 35% ranging from 41 to 65, 7% ranging from 65–91, while 9% of the study area has fault density greater than 91. The map indicates that the northeastern edges of Islamabad consist of lower-density faults than the northwestern region, where there are more mountains. The maximum land use is towards the internal regions with no major geological faults. Previous studies have linked fault-dense regions with higher groundwater recharge potential [10]. Thus, there is a higher recharge potential in the northwestern areas of Islamabad.

Figure 3o shows the thematic layer of drainage density for Islamabad that highlights the northeastern regions to have streams or rivers with relatively long lengths. This ensures a deep-water flow towards the inner edges of Islamabad. Thus, the northeastern region contributes majorly towards the groundwater flow in the areas of highest land use. Further, a flow from the northern to the southern edge is seen with the major contribution from the northeastern region. A previous study showed that high-density drainage regions have greater groundwater recharge potential [122]. This is in line with the current study where major water sources contribute to the water recharge. The same has been highlighted by the rainfall thematic map, where the northeastern region receives most of the rainfall and has a high drainage density, thus contributing to the groundwater flow and recharge.

Figure 3p shows the thematic layer of aspect data for Islamabad that categorizes regions based on their compass directions. The aspect map lists out the front-facing direction of regions along with the compass. For example, the major constituting region in the northeast contains southeastern front-facing regions that align with thematic maps of land use and wells in the study region. The dense regions with most residential and commercial zones are in the southeast. The flow from the north region is ensured towards the southeast region. The southeastern compass front directions of the geological regions act as a gentle slope that promotes groundwater recharge in Islamabad [116].

These influencing factors were considered based on a literature review and classified based on their impact on groundwater recharge contribution, i.e., the class at which lesser the groundwater recharge potential would rank lower and vice versa. For example, a higher slope would have lesser groundwater potential, or a lower TWI would mean low water moisture and low groundwater recharge potential; hence, these classes would have lesser weightage.

5.2. Weightage Calculation for Influence Factor (IF) Techniques

After obtaining the individual thematic maps for each of the contributing factors, these factors were integrated to obtain a potential holistic map that highlights the recharge potential of Islamabad. Accordingly, weights and rates were assigned to the 15 key factors. For incorporating the mutual influence of the factors, rate values were assigned to them. Two points were given for every major effect, while one point was given to the corresponding factor for each minor effect. The cumulative weightage of both major and minor effects was considered for calculating the relative rate, as shown in Table 4. Table 4 shows that factors such as lithology influence six of its fellow factors majorly. It has a noticeable impact on the lineament, drainage, land/use, slope, and soil types. Thus, it has been assigned a value of 2 six times (2 × 6 factors).

Table 4. Relative rates and scores for each potential factor.

Factors	Major Effect (*A*)	Minor Effect (*B*)	Proposed Relative Rates (*A* + *B*)	Normalized Relative Rates (*Y*) in %
Distance to Faults	2 + 2 + 2 + 2	1 + 1 + 1	11	**6.875**
Land use/Land cover	2 + 2 + 2 + 2 + 2 + 2	1 + 1 + 1 + 1	16	**10.000**
Lithology	2 + 2 + 2 + 2 + 2 + 2	1 + 1 + 1 + 1	16	**10.000**
Drainage Density	2 + 2 + 2 + 2 + 2 + 2	1 + 1 + 1	15	**9.375**
Slope	2 + 2 + 2 + 2 + 2	1 + 1 + 1	13	**8.125**
Soil	2 + 2 + 2 + 2 + 2	1 + 1 + 1	13	**8.125**
Rainfall	2 + 2 + 2 + 2 + 2	1 + 1 + 1	13	**8.125**
Plan Curvature	2 + 2	1	5	**3.125**
Fault Density	2 + 2 + 2 + 2	1 + 1	10	**6.250**
Profile Curvature	2 + 2	1	5	**3.125**
TWI	2 + 2 + 2	1 + 1	8	**5.000**
Elevation	2 + 2 + 2 + 2	1 + 1 + 1	11	**6.875**
Aspect	2 + 2 + 2	1 + 1	8	**5.000**
Drainage Distance	2 + 2 + 2 + 2	1 + 1	10	**6.250**
Slope Length	2 + 2	1 + 1	6	**3.750**
			Σ = 160	Σ = 100
			160	100

Similarly, other factors have also been assigned their respective rate values using the same approach. Overall, the major effect (*A*) and minor effect (*B*) are summed for all factors, and their cumulative sums are calculated for each factor to get the proposed relative rates. The cumulative proposed relative rates sum up to 160. Using this value, the normalized relative rates are calculated, where the proposed relative rate of each factor is divided by the cumulative proposed related rates and multiplied by 100 using Equation (1). The values are rounded off to the nearest integer.

$$Normalized\ relative\ rates\ (Y) = \frac{Proposed\ relative\ rates\ (A+B)}{Commulative\ Proposed\ relative\ rates\ (\Sigma(A+B))} \times 100 \quad (1)$$

After the assignment of rate values, the next step is to assign weights. In this process, five major descriptive levels are plotted for each factor ranging from very high to very low, including some interrelated levels as shown in Table 5. Factors contributing majorly, such as rainfall, can be seen as very dominant in relevant studies [108,109] and in abundance in the southeastern regions of the study area and thus were assigned higher weights. In contrast, factors such as profile curvature were assigned a lower weightage, as the area followed a rhythmic curvature, and the influence of curvature was not dominant in terms of groundwater flow, as evident from Figure 3 (previously shown).

With a weightage of 8.1%, rainfall is a dominant factor in the southeastern parts of the study area. Plan curvature data indicate a slight shift in curvature as seen from the thematic map and thus were assigned a weightage of 3.1%. The higher curvature would result in a greater flow of water beneath the surface [66]. Soil is the primary factor that controls seepage and the associated groundwater recharge [117]. Thereby, it was assigned the highest weightage (8.1%). Likewise, faults being the primary indicator of geographical movement (earthquakes or tectonic) over the years indicate a weaker and vulnerable zone suspectable to the greater flow of groundwater channels beneath the surface. It adds greatly to the groundwater recharge and was hence assigned a weightage of 6.8%. Drainage distance, profile curvature, and TWI were assigned weights of 6.2%, 3.1%, and 5%, respectively.

The data obtained from thematic maps do not indicate an abrupt or dominant effect of these considered geographical features (key factors) over the study area, thus acquiring a lower weightage in our study area. The slope indicating the natural flow of water towards the lower altitude area was assigned a weightage of 8.1%. Elevation and slope length were assigned the weightage of 6.8% and 3.7%, indicating the flow towards lower-elevated areas and the flow speed. Accordingly, the lower the speed, the greater the infiltration and vice versa [122]. Lithology has been assigned a weightage of 10%. It indicates the rock characteristics that dictate the water flow beneath the surface in channels and streams. Land use is another primary factor that was assigned 10% weightage. It has been utilized by several related studies [60,66]. Finally, fault and drainage densities and aspects indicated the magnitude of faults, drainage networks, and front-facing direction of slopes signifying the flow of groundwater beneath the surface and were assigned weights of 6.2%, 9.3%, and 5%, respectively, in this study.

After the assignment of rates and weights, the % influencing score was calculated using Equation (2). The % influencing score is defined as the percentage of factor effect on recharge potential (%) and is shown in Table 5 for each factor, where X is the normalized weight from 1 to 10, and Y is the rate from 1 to 10.

$$\%\ influencing\ score = \frac{Total\ Weightage\ \Sigma(X \times Y)}{Grand\ Total\ Weight\ (GTW)} \times 100 \quad (2)$$

Table 5. Weight evaluations of factors influencing potential recharge capacity.

Factors	Categories	Effect	Normalized Weight (X) (1–10)	Normalized Relative Rates (Y) Based on Table 4 (1–10)	Weighted Rating ($X \times Y$)	Total Weightage max(X) $\times$ Y	MAX Effect on Recharge Potential (%)
Rainfall	<882	Very Low	2	8.125	16.25	81.25	**8.125**
	882–999	Low	4		32.50		
	999–1116	Medium	6		48.75		
	1116–1233	High	8		65.00		
	1233–1350	Very High	10		81.25		
Plan curvature	<−2.34	Very Low	2	3.125	6.25	31.25	**3.125**
	−1.74	Low	4		12.50		
	−1.2	Medium	6		18.75		
	0.6–2.6	High	8		25.00		
	>2.6	Very High	10		31.25		
Soil	Calcareous Loamy Soil Piedmont	Very High	10	8.125	81.25	81.25	**8.125**
	Calcareous Silty Soil Gullied Land Complex	High	8		65.00		
	Rough Broken Land	High	8		65.00		
	Mountainous land with nearly continuous soil	Medium	6		48.75		
	Mountainous land with patchy soil	Medium	6		48.75		
	Urban	Very Low	4		32.50		
	Calcareous Loamy Soil	Low	2		16.25		
Distance to fault	<100	Very High	10	6.875	68.75	68.75	**6.875**
	100–200	High	8		55.00		
	200–500	High	8		55.00		
	500–1000	Medium	6		41.25		
	1000–2000	Medium	6		41.25		
	2000–5000	Very Low	4		27.50		
	>5000	Low	2		13.75		

Table 5. *Cont.*

Factors	Categories	Effect	Normalized Weight (X)	Normalized Relative Rates (Y) Based on Table 4	Weighted Rating	Total Weightage max(X) × Y	MAX Effect on Recharge Potential (%)
			(1–10)	(1–10)	(X × Y)		
Drainage distance	<100	Very High	10	6.250	62.50	62.50	**6.250**
	100–200	High	8		50.00		
	200–500	Medium	6		37.50		
	>500	Low	4		25.00		
Profile Curvature	<−3.14	Very Low	2	3.125	6.25	31.25	**3.125**
	−2.24	Low	4		12.50		
	−1.8	Medium	6		18.75		
	0.7–2.9	High	8		25.00		
	>2.9	Very High	10		31.25		
TWI	<2	Very Low	2	5.000	10.00	50.00	**5.000**
	2–4	Low	4		20.00		
	4–6	Medium	6		30.00		
	>6	High or very high	10		50.00		
Slope	<12	Very High	10	8.125	81.25	81.25	**8.125**
	12–24	High	8		65.00		
	24–36	Medium	6		48.75		
	36–48	Low	4		32.50		
	>48	Very Low	2		16.25		
Elevation	<531	Very High	10	6.875	68.75	68.75	**6.875**
	531–617	High	8		55.00		
	617–756	Medium	6		41.25		
	756–994	Low	4		27.50		
	>994	Very Low	2		13.75		
Slope length	<10	Very High	10	3.750	37.50	37.50	**3.750**
	10–20	High	8		30.00		
	20–30	Medium	6		22.50		
	30–40	Low	4		15.00		
	>40	Very Low	2		7.50		
Lithology	Limestone	High	10	10.000	100.00	100.00	**10.000**
	Sandstone	Medium	6		60.00		
	Unconsolidated deposit	Low	4		40.00		

Table 5. *Cont.*

Factors	Categories	Effect	Normalized Weight (X)	Normalized Relative Rates (Y) Based on Table 4	Weighted Rating	Total Weightage max(X) × Y	MAX Effect on Recharge Potential (%)
			(1–10)	(1–10)	(X × Y)		
Land use	Bare area	Low	4	10.000	40.00	100.00	**10.000**
	Vegetation	High	8		80.00		
	Water	Very High	10		100.00		
	Built-up	Medium	6		60.00		
Fault density	<15	Very Low	2	6.250	12.50	62.50	**6.250**
	15–41	Low	4		25.00		
	41–65	Medium	6		37.50		
	65–91	High	8		50.00		
	>91	Very High	10		62.50		
Drainage density	<1	Very Low	2	9.375	18.75	93.75	**9.375**
	01/01/1984	Low	4		37.50		
	1.84–2.68	Medium	6		56.25		
	2.68–3.53	High	8		75.00		
	>3.53	Very High	10		93.75		
Aspect	Flat	Very High	10	5.000	50.00	50.00	**5.000**
	North	Very High	10		50.00		
	Northeast	High	8		40.00		
	East	High	8		40.00		
	Southwest	Medium	6		30.00		
	Southeast	Medium	6		30.00		
	South	Low	4		20.00		
	West	Low	4		20.00		
	Northwest	Very Low	2		10.00		
						GTW: Σ = 3118	Σ = 100

5.3. Final Combined Recharge Potential Map

After considering rate assessment, different layers of recharge potential were superimposed in the ArcGIS tool. As a result of the integration of the 15 contributing factors, the final combined potential map was generated, which highlights the overall recharge potential of Islamabad, as shown in Figure 4. The resulting map generated with the help of influencing factors' relative rates categorizes the region into five descriptive levels based on the rechargeability. These descriptive levels include "best", "high", "medium", "low", and "poor", each with a distinctive color.

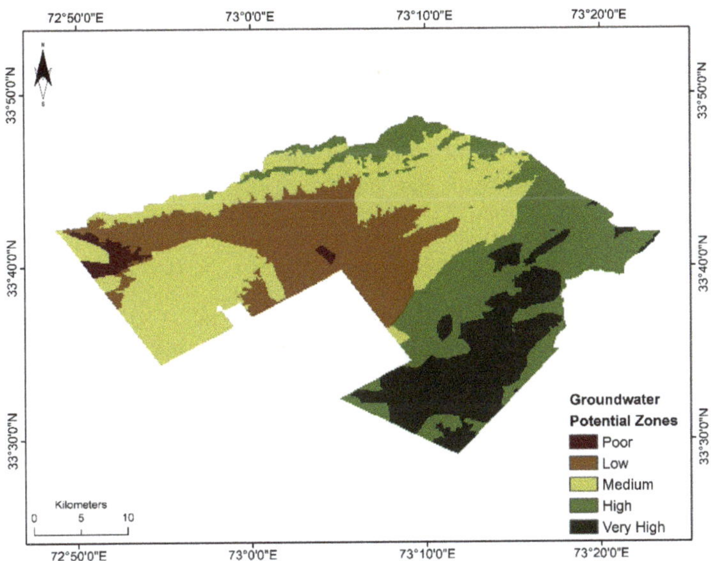

Figure 4. Potential groundwater recharge zones in the study area.

From the output thematic map (Figure 4), it is evident that the eastern region of the study area is the most suitable for groundwater recharge. Accordingly, it is highlighted to be the "best" region. This region received the highest rainfall as per the previously presented maps. This is in line with previous studies that argued that the higher the rainfall, the greater the groundwater recharge and vice versa [116,123]. Moreover, it can be observed from Figure 4 that the groundwater recharge potential decreases as we head towards the western side of Islamabad. A decreasing pattern for groundwater recharge is seen as we move from east to west in the study area. Most of the mountainous region is located towards the northeast of Islamabad, receiving the highest rainfall and having higher slopes, inducing rapid runoff. Towards the center and to the west, the slope length decreases, thus indicating a higher recharge potential, as gentle slopes were attributed to higher recharge potential [122].

Table 6 presents the data of each category shown in graphical form in Figure 4 and gives the exact portions of the study area having best to worst recharge capability. It shows that the area labeled under the "best" comprises 136.8 km^2, covering 15% of the study area. Similarly, an area of 191.52 km^2 falls under our map's "high" classification, covering 21% of the study area. Another 35% of the region collectively serves as a competent region (preferred) for groundwater recharge. The moderate zone covers 218.88 km^2 of area, covering 24% of the study area. In contrast, the potentially poor and low zones make up 13% and 27% of the area, i.e., 118.56 km^2 and 246.24 km^2, respectively.

Table 6. Classification of potential recharge areas.

Recharge Potential Category	Average %	Area Extant (km^2)
Very High	15%	136.8
High	21%	191.52
Medium	24%	218.88
Low	27%	246.24
Poor	13%	118.56

The results show that around more than half (51%) of the total area of Islamabad does not have sufficient recharge capability, and the city is dependent on only 35% of the total area to fulfill the city's demand for groundwater for daily life usage. This can be taken into consideration by local authorities when planning to meet the local water requirements and groundwater recharge. The city planners and policymakers should take mitigation steps and devise strategies to preserve most of this 35% of the land to avoid any further damage to the already fragile water condition of the city. The information devised from this final groundwater potential zones map can help resolve the long due water shortage issues in various sectors of Islamabad and nearby areas through efficient management and preservation of groundwater resources in the area. Compared to the previous studies [36–42], this study addresses the research gap of applying this methodology in a non-coastal region and modifies it by using thematic maps of larger spatial scale and the DEM data of smaller resolution to refine the accuracy of the process. All the previous published research used the one-time dataset and map the output. However, these do not depict the true representation of the groundwater recharge. This is because the considered datasets may change temporally, needing more datasets to overcome this limitation. Hence, this study used the annual mean for all datasets, which change with respect to season or time. Secondly, previously published research used limitedly influencing datasets that might not present the actual situation of the study. In the current research, all the contributing factors were analyzed and used to consider the entire situation. Accordingly, the model gives reliable actual output. Moreover, the study also considers more contributing factors than the previous studies to further enhance the accuracy of the output. The research presents a holistic approach that gives comparatively improved results and can be applied to other regions as and when required.

6. Conclusions

Considering the constant increase in groundwater demand in Islamabad with increasing population growth, the decreasing groundwater level has become a matter of concern for the local authorities. This study attempts to develop a groundwater potential recharge zone map of the study area of Islamabad, Pakistan, to help the policymakers devise efficient policies for mitigating this problem.

The methodology involves the integration of RS and GIS to develop a map that highlights the groundwater recharge potential in the study area. In our scenario, 15 key factors were selected based on their contribution to the recharge. These include soil, land use/land cover, drainage distance, slope, rainfall, plan curvature, distance to faults, profile curvature, TWI, elevation, slope length, lithology, fault density, drainage density, and aspect. Thematic maps were generated and overlayed using GIS. A holistic map was devised at the end, comprising input from 15 of the influencing factors and their weights to produce a weighted map. The resulting map categorizes the region into five different descriptive levels, namely poor, low, medium, high, and best, based on the groundwater recharge potential. The results showed that 13% of the area falls in the poor-recharge-potential category, 27% area has a low potential, 24% has medium potential, 21% has high potential, and 15% has the best chance of recharging the groundwater table. Overall, around 35% of the study area is suitable for groundwater recharge, and more than half is unsuitable for such purposes.

This study provides a holistic model with more accurate results than the previous studies by introducing a comparatively greater number of factors and employing the thematic maps of larger spatial scale and DEM data of a smaller resolution. The current study paves the way for future infrastructure development by the concerned authorities to meet the water demand of Islamabad and preserve the precious natural terrain with high recharge potential.

The study is limited in terms of the factors considered. Further, it is restricted to a single region in a developing country for testing purposes. Moreover, considering that this study was limited in terms of the unavailability of geophysical data for the case study area, future researchers can conduct further research by including the geophysical and field data from multiple regions. This can help in carrying out the subsurface groundwater modeling as well as 3D modeling of the targeted study area. Further, similar studies can be conducted for larger nearby regions and developed countries to help move toward global sustainability goals and tackle climate change effects. The effects of vegetation on recharge can also be investigated in the future.

Author Contributions: Conceptualization, A.M., B.A. and F.U.; methodology, A.M., B.A., N.K. and F.U.; software, A.M., B.A., N.K. and F.U; validation, A.M., B.A., N.K., F.U., H.A., E.E.H. and A.H.A.; formal analysis, A.M., B.A. and N.K.; investigation, A.M., B.A., N.K. and F.U.; resources, A.M., B.A., N.K., H.A., E.E.H. and A.H.A.; data curation, A.M., B.A., N.K. and F.U.; writing—original draft preparation, A.M., N.K., B.A. and F.U; writing—review and editing, F.U., H.A., E.E.H., A.A.A. and A.H.A.; visualization, A.M., B.A., N.K., F.U. and A.A.A.; supervision, A.M., B.A., F.U. and A.H.A.; project administration, A.M., B.A., F.U. and A.H.A.; funding acquisition, A.H.A. and H.A. All authors have read and agreed to the published version of the manuscript.

Funding: This research was funded by Taif University Researchers Supporting Project number TURSP 2020/252, Taif University, Taif, Saudi Arabia.

Institutional Review Board Statement: Not applicable.

Informed Consent Statement: Not applicable.

Data Availability Statement: Data can be shared upon reasonable request.

Acknowledgments: The authors appreciate Taif University Researchers Supporting Project number TURSP 2020/252, Taif University, Taif, Saudi Arabia for supporting this work.

Conflicts of Interest: The authors declare no conflict of interest.

References

1. Megdal, S.B. Invisible Water: The Importance of Good Groundwater Governance and Management. *NPJ Clean Water* **2018**, *1*, 15. [CrossRef]
2. Ganapuram, S.; Kumar, G.V.; Krishna, I.M.; Kahya, E.; Demirel, M.C. Mapping of Groundwater Potential Zones in the Musi Basin Using Remote Sensing Data and Gis. *Adv. Eng. Softw.* **2009**, *40*, 506–518. [CrossRef]
3. Singh, S.K.; Zeddies, M.; Shankar, U.; Griffiths, G.A. Potential Groundwater Recharge Zones within New Zealand. *Geosci. Front.* **2019**, *10*, 1065–1072. [CrossRef]
4. Bear, J. *Hydraulics of Groundwater*; Courier Corporation: North Chelmsford, MA, USA, 2012.
5. Todd, D.K.; Mays, L.W. *Groundwater Hydrology*; John Wiley & Sons: Hoboken, NJ, USA, 2004.
6. Thakur, D.; Bartarya, S.K.; Nainwal, H.C. Mapping Groundwater Prospect Zones in an Intermontane Basin of the Outer Himalaya in India Using Gis and Remote Sensing Techniques. *Environ. Earth Sci.* **2018**, *77*, 368. [CrossRef]
7. Wang, X.J.; Zhang, J.Y.; Shahid, S.; Guan, E.H.; Wu, Y.X.; Gao, J.; He, R.M. Adaptation to Climate Change Impacts on Water Demand. *Mitig. Adapt. Strateg. Glob. Chang.* **2016**, *21*, 81–99. [CrossRef]
8. Rao, Y.S.; Jugran, D.K. Delineation of Groundwater Potential Zones and Zones of Groundwater Quality Suitable for Domestic Purposes Using Remote Sensing and Gis. *Hydrol. Sci. J.* **2003**, *48*, 821–833.
9. Cherry, J.A.; Freeze, R.A. *Groundwater*; Prentice-Hall: Hoboken, NJ, USA, 1979.
10. Shao, Z.; Huq, M.E.; Cai, B.; Altan, O.; Li, Y. Integrated Remote Sensing and Gis Approach Using Fuzzy-Ahp to Delineate and Identify Groundwater Potential Zones in Semi-Arid Shanxi Province, China. *Environ. Model. Softw.* **2020**, *134*, 104868. [CrossRef]
11. Huang, T.; Ma, B.; Pang, Z.; Li, Z.; Li, Z.; Long, Y. How Does Precipitation Recharge Groundwater in Loess Aquifers? Evidence from Multiple Environmental Tracers. *J. Hydrol.* **2020**, *583*, 124532. [CrossRef]

12. Aslam, B.; Maqsoom, A.; Khalid, N.; Ullah, F.; Sepasgozar, S. Urban Overheating Assessment through Prediction of Surface Temperatures: A Case Study of Karachi, Pakistan. *ISPRS Int. J. Geo-Inf.* **2021**, *10*, 539. [CrossRef]
13. Okello, C.; Tomasello, B.; Greggio, N.; Wambiji, N.; Antonellini, M. Impact of Population Growth and Climate Change on the Freshwater Resources of Lamu Island, Kenya. *Water* **2015**, *7*, 1264–1290. [CrossRef]
14. Atif, S.; Umar, M.; Ullah, F. Investigating the Flood Damages in Lower Indus Basin since 2000: Spatiotemporal Analyses of the Major Flood Events. *Nat. Hazards* **2021**, *108*, 2357–2383. [CrossRef]
15. Page, D.; Bekele, E.; Vanderzalm, J.; Sidhu, J. Managed Aquifer Recharge (Mar) in Sustainable Urban Water Management. *Water* **2018**, *10*, 239. [CrossRef]
16. Dillon, P.; Toze, S.; Page, D.; Vanderzalm, J.; Bekele, E.; Sidhu, J.; Rinck-Pfeiffer, S. Managed Aquifer Recharge: Rediscovering Nature as a Leading Edge Technology. *Water Sci. Technol.* **2010**, *62*, 2338–2345. [CrossRef]
17. Bhatti, M.T.; Anwar, A.A.; Aslam, M. Groundwater Monitoring and Management: Status and Options in Pakistan. *Comput. Electron. Agric.* **2017**, *135*, 143–153. [CrossRef]
18. Subhadra, B. Water: Halt India's Groundwater Loss. *Nature* **2015**, *521*, 289. [CrossRef]
19. Qureshi, A.S. Improving Food Security and Livelihood Resilience through Groundwater Management in Pakistan. *Glob. Adv. Res. J. Agric. Sci.* **2015**, *4*, 687–710.
20. Siebert, S.; Kummu, M.; Porkka, M.; Döll, P.; Ramankutty, N.; Scanlon, B.R. A Global Data Set of the Extent of Irrigated Land from 1900 to 2005. *Hydrol. Earth Syst. Sci.* **2015**, *19*, 1521–1545. [CrossRef]
21. Chindarkar, N.; Grafton, R.Q. India's Depleting Groundwater: When Science Meets Policy. *Asia Pac. Policy Stud.* **2019**, *6*, 108–124. [CrossRef]
22. Mancosu, N.; Snyder, R.L.; Kyriakakis, G.; Spano, D. Water Scarcity and Future Challenges for Food Production. *Water* **2015**, *7*, 975–992. [CrossRef]
23. Schneider, U.; Havlík, P.; Schmid, E.; Valin, H.; Mosnier, A.; Obersteiner, M.; Böttcher, H.; Skalský, R.; Balkovič, J.; Sauer, T.; et al. Impacts of Population Growth, Economic Development, and Technical Change on Global Food Production and Consumption. *Agric. Syst.* **2011**, *104*, 204–215. [CrossRef]
24. Iqbal, N.; Hossain, F.; Lee, H.; Akhter, G. Satellite Gravimetric Estimation of Groundwater Storage Variations over Indus Basin in Pakistan. *IEEE J. Sel. Top. Appl. Earth Obs. Remote Sens.* **2016**, *9*, 3524–3534. [CrossRef]
25. Van Steenbergen, F.; Kaisarani, A.B.; Khan, N.U.; Gohar, M.S. A Case of Groundwater Depletion in Balochistan, Pakistan: Enter into the Void. *J. Hydrol. Reg. Stud.* **2015**, *4*, 36–47. [CrossRef]
26. Nag, S.K. Application of Lineament Density and Hydrogeomorphology to Delineate Groundwater Potential Zones of Baghmundi Block in Purulia District, West Bengal. *J. Indian Soc. Remote Sens.* **2005**, *33*, 521–529. [CrossRef]
27. Ravindran, A.A.; Selvam, S. Coastal Disaster Damage and Neotectonic Subsidence Study Using 2d Eri Technique in Dhanushkodi, Rameshwaram Island, Tamilnadu, India. *Middle-East J. Sci. Res.* **2014**, *19*, 1117–1122.
28. Jasrotia, A.S.; Kumar, R.; Saraf, A.K. Delineation of Groundwater Recharge Sites Using Integrated Remote Sensing and Gis in Jammu District, India. *Int. J. Remote Sens.* **2007**, *28*, 5019–5036. [CrossRef]
29. Kumar, P.K.D.; Gopinath, G.; Seralathan, P. Application of Remote Sensing and Gis for the Demarcation of Groundwater Potential Zones of a River Basin in Kerala, Southwest Coast of India. *Int. J. Remote Sens.* **2007**, *28*, 5583–5601. [CrossRef]
30. Rodell, M.; Velicogna, I.; Famiglietti, J.S. Satellite-Based Estimates of Groundwater Depletion in India. *Nature* **2009**, *460*, 999–1002. [CrossRef]
31. Srivastava, P.K.; Bhattacharya, A.K. Groundwater Assessment through an Integrated Approach Using Remote Sensing, Gis and Resistivity Techniques: A Case Study from a Hard Rock Terrain. *Int. J. Remote Sens.* **2006**, *27*, 4599–4620. [CrossRef]
32. Selvam, S.; Sivasubramanian, P. Groundwater Potential Zone Identification Using Geoelectrical Survey: A Case Study from Medak District, Andhra Pradesh, India. *Int. J. Geomat. Geosci.* **2012**, *3*, 55–62.
33. Selvam, S. Use of Remote Sensing and Gis Techniques for Land Use and Land Cover Mapping of Tuticorin Coast, Tamilnadu. *Univers. J. Environ. Res. Technol.* **2012**, *2*, 233–241.
34. Saraf, A.K.; Choudhury, P.R. Integrated Remote Sensing and Gis for Groundwater Exploration and Identification of Artificial Recharge Sites. *Int. J. Remote Sens.* **1998**, *19*, 1825–1841. [CrossRef]
35. Chenini, I.; Ben Mammou, A.; El May, M. Groundwater Recharge Zone Mapping Using Gis-Based Multi-Criteria Analysis: A Case Study in Central Tunisia (Maknassy Basin). *Water Resour. Manag.* **2010**, *24*, 921–939. [CrossRef]
36. Kamal, A.M.; Midorikawa, S. Gis-Based Geomorphological Mapping Using Remote Sensing Data and Supplementary Geoinformation: A Case Study of the Dhaka City Area, Bangladesh. *Int. J. Appl. Earth Obs. Geoinf.* **2004**, *6*, 111–125.
37. Gustavsson, M.; Kolstrup, E.; Seijmonsbergen, H. A New Symbol-and-Gis Based Detailed Geomorphological Mapping System: Renewal of a Scientific Discipline for Understanding Landscape Development. *Geomorphology* **2006**, *77*, 90–111. [CrossRef]
38. Singh, A.K.; Parkash, B.; Choudhury, P.R. Integrated Use of Srm, Landsat Etm+ Data and 3d Perspective Views to Identify the Tectonic Geomorphology of Dehradun Valley, India. *Int. J. Remote Sens.* **2007**, *28*, 2403–2414. [CrossRef]
39. Selvam, S.I.J.D.; Mala, R.I.J.D.; Muthukakshmi, V. A Hydrochemical Analysis and Evaluation of Groundwater Quality Index in Thoothukudi District, Tamilnadu, South India. *Int. J. Adv. Eng. Appl.* **2013**, *2*, 25–37.
40. Selvam, S.; Manimaran, G.; Sivasubramanian, P. Hydrochemical Characteristics and Gis-Based Assessment of Groundwater Quality in the Coastal Aquifers of Tuticorin Corporation, Tamilnadu, India. *Appl. Water Sci.* **2013**, *3*, 145–159. [CrossRef]

41. Selvam, S.; Manimaran, G.; Sivasubramanian, P. Cumulative Effects of Septic System Disposal and Evolution of Nitrate Contamination Impact on Coastal Groundwater in Tuticorin, South Tamilnadu, India. *Res. J. Pharm. Biol. Chem. Sci.* **2013**, *4*, 1207–1218.
42. Singaraja, C.; Chidambaram, S.; Srinivasamoorthy, K.; Anandhan, P.; Selvam, S. A Study on Assessment of Credible Sources of Heavy Metal Pollution Vulnerability in Groundwater of Thoothukudi Districts, Tamilnadu, India. *Water Qual. Expo. Health* **2015**, *7*, 459–467. [CrossRef]
43. Machiwal, D.; Jha, M.K.; Mal, B.C. Assessment of Groundwater Potential in a Semi-Arid Region of India Using Remote Sensing, Gis and Mcdm Techniques. *Water Resour. Manag.* **2011**, *25*, 1359–1386. [CrossRef]
44. Saraf, A.K.; Jain, S.K. Integrated Use of Remote Sensing and Geographical Information System Methods for Groundwater Exploration in Parts of Lalitpur District, Up. In Proceedings of the International Conference on Hydrology and Water Resources, New Delhi, India, 20–22 December 1993; Kluwer Academic Publishers: Dordrecht, The Netherlands, 1994.
45. Krishnamurthy, J.; Srinivas, G. Role of Geological and Geomorphological Factors in Ground Water Exploration: A Study Using Irs Liss Data. *Int. J. Remote Sens.* **1995**, *16*, 2595–2618. [CrossRef]
46. Krishnamurthy, J.; Kumar, N.V.; Jayaraman, V.; Manivel, M. An Approach to Demarcate Ground Water Potential Zones through Remote Sensing and a Geographical Information System. *Int. J. Remote Sens.* **1996**, *17*, 1867–1884. [CrossRef]
47. Kamaraju, M.; Bhattacharya, A.; Reddy, G.S.; Rao, G.C.; Murthy, G.S.; Rao, T.C.M. Ground-Water Potential Evaluation of West Godavari District, Andhra Pradesh State, India—a Gis Approach. *Groundwater* **1996**, *34*, 318–325. [CrossRef]
48. Ravindran, K.V. Drainage Morphometry Analysis and Its Correlation with Geology, Geomorphology and Groundwater Prospects in Zuvari Basin, South Goa: Using Remote Sensing and Gis. In Proceedings of the National Symposium on Remote Sensing for Natural Resources with Special Emphasis on Water Management, Pune, India, 4–6 December 1996.
49. Kumar, A. Sustainable Utilisation of Water Resource in Watershed Perspective-a Case Study in Alaunja Watershed, Hazaribagh, Bihar. *J. Indian Soc. Remote Sens.* **1999**, *27*, 13–22. [CrossRef]
50. Krishnamurthy, J.; Mani, A.; Jayaraman, V.; Manivel, M. Groundwater Resources Development in Hard Rock Terrain-an Approach Using Remote Sensing and Gis Techniques. *Int. J. Appl. Earth Obs. Geoinf.* **2000**, *2*, 204–215. [CrossRef]
51. Srivastava, P.K.; Bhattacharya, A.K. Delineation of Ground Water Potential Zones in a Hard Rock Terrain of Bargarh District, Orissa Using Irs Data. *J. Indian Soc. Remote Sens.* **2000**, *28*, 129–140. [CrossRef]
52. Shahid, S.; Nath, S.K.; Roy, J. Groundwater Potential Modelling in a Soft Rock Area Using a Gis. *Int. J. Remote Sens.* **2000**, *21*, 1919–1924. [CrossRef]
53. Khan, M.A.; Moharana, P.C. Use of Remote Sensing and Geographical Information System in the Delineation and Characterization of Ground Water Prospect Zones. *J. Indian Soc. Remote Sens.* **2002**, *30*, 131–141. [CrossRef]
54. Karanth, K.R.; Seshubabu, K. Identification of Major Lineaments on Satellite Imagery and on Aerial Photographs for Delineation for Possible Potential Groundwater Zones in Penukonda and Dharmavaram Taluks of Anantapur Ditrict. In Proceedings of the Joint Indo-US Workshop on Remote Sensing of Water Resources, National Remote Sensing Agency (NRSA), Hyderabad, India, 10–14 April 1978.
55. Raju, K.C.B.; Rao, P.N.; Rao, G.V.K.; Kumar, B.J. Analytical Aspects of Remote Sensing Techniques for Ground Water Prospection in Hard Rocks. In Proceedings of the 6th Asian Conference on Remote Sensing, Hyderabad, India, 21–26 November 1986.
56. Palanivel, S.; Ganesh, A.; Kumaran, T.V. Geohydrological Evaluation of Upper Agniar and Vellar Basins, Tamil Nadu: An Integrated Approach Using Remote Sensing, Geophysical and Well Inventory Data. *J. Indian Soc. Remote Sens.* **1996**, *24*, 153–168. [CrossRef]
57. Sankar, K. Evaluation of Groundwater Potential Zones Using Remote Sensing Data in Upper Vaigai River Basin, Tamil Nadu, India. *J. Indian Soc. Remote Sens.* **2002**, *30*, 119–129. [CrossRef]
58. Devi, P.S.; Srinivasulu, S.; Raju, K.K. Hydrogeomorphological and Groundwater Prospects of the Pageru River Basin by Using Remote Sensing Data. *Environ. Geol.* **2001**, *40*, 1088–1094. [CrossRef]
59. Saraf, A.K.; Gupta, R.P.; Jain, R.K.; Srivastava, N.K. Gis Based Processing and Interpretation of Ground Water Quality Data. In Proceedings of the Regional Workshop on Environmental Aspects of Ground Water Development, Kurukshetra, India, 17–19 October 1994.
60. Achu, A.L.; Thomas, J.; Reghunath, R. Multi-Criteria Decision Analysis for Delineation of Groundwater Potential Zones in a Tropical River Basin Using Remote Sensing, Gis and Analytical Hierarchy Process (Ahp). *Groundw. Sustain. Dev.* **2020**, *10*, 100365. [CrossRef]
61. Das, B.; Pal, S.C.; Malik, S.; Chakrabortty, R. Modeling Groundwater Potential Zones of Puruliya District, West Bengal, India Using Remote Sensing and Gis Techniques. *Geol. Ecol. Landsc.* **2019**, *3*, 223–237. [CrossRef]
62. Lakshmi, S.V.; Reddy, Y.V.K. Identification of Groundwater Potential Zones Using Gis and Remote Sensing. *Int. J. Pure Appl. Math.* **2018**, *119*, 3195–3210.
63. Roy, A.; Keesari, T.; Sinha, U.K.; Sabarathinam, C. Delineating Groundwater Prospect Zones in a Region with Extreme Climatic Conditions Using Gis and Remote Sensing Techniques: A Case Study from Central India. *J. Earth Syst. Sci.* **2019**, *128*, 201. [CrossRef]
64. Kaur, L.; Rishi, M.S.; Singh, G.; Thakur, S.N. Groundwater Potential Assessment of an Alluvial Aquifer in Yamuna Sub-Basin (Panipat Region) Using Remote Sensing and Gis Techniques in Conjunction with Analytical Hierarchy Process (Ahp) and Catastrophe Theory (Ct). *Ecol. Indic.* **2020**, *110*, 105850. [CrossRef]

65. Selvam, S.; Dar, F.A.; Magesh, N.S.; Singaraja, C.; Venkatramanan, S.; Chung, S.Y. Application of Remote Sensing and Gis for Delineating Groundwater Recharge Potential Zones of Kovilpatti Municipality, Tamil Nadu Using If Technique. *Earth Sci. Inform.* **2016**, *9*, 137–150. [CrossRef]
66. Abijith, D.; Saravanan, S.; Singh, L.; Jennifer, J.J.; Saranya, T.; Parthasarathy, K. Gis-Based Multi-Criteria Analysis for Identification of Potential Groundwater Recharge Zones-a Case Study from Ponnaniyaru Watershed, Tamil Nadu, India. *HydroResearch* **2020**, *3*, 1–14. [CrossRef]
67. Kaliraj, S.; Chandrasekar, N.; Magesh, N.S. Identification of Potential Groundwater Recharge Zones in Vaigai Upper Basin, Tamil Nadu, Using Gis-Based Analytical Hierarchical Process (Ahp) Technique. *Arab. J. Geosci.* **2014**, *7*, 1385–1401. [CrossRef]
68. Banks, W.S.; Paylor, R.L.; Hughes, W.B. Using Thermal-Infrared Imagery to Delineate Ground-Water Discharge D. *Groundwater* **1996**, *34*, 434–443. [CrossRef]
69. Aslam, B.; Maqsoom, A.; Tahir, M.D.; Ullah, F.; Rehman, M.S.U.; Albattah, M. Identifying and Ranking Landfill Sites for Municipal Solid Waste Management: An Integrated Remote Sensing and GIS Approach. *Buildings* **2022**, *12*, 605. [CrossRef]
70. Gilani, H.; Ahmad, S.; Qazi, W.A.; Abubakar, S.M.; Khalid, M. Monitoring of Urban Landscape Ecology Dynamics of Islamabad Capital Territory (Ict), Pakistan, over Four Decades (1976–2016). *Land* **2020**, *9*, 123. [CrossRef]
71. Das, S.; Pardeshi, S.D. Integration of Different Influencing Factors in Gis to Delineate Groundwater Potential Areas Using If and Fr Techniques: A Study of Pravara Basin, Maharashtra, India. *Appl. Water Sci.* **2018**, *8*, 197. [CrossRef]
72. Juandi, M.; Syahril, S. Empirical Relationship between Soil Permeability and Resistivity, and Its Application for Determining the Groundwater Gross Recharge in Marpoyan Damai, Pekanbaru, Indonesia. *Water Pract. Technol.* **2017**, *12*, 660–666. [CrossRef]
73. Arkoprovo, B.; Adarsa, J.; Animesh, M. Application of Remote Sensing, Gis and Mif Technique for Elucidation of Groundwater Potential Zones from a Part of Orissa Coastal Tract, Eastern India. *Res. J. Recent Sci.* **2013**, *2277*, 2502.
74. Butt, M.J.; Waqas, A.; Iqbal, M.F.; Muhammad, G.; Lodhi, M.A.K. Assessment of Urban Sprawl of Islamabad Metropolitan Area Using Multi-Sensor and Multi-Temporal Satellite Data. *Arab. J. Sci. Eng.* **2012**, *37*, 101–114. [CrossRef]
75. Shaheen, N.; Baig, M.A.; Mahboob, M.A.; Akbar, S.; Khokar, M.F.; Wwf-Pakistan, G. Application of Remote Sensing Technologies to Detect the Vegetation Changes During Past Two Decades in Islamabad, Pakistan. *J. Soc. Sci.* **2015**, *4*, 886–900. [CrossRef]
76. Hassan, Z.; Shabbir, R.; Ahmad, S.S.; Malik, A.H.; Aziz, N.; Butt, A.; Erum, S. Dynamics of Land Use and Land Cover Change (Lulcc) Using Geospatial Techniques: A Case Study of Islamabad Pakistan. *SpringerPlus* **2016**, *5*, 812. [CrossRef]
77. Butt, A.; Shabbir, R.; Ahmad, S.S.; Aziz, N. Land Use Change Mapping and Analysis Using Remote Sensing and Gis: A Case Study of Simly Watershed, Islamabad, Pakistan. *Egypt. J. Remote Sens. Space Sci.* **2015**, *18*, 251–259. [CrossRef]
78. Shabbir, R.; Ahmad, S.S. Water Resource Vulnerability Assessment in Rawalpindi and Islamabad, Pakistan Using Analytic Hierarchy Process (Ahp). *J. King Saud Univ.-Sci.* **2016**, *28*, 293–299. [CrossRef]
79. Shah, A.; Ali, K.; Nizami, S.M. Spatio-Temporal Analysis of Urban Sprawl in Islamabad, Pakistan During 1979–2019, Using Remote Sensing. *GeoJournal* **2021**, 1–14. [CrossRef]
80. Abbasi, K. Islamabad's Groundwater Has Lowered by Five Times over Last Five Years: Minister. *DAWN News*. 20 December 2018. Available online: https://www.dawn.com/news/1452462/islamabads-groundwater-has-lowered-by-five-times-over-last-five-years-minister (accessed on 18 April 2022).
81. Doxiadis, C.A. Islamabad: The Creation of a New Capital. *Town Plan. Rev.* **1965**, *36*, 1–35. [CrossRef]
82. Adeel, M. Methodology for Identifying Urban Growth Potential Using Land Use and Population Data: A Case Study of Islamabad Zone Iv. *Procedia Environ. Sci.* **2010**, *2*, 32–41. [CrossRef]
83. Aslam, A.; Rana, I.A.; Bhatti, S.S. The Spatiotemporal Dynamics of Urbanisation and Local Climate: A Case Study of Islamabad, Pakistan. *Environ. Impact Assess. Rev.* **2021**, *91*, 106666. [CrossRef]
84. Sandhu, G.S.; Naeem, M.A. A Case Study of Innovative Businesses Involved with Efficient Municipal Solid Waste Management in Islamabad, Pakistan. *WIT Trans. Ecol. Environ.* **2017**, *223*, 529–538.
85. Maria, S.I.; Imran, M. Planning of Islamabad and Rawalpindi: What Went Wrong. In Proceedings of the 42nd ISoCaRP Congress, Istanbul, Turkey, 14–18 September 2006.
86. Arribas-Bel, D.; Nijkamp, P.; Scholten, H. Multidimensional Urban Sprawl in Europe: A Self-Organizing Map Approach. *Comput. Environ. Urban Syst.* **2011**, *35*, 263–275. [CrossRef]
87. Liu, Y.; Din, S.U.; Jiang, Y. Urban Growth Sustainability of Islamabad, Pakistan, over the Last 3 Decades: A Perspective Based on Object-Based Backdating Change Detection. *GeoJournal* **2020**, *86*, 2035–2055. [CrossRef]
88. Ghalib, H.; Elkhorazaty, M.T.; Serag, Y. New Capital Cities: A Timeless Mega-Project of Intercontinental Presence. In Proceedings of the IOP Conference Series: Materials Science and Engineering, Chennai, India, 16–17 September 2020.
89. Sohail, M.T.; Mahfooz, Y.; Azam, K.; Yat, Y.; Genfu, L.; Fahad, S. Impacts of Urbanization and Land Cover Dynamics on Underground Water in Islamabad, Pakistan. *Desalin Water Treat.* **2019**, *159*, 402–411. [CrossRef]
90. Yaseen, H. People in Islamabad Facing Extreme Water Crisis. *MM News*, 4 February 2020; p. 1.
91. APP. Water Shortage Puts Islamabad's I-8/1 Residents at Risk. *The Express Tribune*. 13 April 2020. p. 2013. Available online: https://tribune.com.pk/story/2196636/water-shortage-puts-islamabads-81-residents-risk (accessed on 18 April 2022).
92. Bishop, J.M.; Glenn, C.R.; Amato, D.W.; Dulai, H. Effect of Land Use and Groundwater Flow Path on Submarine Groundwater Discharge Nutrient Flux. *J. Hydrol. Reg. Stud.* **2017**, *11*, 194–218. [CrossRef]

93. Kotchoni, D.O.V.; Vouillamoz, J.-M.; Lawson, F.M.A.; Adjomayi, P.; Boukari, M.; Taylor, R.G. Relationships between Rainfall and Groundwater Recharge in Seasonally Humid Benin: A Comparative Analysis of Long-Term Hydrographs in Sedimentary and Crystalline Aquifers. *Hydrogeol. J.* **2019**, *27*, 447–457. [CrossRef]
94. Lei, M.; Liu, W.; Gao, Y.; Zhu, T. Mobile User Behaviors in China. In *Encyclopedia of Mobile Phone Behavior*; IGI Global: Hershey, PA, USA, 2015; pp. 1110–1128.
95. Xu, G.; Li, P.; Lu, K.; Tantai, Z.; Zhang, J.; Ren, Z.; Wang, X.; Yu, K.; Shi, P.; Cheng, Y. Seasonal Changes in Water Quality and Its Main Influencing Factors in the Dan River Basin. *Catena* **2019**, *173*, 131–140. [CrossRef]
96. Mengel, L.; Krauss, H.-W.; Lowke, D. Water Transport through Cracks in Plain and Reinforced Concrete–Influencing Factors and Open Questions. *Constr. Build. Mater.* **2020**, *254*, 118990. [CrossRef]
97. Zhang, M.; Li, J.; Ding, H.; Ding, J.; Jiang, F.; Ding, N.X.; Sun, C. Distribution Characteristics and Influencing Factors of Microplastics in Urban Tap Water and Water Sources in Qingdao, China. *Anal. Lett.* **2020**, *53*, 1312–1327. [CrossRef]
98. Zhao, Y.; Wang, Y.; Wang, Y. Comprehensive Evaluation and Influencing Factors of Urban Agglomeration Water Resources Carrying Capacity. *J. Clean. Prod.* **2021**, *288*, 125097. [CrossRef]
99. Gebreyohannes, T.; De Smedt, F.; Walraevens, K.; Gebresilassie, S.; Hussien, A.; Hagos, M.; Amare, K.; Deckers, J.; Gebrehiwot, K. Regional Groundwater Flow Modeling of the Geba Basin, Northern Ethiopia. *Hydrogeol. J.* **2017**, *25*, 639–655. [CrossRef]
100. Liu, J.; Gao, G.; Wang, S.; Jiao, L.; Wu, X.; Fu, B. The Effects of Vegetation on Runoff and Soil Loss: Multidimensional Structure Analysis and Scale Characteristics. *J. Geogr. Sci.* **2018**, *28*, 59–78. [CrossRef]
101. Raduła, M.W.; Szymura, T.; Szymura, M. Topographic Wetness Index Explains Soil Moisture Better Than Bioindication with Ellenberg's Indicator Values. *Ecol. Indic.* **2018**, *85*, 172–179. [CrossRef]
102. Akinluyi, F.O.; Olorunfemi, M.O.; Bayowa, O.G. Investigation of the Influence of Lineaments, Lineament Intersections and Geology on Groundwater Yield in the Basement Complex Terrain of Ondo State, Southwestern Nigeria. *Appl. Water Sci.* **2018**, *8*, 49. [CrossRef]
103. Akter, A.; Uddin, A.M.H.; Ben Wahid, K.; Ahmed, S. Predicting Groundwater Recharge Potential Zones Using Geospatial Technique. *Sustain. Water Resour. Manag.* **2020**, *6*, 24. [CrossRef]
104. Gnanachandrasamy, G.; Zhou, Y.; Bagyaraj, M.; Venkatramanan, S.; Ramkumar, T.; Wang, S. Remote Sensing and Gis Based Groundwater Potential Zone Mapping in Ariyalur District, Tamil Nadu. *J. Geol. Soc. India* **2018**, *92*, 484–490. [CrossRef]
105. Jerbi, H.; Massuel, S.; Leduc, C.; Tarhouni, J. Assessing Groundwater Storage in the Kairouan Plain Aquifer Using a 3d Lithology Model (Central Tunisia). *Arab. J. Geosci.* **2018**, *11*, 236. [CrossRef]
106. Torabi, A.; Alaei, B.; Ellingsen, T. Faults and Fractures in Basement Rocks, Their Architecture, Petrophysical and Mechanical Properties. *J. Struct. Geol.* **2018**, *117*, 256–263. [CrossRef]
107. Daryono, M.R.; Natawidjaja, D.H.; Sapiie, B.; Cummins, P. Earthquake Geology of the Lembang Fault, West Java, Indonesia. *Tectonophysics* **2019**, *751*, 180–191. [CrossRef]
108. Ahmed, R.; Sajjad, H. Analyzing Factors of Groundwater Potential and Its Relation with Population in the Lower Barpani Watershed, Assam, India. *Nat. Resour. Res.* **2018**, *27*, 503–515. [CrossRef]
109. Khan, A.; Govil, H.; Taloor, A.K.; Kumar, G. Identification of Artificial Groundwater Recharge Sites in Parts of Yamuna River Basin India Based on Remote Sensing and Geographical Information System. *Groundw. Sustain. Dev.* **2020**, *11*, 100415. [CrossRef]
110. Thapa, R.; Gupta, S.; Gupta, A.; Reddy, D.V.; Kaur, H. Use of Geospatial Technology for Delineating Groundwater Potential Zones with an Emphasis on Water-Table Analysis in Dwarka River Basin, Birbhum, India. *Hydrogeol. J.* **2018**, *26*, 899–922. [CrossRef]
111. Magesh, N.S.; Chandrasekar, N.; Soundranayagam, J.P. Delineation of Groundwater Potential Zones in Theni District, Tamil Nadu, Using Remote Sensing, Gis and Mif Techniques. *Geosci. Front.* **2012**, *3*, 189–196. [CrossRef]
112. Shaban, A.; Khawlie, M.; Abdallah, C. Use of Remote Sensing and Gis to Determine Recharge Potential Zones: The Case of Occidental Lebanon. *Hydrogeol. J.* **2006**, *14*, 433–443. [CrossRef]
113. Ahirwar, S.; Malik, M.S.; Ahirwar, R.; Shukla, J.P. Application of Remote Sensing and Gis for Groundwater Recharge Potential Zone Mapping in Upper Betwa Watershed. *J. Geol. Soc. India* **2020**, *95*, 308–314. [CrossRef]
114. Pakistan. Generalized Soil Map. Soil Survey of Pakistan. Lahore. Available online: https://esdac.jrc.ec.europa.eu/content/pakistan-generalized-soil-map (accessed on 18 April 2022).
115. Bakr, M.U. Geological Map of Pakistan. (East and West Pakistan). Direction of N.M. Khan. Director General. Geological Survey of Pakistan. Available online: https://esdac.jrc.ec.europa.eu/content/geological-map-pakistan-east-and-west-pakistan (accessed on 18 April 2022).
116. Dar, T.; Rai, N.; Bhat, A. Delineation of Potential Groundwater Recharge Zones Using Analytical Hierarchy Process (Ahp). *Geol. Ecol. Landsc.* **2020**, *5*, 292–307. [CrossRef]
117. Lehmann, P.; Berli, M.; Koonce, J.E.; Or, D. Surface Evaporation in Arid Regions: Insights from Lysimeter Decadal Record and Global Application of a Surface Evaporation Capacitor (Sec) Model. *Geophys. Res. Lett.* **2019**, *46*, 9648–9657. [CrossRef]
118. Igwe, O.; Ifediegwu, S.I.; Onwuka, O.S. Determining the Occurrence of Potential Groundwater Zones Using Integrated Hydro-Geomorphic Parameters, Gis and Remote Sensing in Enugu State, Southeastern, Nigeria. *Sustain. Water Resour. Manag.* **2020**, *6*, 39. [CrossRef]
119. Kumar, V.A.; Mondal, N.C.; Ahmed, S. Identification of Groundwater Potential Zones Using Rs, Gis and Ahp Techniques: A Case Study in a Part of Deccan Volcanic Province (Dvp), Maharashtra, India. *J. Indian Soc. Remote Sens.* **2020**, *48*, 497–511. [CrossRef]

120. Moeck, C.; Grech-Cumbo, N.; Podgorski, J.; Bretzler, A.; Gurdak, J.J.; Berg, M.; Schirmer, M. A Global-Scale Dataset of Direct Natural Groundwater Recharge Rates: A Review of Variables, Processes and Relationships. *Sci. Total Environ.* **2020**, *717*, 137042. [CrossRef] [PubMed]
121. Martos-Rosillo, S.; Ruiz-Constán, A.; González-Ramón, A.; Mediavilla, R.; Martín-Civantos, J.M.; Martínez-Moreno, F.J.; Jódar, J.; Marín-Lechado, C.; Medialdea, A.; Galindo-Zaldívar, J.; et al. The Oldest Managed Aquifer Recharge System in Europe: New Insights from the Espino Recharge Channel (Sierra Nevada, Southern Spain). *J. Hydrol.* **2019**, *578*, 124047. [CrossRef]
122. Lentswe, G.B.; Molwalefhe, L. Delineation of Potential Groundwater Recharge Zones Using Analytic Hierarchy Process-Guided Gis in the Semi-Arid Motloutse Watershed, Eastern Botswana. *J. Hydrol. Reg. Stud.* **2020**, *28*, 100674. [CrossRef]
123. Kolli, M.K.; Opp, C.; Groll, M. Mapping of Potential Groundwater Recharge Zones in the Kolleru Lake Catchment, India, by Using Remote Sensing and Gis Techniques. *Nat. Resour.* **2020**, *11*, 127. [CrossRef]

Article

Ecological Compensation Mechanism in a Trans-Provincial River Basin: A Hydrological/Water-Quality Modeling-Based Analysis

Wenhua Wan [1], Hang Zheng [1,*], Yueyi Liu [1], Jianshi Zhao [2], Yingqi Fan [1] and Hongbo Fan [1]

[1] School of Environment and Civil Engineering, Dongguan University of Technology, Dongguan 523808, China
[2] State Key Laboratory of Hydro-Science and Engineering, Department of Hydraulic Engineering, Tsinghua University, Beijing 100084, China
* Correspondence: zhenghang00@163.com

Abstract: Ecological compensation is an important economic means of water pollution control and quality management, especially for trans-regional rivers with unbalanced economic and social development between upstream and downstream. The Tangbai River Basin (TRB), a watershed crossing Henan province and Hubei province, China, forms one of the nation's most productive agricultural regions. The TRB has been exposed to high doses of fertilizers for a long time. This study simulates hydrologic and nutrient cycling in the TRB using Soil and Water Assessment Tool (SWAT) with limited data available. The results indicate that dryland fields, which constitute 62% of the basin area, produce 80% of total nitrogen (TN) and 85% of total phosphorus (TP) yields of the whole river basin. The water quality of river sections at the provincial boundary shows that only 29% of the time from 2000 to 2019 met the Class III standard regarding TN and TP concentrations, and the concentrations in the spring flood season are approximately three times the mean in the non-flood season. The Grain for Green ecological restoration measure in Henan province shows that restoration of non-flat drylands can reduce nutrient loads at trans-provincial sections by 3.5 times compared to that of slope-independent drylands; however, the water quality compliance rate remains similar. The value of ecological compensation can also vary widely depending on different quantitative criteria. The SWAT-based pollutant quantification method adopted in this study could have implications for ecological compensation in trans-regional rivers.

Keywords: ecological compensation; trans-regional river; non-point-source pollution; grain for green; SWAT

Citation: Wan, W.; Zheng, H.; Liu, Y.; Zhao, J.; Fan, Y.; Fan, H. Ecological Compensation Mechanism in a Trans-Provincial River Basin: A Hydrological/Water-Quality Modeling-Based Analysis. *Water* **2022**, *14*, 2542. https://doi.org/10.3390/w14162542

Academic Editors: Dengfeng Liu, Hui Liu and Xianmeng Meng

Received: 4 July 2022
Accepted: 15 August 2022
Published: 18 August 2022

1. Introduction

Water pollution control and sustainable development of international river basins have been a major challenge in the current ecological research. Yet, due to the difficulty of data acquisition, reluctant cooperation across boundaries, and the absence of exchange mechanisms [1,2], there are few successful cases. Also arising from unbalanced development along rivers, the trans-provincial river management is a regional form of trans-boundary problem. However, provinces (or states) are more willing to cooperate in controlling pollution than countries given sovereignty and territorial integrity. Successful cases include the Murray–Darling Basin in Australia [3], Delaware River in the United States [4], and Xin'an River in China [5]. Therefore, the management of trans-provincial river basins in water ecology can provide valuable information for solving trans-boundary ecological issues.

Ecological compensation (hereinafter "eco-compensation") has long been considered an effective economic instrument for controlling water pollution in trans-regional basins [6]. Eco-compensation, also known as "payment for environmental services", was proposed in the late 1990s for watershed management to address the environmental problems caused by the Industrial Revolution [7]. At present, eco-compensation has captured increasing attention nationwide and is seen as a promising complementary method to alleviate the identified contradiction among stakeholders [8]. The form of eco-compensation measures varies by country. In China, the central government has launched a series of explorations and research projects for eco-compensation since 1999. One typical scheme is the national-scale Grain for Green Program, which compensates rural households for converting sloping croplands to forests or grasslands to reduce soil erosion [9,10]. At the local administrative level, the first trans-provincial eco-compensation pilot scheme is the Xin'an River scheme, launched in 2011. This is an eco-compensation agreement between Anhui and Zhejiang provinces, targeting pollution control from the upper regions in Huangshan city, Anhui, and thereby maintaining high water quality in the lower reaches, Qiandao Lake in Hangzhou city, Zhejiang [11]. The list of other pilot eco-compensation schemes in China is found in Wang et al. [12]. These studies are crucial in clarifying the implications of watershed eco-compensation and establishing typical methods and standards. How to reasonably account for compensation and what measures should be taken to effectively protect trans-regional watersheds from water pollution depend largely on the accurate mapping of compensation stakeholders [13]. A solution generally lies in addressing the four questions below. First, what are the major water pollutants and where do they come from? Second, how do pollutants change throughout their journey from upstream to downstream, and are they accumulating or separating? Third, what actions could be taken to reduce the load across regional boundaries? Last, but not least, what is the cost of pollution control and how should it be compensated?

Despite a number of studies on these questions, there is still a lack of systematic quantitative analysis on how eco-compensation responds to the dynamic change in pollutants. Previous scholars from different disciplines have investigated the effectiveness of eco-compensation and measures of reducing water pollution from different perspectives. For research on eco-compensation, current studies primarily focus on policy analysis involving game theory [6] and water-related regulations [11]. However, such methods are mostly static and disregard the responses to hydrology and water quality and, thus, cannot adequately address non-point-source pollution.

In pollution control, the wide range of substances that may pollute water bodies leads to a variety of options for reducing pollution [14,15]. In China, more than 2600 lakes, including the nation's largest freshwater lake, Poyang Lake, have been subjected to high loads of nitrogen (N) and phosphorus (P) [16,17]. Agricultural operations, including crop fertilization and livestock farming, are one of the major sources of N and P pollution to surface water. For example, Boesch et al. [18] and Reckhow et al. [14] found that high nutrient leaching from farmland into the US Chesapeake Bay is the foremost water quality concern for the waterway, and thus nearby agricultural land needs to implement the best management practices. Successful tracking of pollution control strategies relies upon the careful modeling of non-point nutrient fluxes, and distributed process-based models are then introduced [19–21]. These models include, but are not limited to, Soil and Water Assessment Tool (SWAT) [22], Integrated Valuation of Environmental Services and Tradeoffs (InVEST) [23], and Annualized AGricultural Non-point Source (AnnAGNPS) [24]. Among these, SWAT is arguably the most widely used model, especially at the watershed scale [25,26]. In summary, the above scattered studies reveal a fragmented and insufficient link between non-point-source simulation and eco-compensation, thereby failing to support effective policy formulation.

To help decision-makers understand trans-regional non-point-source pollution issues from both environmental and economic perspectives, this study utilized the SWAT hydrological/water-quality model to simulate nitrogen/phosphorus cycling and their responses to eco-compensation strategies for an agricultural watershed, the Tangbai River Basin (TRB). As the TRB receives the most runoff in Henan province and joins the Han River, the longest tributary of the Yangtze River, in Hubei province, it is a representative trans-provincial river basin. A watershed eco-compensation mechanism is selected as a water pollution control measure to provide a reference for trans-boundary water quality management.

2. Data and Methodology

2.1. Study Area

The Tangbai River Basin (TRB, 31°38′–33°43′ N, 111°34′–113°40′ E, 24,190 km^2, Figure 1) is a subbasin of the Han River Basin, which is the source of the central route of the South-to-North Water Transfer Project. The Tangbai River is formed by the convergence of two tributaries, the Tang River and the Bai River, flowing through Henan province and Hubei province, China (Figure 1b). The average flow rate at the basin outlet is 323 m^3/s (1980–2012 time series). The two tributaries both originate from Nanyang city, the south-western part of Henan province, and then flow into Xiangyang city of Hubei province and finally form the Tangbai River, which later joins the Han River.

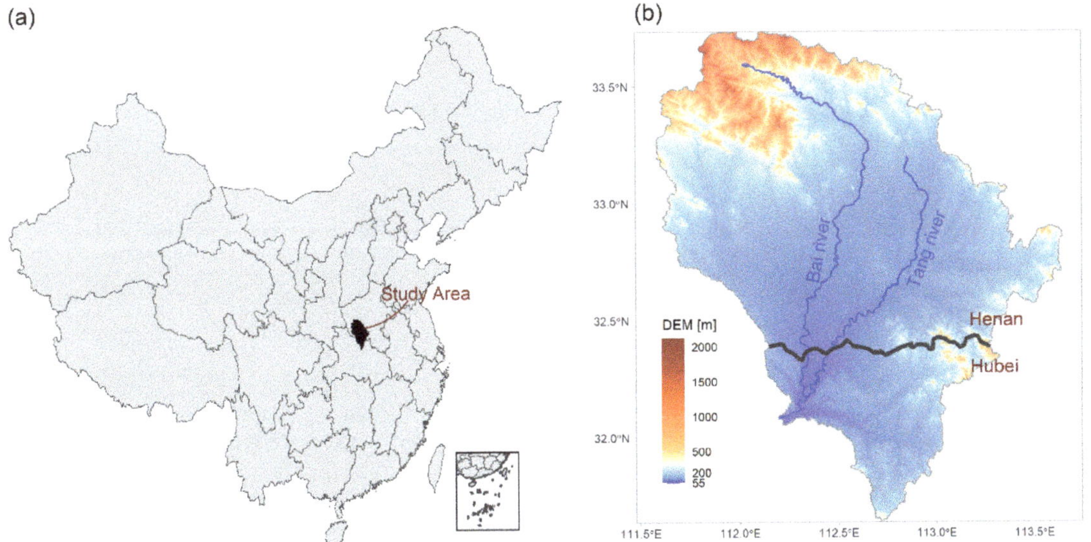

Figure 1. Geographic location (**a**) and 90 m digital elevation map (DEM, **b**) of the Tangbai River Basin, a shared river basin of Henan province and Hubei province. The curved path in (**b**) is the provincial boundary, and the blue lines are the two major tributaries constituting the Tangbai River.

The TRB is characterized by a dense population. Around 90,000 people live in the middle and upper reaches of the watershed, which is the famous Nanyang Basin, the nation's most productive agricultural regions. The excess use of fertilizers, traditional inadequate farming techniques, and dramatic growth of industrial and decreasing runoff during the 1990s due to climate change have led to serious pollution and environmental degradation of the Tangbai River [27]. In recent years, the surface water quality has improved to a common Class IV status due to the long-term regional cooperative actions and joint management between Nanyang (in Henan province) and Xiangyang (in Hubei province) [28]. However, such pollution control mainly focuses on point-source pollution from manufacturing and runoff pollution from urban areas [29]. The agricultural non-point-source pollution in the rural areas of Nanyang city is still prominent. The rainfall- and snowmelt-runoff processes affect the transfer of non-point-source pollution, and thereby disturb the availability of water resources between Nanyang and Xiangyang. As a typical trans-provincial watershed, managing non-point-source pollution in TRB remains complicated and difficult.

2.2. SWAT Model and Inputs

The SWAT model is a typical semi-distributed, physically-based hydrological model. It splits the watershed into subbasins connected by a stream network and further delineates each subbasin into hydrologic response units (HRUs), which comprise unique combinations of land use, soil type, and slope. The HRU aggregates water and nutrient fluxes to the subbasin level, and they are then routed through the stream network to the watershed outlet. Therefore, the major model inputs consist of topography, land-cover type, soil property, climate data, and land management practices. The resolution and sources of all data inputs in this study are listed in Table 1.

Table 1. Input and validation datasets, sources, and main attributes.

Data Type	Resolution	Year	Data Source
1. Digital elevation model	90 m	2006	SRTM Digital Elevation Database v4. 1 (https://bigdata.cgiar.org/srtm-90m-digital-elevation-database/, accessed on 1 February 2021)
2. Land use land cover	1 km	2000, 2005, 2010, 2015	Resource and Environment Science and Data Center (https://www.resdc.cn/, accessed on 1 February 2021)
3. Soil map	1 km	1995	Resource and Environment Data Cloud Platform (https://www.resdc.cn/, accessed on 1 March 2021); China Soil Science Database (http://vdb3.soil.csdb.cn/, accessed on 1 May 2021); SPAW software (https://hrsl.ba.ars.usda.gov/SPAW/, accessed on 1 May 2021)
4. Climate observation (i.e., station location, precipitation, temperature, solar, relative humidity, solar radiation, wind speed)	daily	1960–2019	National Meteorological Information Centre (http://data.cma.cn/, accessed on 1 February 2021); Angstrom-Prescott radiation model
5. Crop management (i.e., land cover, type, and mode)	date	2019	Field survey
6. Fertilization application (i.e., type, date, and quantity)	date	2005	Field survey; literature review [1]
7. Livestock manure production	daily	2011	Literature review
8. Streamflow data (i.e., station location, flow rate)	monthly	1997–2001; 2007–2012	Changjiang Water Resources Commission of the Ministry of Water Resources
9. Water quality data (i.e., monitoring location, and total nitrogen and phosphorus)	date	2000–2019	Literature review; Monthly Report on Water Quality of Han River (http://sthjj.xiangyang.gov.cn/hjxx/tjsj/hjszyb/, accessed on 1 May 2021); Xiangyang Municipal Ecological Environment Bureau

Note: [1] Namely, statistical analyses of the results of a government-led survey conducted in 2005 on the agricultural pollution along the central route of the South-to-North Water Transfer Project.

In this study, we used the ArcSWAT interface (version 2012) for ArcGIS. The simulation was performed with a monthly time step from 1998–2019, in which the first two years were excluded from the analysis as they were used as the warmup period.

2.2.1. Datasets for SWAT Model Building

A digital elevation model (DEM, 90 m × 90 m, Figure 1b) was used to delineate the watershed. We obtained a DEM from the SRTM 90 m Digital Elevation Database (https://bigdata.cgiar.org/, accessed on 1 February 2021) and clipped it into the area that covers the whole TRB. Note that additional subbasin outlets were manually defined at the locations with hydrological gauging stations (Figure 2d) in addition to the autogenerated subbasins, allowing us to compare the simulated results with observations at any location of interest.

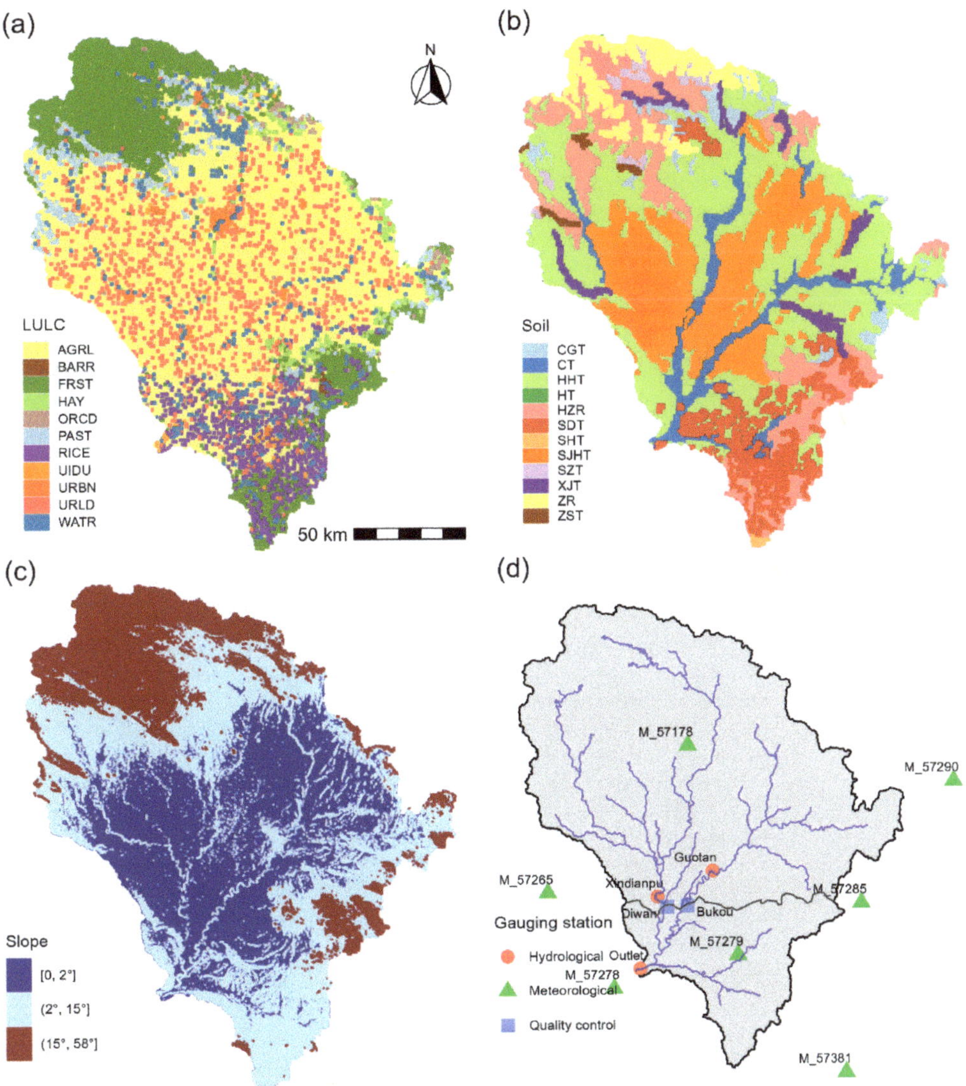

Figure 2. The spatial distribution of the land-use/land-cover (LULC) classes in 2015 (**a**), soil types (**b**), slope classes (**c**), and (**d**) gauging stations in the SWAT-generated stream network, distinguishing meteorological stations used for building SWAT (triangle), hydrological stations used for model validation (circle), and water quality monitoring sections at the provincial boundary (square). Abbreviations of land-cover classes and soil types are found in the main text.

Approximately 66% of the study area is in agricultural use (AGRL and RICE). The remaining area is covered by forest (FRST), grassland (PAST and HAY), residential zones (URLD and URBN), and others. As the area under cultivation changed only slightly from 2000 to 2015, with the dryland (AGRL) decreasing from 62.4% to 61.6% and the paddy field (RICE) increasing from 4.4% to 4.5%, we therefore categorized land-use/land-cover classes based on the conditions in 2015 (Figure 2a). A total of 18.4% of the land-cover area was considered to be mixed forest (FRST), and 6.8% was considered to be low density residential zones. Both the land-cover map and soil map hold a resolution of 1 km $\times$ 1 km. The soil characteristics for preparing the user-defined soil database include soil component parameters (e.g., soil name, hydrological soil group, and number of soil layers) and soil layer parameters (e.g., soil depth, salinity, organic matter content, and particle size distribution). All characteristics, except for the soil name, were extracted from the China Soil Science Database (http://vdb3.soil.csdb.cn/, accessed on 1 May 2021) for central China, with some further calculations using SPAW software (Table 1). The most dominant soil types are yellow-cinnamon soil (HHT, 37.4%), shajiang black soil (SJHT, 19.3%), and yellow-brown earth (HZR, 15.2%). The remaining soil types include skeletal soil (CGT), fluvo-aquic soil (CT), cinnamon soil (HT), paddy soil (SDT), lime soil (SHT), litho soil (SZT), alluvial soil (XJT), brown earth (ZR), and purplish soil (ZST). Considering the large span of slope in TRB (i.e., 0–163%, 0–58°), it is necessary to categorize the slope into separate classes. Referring to the soil erosion condition on the steepness of cultivated fields [30], we divided the whole TRB into three slope classes: 0–3.5% (i.e., 0–2°, flat land with no soil erosion), 3.5%–26.8% (i.e., 2–15°, terraced land with a low soil erosion rate), and 26.8–9999% (i.e., 15–58°, sloping land with a high erosion rate) (Figure 2c).

The meteorological observations used were obtained from the National Meteorological Information Centre (Table 1) and consist of rainfall, temperature, relative humidity, solar radiation, and wind speed from seven stations (Figure 2d). Solar radiation was calculated based on sunshine observations using the Angstrom–Prescott equation. Stations outside the watershed were also included for more accurate spatial interpolation of climate data, particularly near the border.

2.2.2. Management Practices for the SWAT Model

To obtain a relatively accurate estimation of crop yields and nutrient transformations in the TRB, we investigated crop rotation, fertilization, and livestock breeding practices based on field surveys and literature review (Tables 1 and 2). Specifically, these management activities were established for three types of land cover: AGRL, RICE, and URLD. As listed in Table 2, the main crops in the AGRL field are rainfed grains (winter wheat, summer corn) with rotation [31], while the RICE field is used for growing rice [32]. The nitrogen fertilizer applied was urea, and the phosphorus fertilizer was phosphorus pentoxide. The amount and fertilization period were set following Lan [33] and local crop guidance [34]. The irrigation schedule was set as default. As livestock and poultry breeding is mainly concentrated in Nanyang city and much less in Xiangyang city, we assumed that in the model all animal breeding occurs in Nanyang in low-density residential zones (URLD). The total amount of fresh manure discharge F_{manure} is then estimated based on the statistics of animal breeding [35], with

$$F_{manure} = \mu \sum_{i} n_i X_i \text{Area}_{TRB} / \text{Area}_{URLD} \quad (1)$$

where μ is the fraction of livestock and poultry manure, which is set to 0.9 according to Cai [32], n_i is the number of the ith specific livestock or poultry breeding, and X_i is manure produced per unit area (kg/ha·d); Area_{TRB} and Area_{URLD} are the area of TRB (ha) and area covered by URLD in TRB (ha).

Three variants of fertilization were considered depending on land-cover classes: (1) for crop fields (i.e., AGRL and RICE) listed in Table 2, fertilizer was applied according to the calendar; (2) for animal breeding areas (i.e., URLD), compound fresh manure was applied once every day in the form of continuous fertilization; (3) for the remaining land-cover types, an auto-fertilization mode was adopted based on plant nitrogen stress.

Table 2. Settings on non-point-source pollution from planting in terms of crop cycle and management practices, and livestock breeding in terms of manure discharge in rural areas based on land cover in 2015.

LULC	Crop	Cycle	Fertilization	Urea Applied (kg/ha)	P_2O_5 [2] Applied (kg/ha)	Fertilizer on Surface
AGRL [1]	Winter wheat	15 October–15 June	15 October	100.38	102.60	0.2
			1 April	43.02	-	0.2
	Summer corn	1 July–15 September	1 August	191.25	59.25	0.2
RICE [1]	Rice	15 April–1 October	1 May	82.80	114.75	0.5
			1 July	49.68	-	0.5
			1 August	33.12	-	0.5
LULC	**Livestock**	**Duration (days)**	**Application frequency**	**Manure [3] applied (kg/ha)**		**Heat Unit**
URLD	Swine/beef/broiler	365	1 (daily)	6.38		0

Note: [1] The auto-fertilization based on heat unit in AGRL and RICE fields were removed. [2] P_2O_5 is a self-defined P fertilizer, in which the fraction of mineral P is 0.43. [3] Compound manure is a self-defined compound manure, in which nutrient fractions are proportional to that of the fresh manure produced by different livestock/poultry species.

Data used for validation include recorded time series of streamflow and water quality. The monthly streamflow was available at three stations on two tributaries: Guotan station (32°32′ N, 112°36′ E) on the Tang River, Xindianpu station (32°25′ N, 112°18′ E) on the Bai River, and at the watershed outlet (32°5′ N, 112°12′ E) (Figure 2d). The streamflow time series span from 1997 to 2012, with 2001–2006 missing. Regarding water quality data, we did not find any long-term observations. We therefore decided to use data from articles and online reports, e.g., [28,36,37], in which nutrients (N and P) are presented in concentration. Data on TN and TP concentrations were collected on different dates, at eight different sites (Figure 3d), including the two river sections at the provincial boundary (Bukou section on the Tang River and Diwan section on the Bai River, Figure 2d).

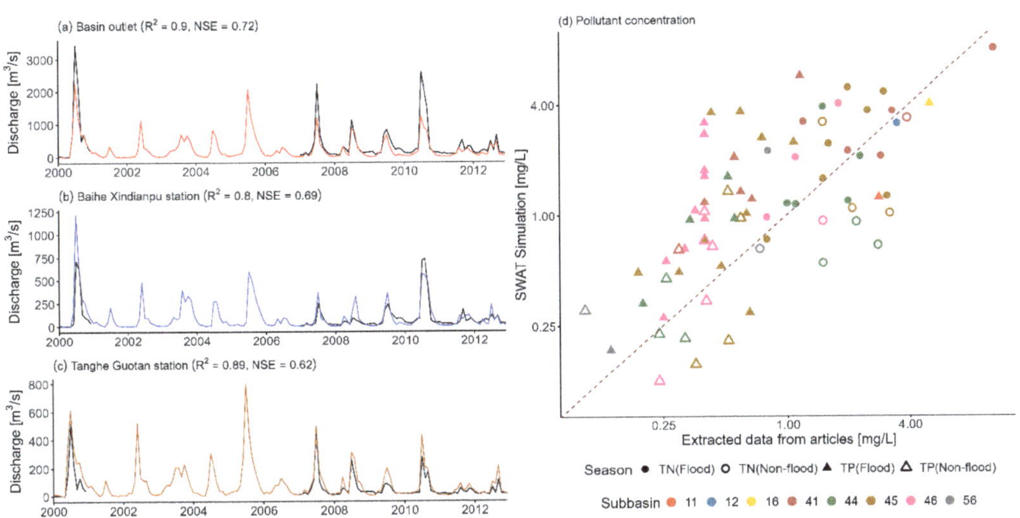

Figure 3. Comparison between SWAT simulation and record extracted from articles with respect to monthly streamflow (**a**–**c**) and TN and TP concentrations (mg/L) (**d**). The flood season refers to the months of April to September, and the non-flood season refers to the remaining months. The distribution of the subbasin index is found in the following Figure 4.

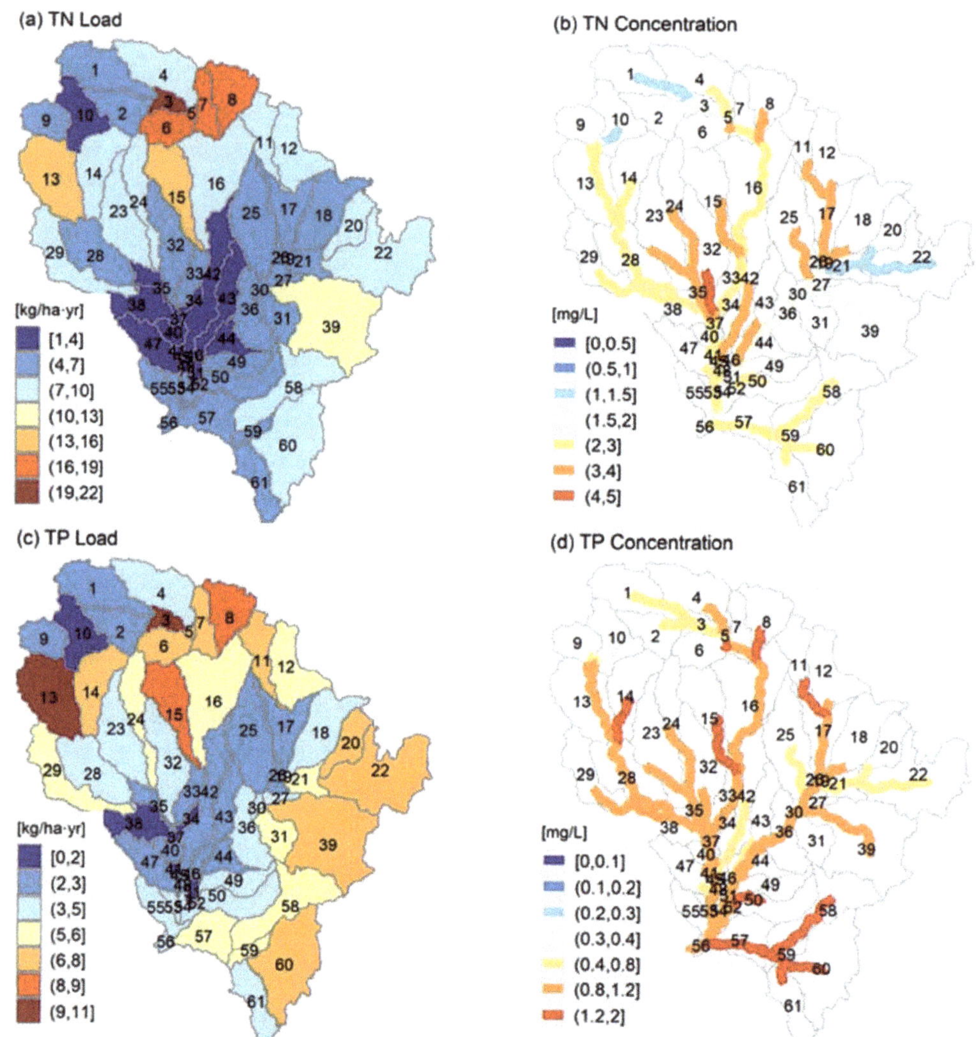

Figure 4. Spatial distribution of simulated TN (**a**,**b**) and TP (**c**,**d**) averaged over 2000–2019. Panels (**a**,**c**) are the annual mean load into reach from each subbasin and (**b**,**d**) are the mean monthly pollutant concentrations in the reach. The blue palette in (**b**,**d**) indicates concentrations of water quality of Class I (applicable to source water and national nature reserves, with TN $\leq$ 0.2 mg/L and TP $\leq$ 0.02 mg/L), Class II (primary protection zone for centralized drinking water, with TN $\leq$ 0.5 mg/L and TP $\leq$ 0.1 mg/L), Class III (secondary protection zone for centralized drinking water, with TN $\leq$ 1 mg/L and TP $\leq$ 0.2 mg/L), Class IV (undrinkable, applicable to general industrial water, with TN $\leq$ 1.5 mg/L and TP $\leq$ 0.3 mg/L), and Class V (undrinkable, applicable to agricultural water, with TN $\leq$ 2 mg/L and TP $\leq$ 0.4 mg/L), respectively, according to Environmental Quality Standards for Surface Water (GB3838-2002). The red palette indicates water quality inferior to Class V (i.e., TN > 2 mg/L and TP > 0.4 mg/L).

SWAT simulated nutrient yields are expressed as load. The term was converted from load to concentration to facilitate comparison between simulation, record, and Environmental Quality Standards for Surface Water (GB3838-2002). The total N/P load in surface water is TOT_X (X represents nitrogen or phosphorus), while that leaching from HRU or subbasin into the reach during the time step is the sum of N/P in different forms (Equations (2) and (3). The corresponding concentration C_X (mg/L) in surface water is the total load divided by the mass flow rate (Equation (4)).

$$TOT_N = ORGN + NUSRQ \tag{2}$$

$$TOT_P = ORGP + SEDP + SOLP \tag{3}$$

$$C_X = \frac{TOT_X}{FLOW_{out}} \times \eta \tag{4}$$

where TOT_N and TOT_P are total nitrogen TN (kg) and total phosphorus TP (kg) in the reach. $ORGN$ and $NUSRQ$ are organic N yield (kg N/ha) and nitrate NO_3^- transport with surface runoff into the reach (kg N/ha). $ORGP$, $SEDP$, and $SOLP$ are organic P yield (kg P/ha), mineral P in sediment transported into reach (kg P/ha), and soluble mineral forms of P transport by surface runoff (kg P/ha), respectively. $FLOW_{out}$ is the streamflow out of the reach during time step (m^3/s), and η is the unit conversion factor.

2.3. Ecological Compensation Implementation

Understanding the upstream–downstream linkage in hydrological processes and nutrient (N and P) pollution is essential. Generally, the economy of the upstream basin is relatively backwards compared to the downstream basin; thus, the conflict between economic development and water ecological protection is more pronounced. Protection of water resources in the upper regions is likely to be much more beneficial for downstream areas, especially in terms of urban landscape and ecology. For TRB, Nanyang city (in upstream Henan province) had a per capita GDP of CNY 38,064 (note: CNY 1 = USD 0.15), while Xiangyang city (in downstream Hubei province) had a GDP of 2.3 times the upstream GDP (CNY 84,815) in 2019. It is necessary to establish water pollution ecological compensation between downstream beneficiaries and upstream water protectors to achieve ecological sustainability for the entire basin.

Common river restoration measures include the collection of pollution taxes for individual polluters [38], payments for environmental services (e.g., agricultural practices) [39], conservation reserve programs [40], and conversion of cropland to forest (i.e., Grain for Green) [41]. Among them, the Grain for Green (hereinafter "GFG") measure is a widely adopted ecological restoration measure in China that is applicable to agricultural non-point-source pollution. In this study, we established four sets of GFG measures applied to AGRL lands in Henan province, thereby adjusting the economic structure and promoting industrial upgrades using funds from eco-compensation. (1) Converting all sloping drylands (AGRL, slope > 15°) in Henan province, comprising 1.5% of dryland areas in Henan, to forest (FRST); (2) converting all non-flat drylands (AGRL with slope > 2°), constituting 21.8% of drylands in Henan, to FRST; (3) converting all drylands in Henan province to FRST; and (4) performing conversions based on the AGRL area ratio (without distinguishing slopes), 10%, 20%, 40%, 60%, and 80% of the AGRL land in Henan province to FRST. For each area ratio, we generated 10 scenarios by randomly selecting HRUs of unique combinations of SOIL-LULC until the threshold area was reached. A total of 53 AGRL-to-FRST scenarios were generated, and each was used as LULC input to rerun SWAT. The corresponding nutrient (N and P) and crop yield outputs were then simulated.

The eco-compensation quantification, which is the key for implementing compensation, is always measured by monetary values. The ecosystem services were estimated based on the opportunity cost of land use. Specifically, we added up the cost of planting trees and the annual mean value of crops lost because of the GFG activities. The tree planting fees were calculated based on the annual compensation standard for state-owned forests, which is CNY 75 per hectare per year. The value of wheat was assumed to be the average purchase price in Henan in 2020, which was CNY 2.24/kg, while corn was CNY 2.62/kg. This total cost then served as the upper boundary of the compensation standard. Additionally, the lower boundary of the compensation standard was estimated according to the up-to-date national policy of the GFG program, which is subsiding CNY 4800 per hectare per year, for a total of 5 years.

3. Results and Discussion

According to the input data, the watershed is discretized into 61 subbasins and divided into 2298 HRUs. The watershed outlet (Zhangwan section on the Tangbai River–Han River intersection) is located in subbasin 56, which drains an area of 24,200 km^2. The river sections lying at the provincial boundary, the Diwan section on the Bai River and the Bukou section on the Tang River (Figure 2d), are located in subbasins 45 and 46, with drainage areas of 11,780 km^2 and 7834 km^2, respectively.

3.1. Suitability of SWAT for Simulating Streamflow and Nutrients

SWAT model performance in simulating hydrological processes is typically determined by the fitness of the streamflow time series at the watershed outlet. According to the existing literature, when the Nash–Sutcliffe efficiency (NSE) exceeds 0.5 and the coefficient of determination (R^2) exceeds 0.6, the simulation results are deemed satisfactory [42]. Before calibration, the NSE and R^2 values for monthly streamflow at the outlet of TRB are already 0.9 and 0.72, respectively (Figure 3a). The streamflow in high-flow seasons is underestimated at the outlet. This could be related to the interpolation and aggregation of daily rainfall. The NSEs for the reaches on the tributaries also exceed 0.6, while R^2s exceed 0.8 (Figure 3b,c). According to the statistical indicators, it appears that the SWAT model has satisfactory applicability to TRB. This may be partly due to the relatively uniform distribution of meteorological inputs (Figure 2d). In contrast, simulated streamflow does not improve much after 2000 auto-calibrations using SWATCUP, a phenomenon which was also observed in another Yangtze River subbasin study [43]. Therefore, the rest of this study used the default parameters and settings, instead of the calibrated parameters. Literature discussions are available for the SWAT model calibration method, sensitivity analysis, and the possible range of parameters [44,45].

Water quality comparison in terms of concentrations of TN and TP is shown in Figure 3d. Because only a monthly step was adopted when running SWAT, the impact of daily step storm scour on nutrients cannot be considered; thus, the hydrologic condition when water was sampled (i.e., water quality data obtained from the literature) may differ from that we simulated. Most TN concentrations are greater than 1.0 mg/L and the overestimation of the SWAT model usually occurs during the flood season (April to September). TP concentrations are mostly within 1.0 mg/L but overestimation of TP is more pervasive. These frequent overestimations of N and P may likely be the result of uncalibrated parameters impacting sediment and nutrient transportation simulations. Nevertheless, we failed to calibrate the simulations as calibration requires observations of nutrient loads (kg), whereas the collected measurements are in the form of concentration (kg/L) and, more importantly, include both point and non-point sources of pollution at distinct locations. Due to the difficulty of collecting point-source pollution data, the nutrient emissions caused by industrial waste and residential sewage were not considered in this study.

In general, despite the dispersion of nutrient concentrations, the estimated values mostly fall within acceptable ranges according to available data for the study area. We claim that the SWAT model can essentially convey the hydrological processes and nutrient pollution in TRB.

3.2. Spatial and Temporal Distribution of Nutrient Pollution

3.2.1. Spatial Variation in Nutrients

As an important component in water pollution, SWAT simulates the transformation and transportation processes of N and P nutrients. Figure 4 shows the simulated spatial distribution of nutrient (TN and TP) leaching loads from each subbasin and the concentration in each river reach. The nutrients in Figure 4a,c are illustrated in terms of load per unit area (kg/ha) to make the value comparable among different subbasins with distinct total areas. The spatial distribution of non-point-source pollution strongly depends on the heterogeneities of soil properties, crop types, vegetation types, farming activities, etc. [46]. The TN load is mainly concentrated in the mountainous cultivated field of the upstream areas of the Bai River, such as subbasins 3 (22 kg/ha·y), 5–8, whereas TN load to the Tang River is lower, ranging from 2.9 kg/ha·y (subbasin 44) to 11.8 kg/ha·month (subbasin 39). The intensity of TP load is about half that of TN load. Again, subbasin 3 leached the most TP into the river, at a rate of 10 kg/ha per year. In addition to the upper Bai River regions, TP load is also high in the southeast TRB, including the upper Tang River regions and the other tributary of the Tangbai River, the Gun River in Hubei Province. In the middle flat land, only small amounts of N and P loads are observed. This may be attributed to (1) the low altitude and (2) the occupation of shajiang black soil (Figure 2b), which belongs to a clayey soil (40–54% of clay among four layers), thereby retaining nutrients much longer and eroding less [47].

The mean monthly TN and TP concentrations at the watershed outlet are 1.9 mg/L and 1.1 mg/L, respectively. Regarding the variation along river reaches (Figure 4b,d), none have mean TN or TP concentration that met the Class III standard (i.e., $\leq$1.0 mg/L for TN and $\leq$0.2 mg/L for TP). Only 441 km (30.5%) of the river total length had TN concentrations belonging to Classes IV–V. The Bai River is more polluted with N than the Tang River. Except for reaches flowing through the northern mountainous forests, all reaches of the Bai River have TN concentrations inferior to Class V (Figure 4b). The Tang River, in contrast to the Bai River, has TN concentration in the mainstem within Class V and is only inferior to Class V in three tributaries. The magnitude of TP concentration on average is lower than that of TN, but its quality class is significantly worse. Only 18 km of river reaches are of Classes I–V. The remaining 98.8% of the rivers have TP concentrations that are inferior to Class V (Figure 4d).

Table 3 lists the nutrient inputs through fertilization and leaching for different land covers. The average N application for the whole TRB is estimated to be 114.3 kg/ha·y. The highest N fertilization occurs in hay land (HAY), where auto-fertilization is applied based on nitrogen stress and heat units (Section 2.2.2). The elemental N and P fertilizers applied to generic dryland (AGRL) are large because of chemical fertilizer application, as shown in Table 2. The three land covers with the highest TN leaching per unit area are barren land (BARR), dryland AGRL, and pasture (PAST), corresponding to 14, 10, and 9.7 kg/ha·y, respectively. The highest TP leaching is observed in agricultural fields with values of 8.4 and 7.1 kg/ha·y for RICE and AGRL, respectively. The nutrient loads from fields are dominated by organic N (ORGN in Equation 2, 92% and 86% of TN for AGRL and RICE, not shown) and sediment mineral P (SEDP in Equation 3, 68% and 77% of TP, not shown). SWAT simulates three forms of organic N/P: active organic N/P, stable organic N/P associated with humic substances, and fresh organic N/P associated with plant residues [48]. Such predominant organic N and sediment P may partly due to that (1) SWAT tends to overestimate organic N but underestimate dissolved N [49,50], and (2) the accuracy of organic N and sediment P relies on how sediment is simulated [49,51]. The SWAT model has been found to tend to overestimate organic N but underestimate dissolved N [50]. The high nutrient leaching from fields is due to the loosening surface, ponding fresh water on the soil surface, harvesting activities, and rich content of nutrients in the fields; thus, sediment and nutrients can be easily washed away or infiltrated into streams during heavy rainfall. Organic N accounts for 100% of TN losses in pasture (PAST), barren land (BARR), and forest (FRST), which is caused by rain erosion. Forests leach much less nutrients than other land covers, with 2.7 kg/ha·y for TN and 0.9 kg/ha·y for TP, owing to their complex vegetation coverage and interception effects of higher leaf area. This difference in nutrient loss is consistent with previous studies [52,53].

The annual mean loads of N and P leaching from HRUs to stream for the whole TRB are 18.6 and 12.4 kiloton, respectively (see Table 3), whereas the corresponding loads flowing through the watershed outlet (reach 56) are slightly lower at 16.2 (13% less) and 11.6 (6.5% less) kiloton (see Figure 5). When area factions of different land covers are considered, AGRL land is the predominant source of nutrient pollution in the Tangbai River, constituting 79.7% of TN and 85.2% of TP. Such serious non-point-source pollution should be controlled by strengthening dryland management and emphasizing conservation tillage to reduce sediment and nutrient losses.

Table 3. The type of plant, area and area ratio, elemental nitrogen, and phosphorus fertilizers applied through calendar fertilization and auto-fertilization in HRU (Nfert, Pfert, see Section 2.2.2), and total N and total P transport from a certain HRU into reach (TN, TP) in load and load per unit area per year for individual land-use/land-cover (LULC) classes.

LULC	Plant	Area (km^2)	Area (%)	Nfert (kg/ha·y)	Pfert (kg/ha·y)	TN (kg/ha·y)	TP (kg/ha·y)	TN (ton·y)	TP (ton·y)
AGRL	WWHT/CORN	14,913.9	61.6	153.9	69.6	10.0	7.1	14,850.9	10,566.6
FRST	FRST	4462.2	18.4	-	-	2.7	0.9	1223.9	387.2
URLD	BERM	1638.0	6.8	121.2	-	5.6	1.5	917.0	252.5
RICE	RICE	1083.2	4.5	76.2	49.3	6.3	8.4	680.8	913.5
WATR	-	657.6	2.7	-	-	-	-	-	-
PAST	Panicum	567.3	2.3	-	-	9.7	3.1	550.1	173.8
HAY	Hay	392.0	1.6	321.6	12.5	3.0	1.5	118.3	60.4
URBN	BERM	221.3	0.9	226.2	-	6.6	0.3	147.1	7.5
ORCD	ORCD	142.7	0.6	2.9	-	4.1	1.1	57.9	15.0
UIDU	BERM	83.2	0.3	129.2	-	6.0	1.5	49.6	12.5
BARR	BARR	30.4	0.1	-	-	14.0	4.2	42.6	12.8
Whole basin		24,192	100	114.3	45.3	7.7	5.1	18,638.2	12,401.8

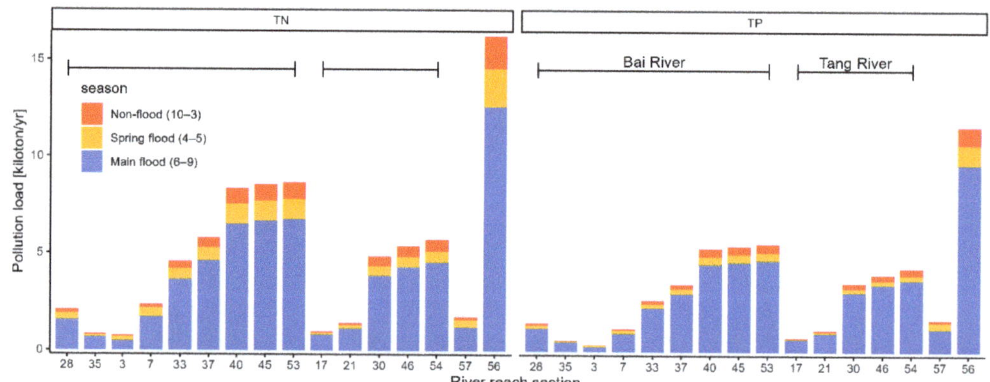

Figure 5. Seasonal mean TN and TP loads transported with water out of the individual river reaches averaged over 2000–2019. Reaches numbered before 53 belong to the Bai River, and those numbered 17–54 belong to the Tang River. Reach 57 is the downstream tributary called the Gun River, and 56 is the watershed outlet of TRB.

3.2.2. Seasonal Variation in Nutrients

We divided the year into three seasons according to the monthly variation in precipitation: the non-flood season (from October to March), spring flood season (April and May), and main flood season (June to September). The respective mean seasonal total precipitation from 2000–2019 are 174.9 mm (21% of the mean annual), 150.6 mm (18%), and 497.4 mm (61%). Figure 5 shows the TN and TP loads at each representative reach for the three seasons. The nutrient loads are overwhelmingly dominant in the main flood season, constituting on average 78% of the annual TN load and 84% of the TP load. Using the watershed outlet (reach 56) as an example, the TN loads in spring flood season and main flood season are approximately 3.5 and 9 times that in the non-flood season, respectively; the TP loads are 3.3 and 16 times that in the non-flood season, respectively. The simulated low nutrient pollution in non-flood seasons is because the pollutants gradually accumulate on the land surface when there is little or no rain, regardless of the point-source pollution emissions. The spring flood season is the first period of concentrated rainfall after the dry winter; thus, it allows the pollutants accumulated in the soil and floating in the air to be washed away into the stream.

Comparing the loads of different reaches, a clear increasing trend of nutrients is observed from upstream to downstream, suggesting relatively lower rate of pollutant degradation than accumulation. For example, except for the two reaches 28 and 35, which are tributaries of the Bai River (Figure 4b), TN and TP loads increase continuously along reaches 3 to 53. The changes among reaches 40, 45, and 53 are limited because the leaching losses from the respective subbasins are low (Figure 4a,c). The TN load in the Bai River (reach 53) is around 1.5 times that of the Tang River (reach 54), accounting for around 53.4% and 35.3% of total nutrient loads at the outlet reach 56. The TP loads in the Tang River and the lower reaches of the Tangbai River contribute more to that in reach 56, as a result of increasing phosphorus loss from paddy fields (RICE, Table 3) which are distributed mainly in the southern part of TRB (Figure 2a).

3.2.3. Nutrients in the Trans-Provincial Key Sections

The trans-provincial analysis focuses on three controlled sections, including the outlet of TRB (Zhangwan section on reach 56) and the junctions of Henan and Hubei province (Diwan section and Bukou section on reaches 45 and 46). Figure 6 illustrates the monthly distribution of nutrient concentrations in these sections. Both the multi-year mean and median results are shown to eliminate the disturbance from large outliers.

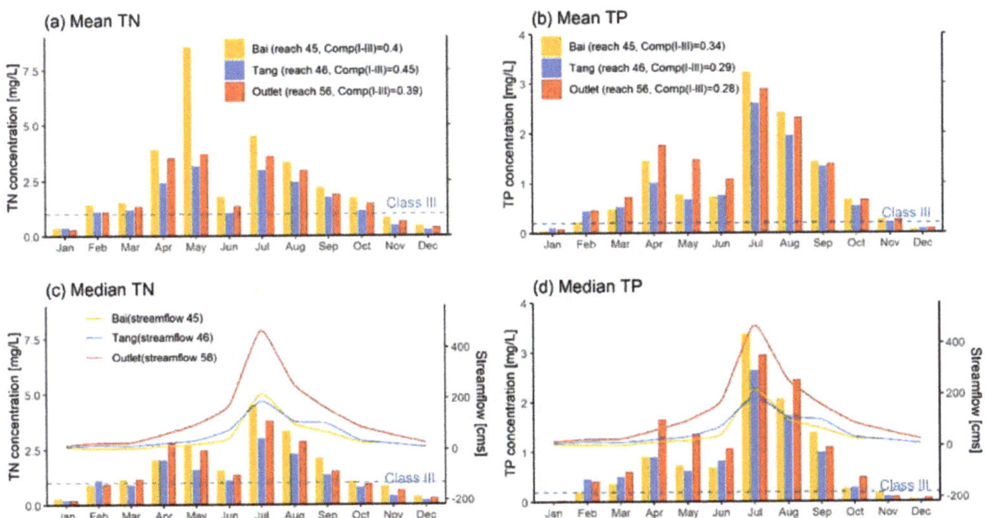

Figure 6. Mean and median values of monthly TN (**a**,**c**) and TP (**b**,**d**) concentrations in reaches 45 (Diwan section) on the Bai River, 46 (Bukou section) on the Tang River, and 56 (Zhangwan section) of the watershed outlet. The dashed horizontal lines are the respective concentrations of water quality Class III. The three curves in (**c**,**d**) are SWAT-simulated mean monthly streamflow for the individual reaches. Comp(I–III) in the legend is the respective compliance rate of Classes I–III water in the period 2000–2019.

Nutrient pollution is concentrated during the flood season (April to September): it increases sharply in April, decreases in June, peaks in July, and then gradually declines with decreasing streamflow. In accordance with Figure 5, the nutrient concentration is high in the spring flood season (April and May) despite the small proportion of annual rainfall and streamflow. The N and P leaching losses under heavy rainfall during the spring flood season far exceed those under weak rainfall during the non-flood season, which is consistent with previous findings [54]. This in turn causes the nutrient concentration to exceed the water quality standard. For the Bai River, the mean TN concentration in the spring flood season can be even higher than that in the main flood season. Interestingly, we find the mean TN (Figure 6a) to be much larger than the median value (Figure 6c), while the difference is insignificant in the remaining months. This indicates that the extremely high TN concentration on the Bai River can occur during spring flood seasons. In contrast, the TP concentration in the spring flood season is only about half of the peak value (in July). This is because paddy fields (RICE), which are the main source of P pollutants (Table 3), lay fallow in the non-flood season (Table 2); therefore, phosphorus accumulation is limited in winter, and the amount of P flushed out from paddy fields is also not high during the spring flood season.

Considering the water quality standard, the Class III water quality compliance rate for monthly TN concentration over 2000–2019 is around 40%, and for TP, it is only 30%. The higher compliance rate of TN than TP is consistent with the spatial distribution results in Figure 4. On average, only months in the non-flood season (October to March) may reach Class III or Class V water. The TN pollution in the Bai River is more serious than that in the Tang River; but the opposite is true for TN pollution.

3.3. Effect and Quantification of Eco-Compensation

Stretching across Henan province (Nanyang city) and Hubei province (Xiangyang city), TRB has a high proportion of traditional agriculture. Nanyang has 86.8% of the total agricultural fields of TRB, while Xiangyang has only 13.2%. The boundary between wheat–corn fields and rice fields almost coincides with the provincial boundary between Henan and Hubei provinces. Nanyang city has been accustomed to growing wheat and corn for thousands of years for geographical reasons. The major soil types in Nanyang are shajiang black soil (SJHT) and yellow-cinnamon soil (HHT, Figure 2b). Both are typical low-yield soils due to their low organic matter content and poor soil structure [55,56]. The wide extent of dryland farming in Henan province, along with intensive human activities, has accelerated soil erosion, leading to serious nutrient losses.

To realize the simultaneous growth of the economy and restoration of water ecosystems in Nanyang city, increasing crop yield, for example, through implementing more scientific crop management (e.g., agricultural industrial agglomeration) and a higher portion of organic fertilizer [57], is necessary but not sufficient. It is an inevitable trend to abandon part of drylands, and transform and upgrade the local industrial structure. This section predicts the effect of converting dryland to forest on water quality in streams across the provincial boundary. Note that we assume that dryland fields (AGRL) in Hubei province and all paddy fields (RICE) remain unchanged because (1) this study investigates trans-boundary eco-compensation, the logic of which is the downstream Hubei province may benefit from the ecological contribution of Nanyang in the upstream and, in turn, compensate for this contribution; (2) the soil type in the paddy fields is mostly paddy soil (SDT), which is one of the three high-yield soils in China; thus the ecological benefits of retiring paddy fields are likely to be much fewer than the economic benefits of retaining farming.

Figure 7 shows the changes in pollutants and reference eco-compensation values at trans-provincial sections under different 53 Grain for Green (GFG) settings (see Section 2.3). Overall, a clearly more noticeable nutrient load decrease is observed compared to the TN/TP compliance rate increase with the increase in GFG area. For instance, up to 60% area achieves only limited improvement (within 10%) for the TP compliance rate while the TP load is reduced by more than 50% (Figure 7b,d). Under the same horizontal axis, there is a disparity in results because the HRUs for GFG are randomly selected, either in the upper regions of the Tang River or Bai River. However, for the whole basin, the different selections in HRUs bring only limited difference in the nutrient pollution (i.e., thin ribbons for the basin outlet shown in Figure 7e). If all dryland fields in TRB in Henan province could be converted to forest (i.e., 100% GFG area), the water compliance rate in the Diwan and Bukou sections would increase by 32%–39%. The months with the worst water quality are mainly concentrated in Spring (March to May, Figure 8e,f). However, the improvement in TN at the outlet section is stronger than that in TP, with maximum increases in the compliance rate in TN of 29.8% and in TP of only 13.2%. Under the 100% GFG, the further accumulation of TN and TP loads in the downstream Hubei province is responsible for 38% and 62% of the N and P transported out of the basin, respectively. This implies that TP pollution is more extensive in Hubei and therefore cannot be controlled by eco-protection measures in Henan province only.

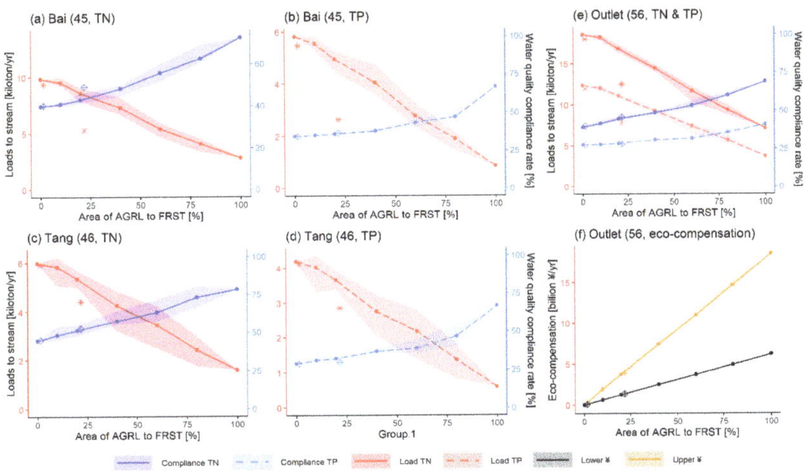

Figure 7. Effects of converting drylands (AGRL) in Henan to forest in terms of TN and TP leaching into reaches from upstream subbasins and compliance rate of water quality (Class III water) in the individual controlled sections (**a–e**), and the value of eco-compensations in CNY based on opportunity cost (upper line, **f**) and GFG compensation standard (lower line, **f**). The ribbons are the range and means of the respective results for GFG based on certain area ratios, and the scatter diamonds and stars are results based on slopes.

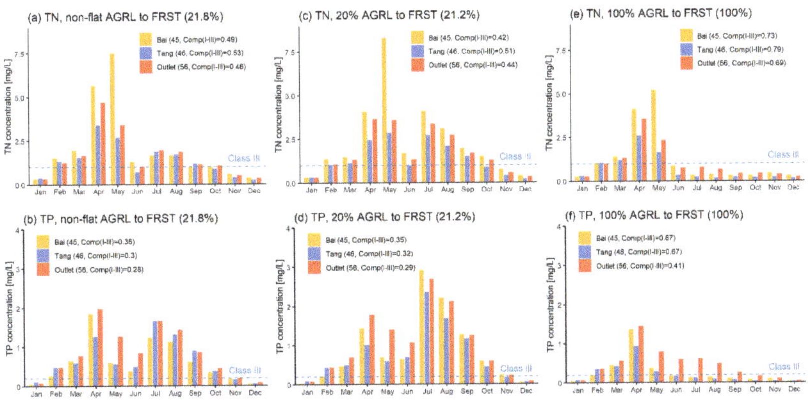

Figure 8. Comparison of mean monthly TN (**a**,**c**,**e**) and TP (**b**,**d**,**f**) concentrations in pollutant control sections between three different sets of GFG scenarios: converting all non-flat drylands (AGRL with slope > 2.5%) in Henan province to FRST (**a**,**b**), one scenario of randomly selecting 20% AGRL to FRST based on HRUs (**c**,**d**), and converting all AGRL land to FRST (**e**,**f**).

As sloping lands are more susceptible to soil erosion, the scatter dots in Figure 7 depict the two slope-specific GFG scenarios. It is evident that GFG of sloping drylands (slope > 15°, area = 1.5%) and non-flat drylands (slope > 2°, area = 21.8%) are more effective in controlling nutrient loss than slope-independent GFG scenarios. Using the TN load at the Diwan section (on the Bai River, Figure 7a) as an example, the TN reduction under the 21.8% non-flat dryland GFG is 3.5 times the mean reduction under the 20% slope-independent GFG; however, water quality compliance rates in this section are not much different from each other. This inconsistency is because although GFG of sloping/non-flat drylands reduces the pollutant load considerably, mainly in the main flood season (June to September), TN or TP concentrations during this period are still worse than Class III (Figure 8a,b and Figure 6a,b).

To summarize, if the water quality standard is the only benchmark, then improving the nutrient concentration at the provincial boundary sections requires control of both the amount of nutrients being washed out and the timing of their washing out. Thus, enhancing the monitoring and management of water quality, along with the ecological operation of water projects, are the key engineering measures that allow water bodies to self-purify more effectively.

The payments of eco-compensation are calculated based on the GFG area and crop yields (Figure 7f). In the baseline scenario without any GFG, the mean annual yields of maize and wheat in TRB within Henan province are 3839.1 and 3740.7 kiloton, respectively. The sum of this is close to, but slightly higher than, the statistic annual grain production of 7105.9 kiloton in Nanyang city in 2019 [58]. The total average annual value of both crops simulated by SWAT is CNY 18.4 billion. The compensation amount considering agricultural opportunity costs increases almost linearly with the increase in GFG area, from CNY 1.81 billion (for 10% GFG) to CNY 18.4 billion (for 100% GFG) (upper line in Figure 7f) per year. When compensation is based on merely the GFG area, the value also increases proportionally, but grows more slowly (lower line in Figure 7f). Under the 100% GFG scenario, the lower value is only one-third of the upper value. This indicates that even under the same ecological restoration strategy, the amount of eco-compensation available to farmers may vary considerably due to the different quantitative criteria adopted.

3.4. Implications for Eco-Compensation

In China, the main factors accounting for eco-compensation include the following three with reference to guidelines in other countries. (1) The opportunity-cost factor: additional consideration must be given to opportunity costs, which are the non-ecological benefits being sacrificed by environmental protectors to protect the environment [59]. (2) The polluter-pays factor: the actors causing the pollution should pay to correct the wrong, thereby limiting their pollution activities and effectively reducing the environmental free-riding behaviors [60]. (3) The beneficiary-pays factor: beneficiaries of ecosystem services should reimburse the upstream providers of water-related environmental services either in full or according to a share of the total [61,62]. Generally, the finance of eco-compensation in China comes from both the government and directly responsible stakeholders [41]. The calculations performed in this study, especially Figure 7f, can provide the bases for estimating opportunity cost. Although the money involved may seem large, it does not reflect the actual amount of funds needed for reforestation but provides reference boundaries of the investment that needs to be paid to the upstream by downstream water users. Local ecological requirements and financial affordability should be considered when accounting for eco-compensation and making decisions.

Funding for watershed protection can be implemented in two ways. The first is paying for ecological projects, such as water project construction, including sewage treatment and ecological restoration plants, and their subsequent maintenance. The second is to directly subsidize the contributors of the affected areas, such as paying out the protection funds to the residents, enterprises, and local governments involved on an annual or quarterly basis to compensate their losses due to changes in crop production and lifestyles. The former project payment can promote regional sustainable development and maintain long-term operation, but fixed, static investments may lack flexibility [63]. In contrast, the latter subsidy, although highly flexible and easily gains trust from local residents, can easily turn into consumption expenditure [64], thereby deviating from the policy objective of agricultural transformation. In this study area, the eco-compensation fund needs to be allocated in a combination of project support and subsidies. That is, to achieve a good water ecosystem status in the downstream TRB, it is necessary not only to carry out integrated management at the watershed level but also to provide cash payments to the upstream government and residents, especially to farmers who return farmland to forest.

The standards of eco-compensation can be determined through negotiation based on project cost analysis and water value assessment [65] or on flexible gambling agreements [66]. The value assessment can be calculated by incorporating both direct cost and opportunity cost, such as the money inputs for pollutant dilution and the lost development opportunities due to ecological protection activities. The gambling agreement is more often utilized for eco-compensation regarding trans-regional water protection, where negotiation is made between upstream and downstream in terms of water quality standards and pollutant thresholds in the controlled river sections. If the water quality at the junctions meets the standards, the co-protection fund should be allocated upstream as compensation for protecting the watershed; conversely, the fund will be given downstream as compensation for purifying water.

4. Conclusions

The Tangbai River Basin (TRB) is a typical trans-provincial watershed in central China that experiences severe non-point-source agricultural pollution. Based on the SWAT model simulations in TRB, the spatial and temporal distributions of nitrogen and phosphorus pollution are described in terms of nutrient load and concentration. Their responses to Grain for Green (GFG) ecological restoration measures are then investigated. With respect to trans-regional eco-compensation, we evaluated the effects of location, area, and the slope-dependency of GFG on the water quality along the provincial boundary sections, using TN and TP as indicators. It appears that GFG measures may be more effective at reducing nutrient loads than increasing the water quality compliance rate. The monetary values of the corresponding eco-compensation are quantified based on crop production changes and the GFG area. We found that the compensation amount using opportunity cost can be as high as three times the amount typically paid based on area. This study could provide suggestions for eco-compensation and offer new ideas for controlling non-point-source pollution of trans-boundary rivers.

Author Contributions: W.W., H.Z. and Y.L. planned the study and prepared the initial data. W.W. performed computations and data analysis, and wrote the initial draft. H.Z. supervised the writing of the paper. Y.F. coordinated a part of computation. H.Z. and H.F. provided financial support. The revised version was written by W.W., with specific contribution from H.Z. and J.Z. All authors have read and agreed to the published version of the manuscript.

Funding: This research was supported by the National Natural Science Foundation of China, (No. 52179009, 51909035 and U2040206).

Institutional Review Board Statement: Not applicable.

Informed Consent Statement: Not applicable.

Data Availability Statement: The public datasets used for SWAT model building have been properly cited in the main text, while the data of streamflow and nutrient concentration adopted for model validation can be found in https://doi.org/10.6084/m9.figshare.20402019.

Conflicts of Interest: The authors declare that they have no known competing financial interests or personal relationships that could have appeared to influence the work reported in this paper.

References

1. Prathumratana, L.; Sthiannopkao, S.; Kim, K.W. The relationship of climatic and hydrological parameters to surface water quality in the lower Mekong River. *Environ. Int.* **2008**, *34*, 860–866. [CrossRef] [PubMed]
2. Zeitoun, M.; Goulden, M.; Tickner, D. Current and future challenges facing transboundary river basin management. *WIREs Clim. Chang.* **2013**, *4*, 331–349. [CrossRef]
3. Connell, D.; Grafton, R.Q. Water reform in the Murray-Darling Basin. *Water Resour. Res.* **2011**, *47*, W00G03. [CrossRef]
4. Isakson, R.S. *Payments for Environmental Services in the Catskills: A Socio-Economic Analysis of the Agricultural Strategy in New York City's Watershed Management Plan*; Report was Elaborated for the "Payment for Environmental Services in the Americas" Project; FORD Foundation and Fundación PRISMA: San Salvador, El Salvador, 2002.
5. Ren, Y.; Lu, L.; Yu, H.; Zhu, D. Game strategies in government-led eco-compensation in the Xin'an River Basin from the perspective of the politics of scale. *J. Geogr. Sci.* **2021**, *31*, 1205–1221. [CrossRef]
6. Jiang, K.; Merrill, R.; You, D.; Pan, P.; Li, Z. Optimal control for transboundary pollution under ecological compensation: A stochastic differential game approach. *J. Clean. Prod.* **2019**, *241*, 118391. [CrossRef]
7. Pagiola, S.; Platais, G. *Payments for Environmental Services*; World Bank: Washington, DC, USA, 2002.
8. Adhikari, B. *Market-Based Approaches to Environmental Management: A Review of Lessons from Payment for Environmental Services in Asia*; Asian Development Bank Institute: Tokyo, Japan, 2009.
9. Sun, X.Z.; Xie, G.D.; Zhen, L. Effects of converting arable land into forest (grassland) and eco-compensation: A case study in Yuanzhou county, Guyuan city of Ningxia Hui Autonomous Region. *Resour. Sci.* **2007**, *29*, 194–200. (In Chinese)
10. Xian, J.; Xia, C.; Cao, S. Cost–benefit analysis for China's Grain for Green Program. *Ecol. Eng.* **2020**, *151*, 105850. [CrossRef]
11. Wang, J.; Tian, R.; Dong, Z.; Shi, G.; Hou, C. *Watershed Eco-Compensation in China: Practice and Review*; Springer: Singapore, 2022; pp. 1–37.
12. Wang, H.; Dong, Z.; Xu, Y.; Ge, C. Eco-compensation for watershed services in China. *Water Int.* **2016**, *41*, 271–289. [CrossRef]
13. Chen, C.; Matzdorf, B.; Zhen, L.; Schroeter, B. Social-Network Analysis of local governance models for China's eco-compensation program. *Ecosyst. Serv.* **2020**, *45*, 101191. [CrossRef]
14. Reckhow, K.H.; Norris, P.E.; Budell, R.J.; Di Toro, D.M.; Galloway, J.N.; Greening, H.; Sharpley, A.N.; Shirmohammadi, A.; Stacey, P.E. *Achieving Nutrient and Sediment Reduction Goals in the Chesapeake Bay: An Evaluation of Program Strategies and Implementation*; National Academies Press: Washington, DC, USA, 2011.
15. Tang, W.; Pei, Y.; Zheng, H.; Zhao, Y.; Shu, L.; Zhang, H. Twenty years of China's water pollution control: Experiences and challenges. *Chemosphere* **2022**, *295*, 133875. [CrossRef] [PubMed]
16. Jin, X.; Xu, Q.; Huang, C. Current status and future tendency of lake eutrophication in China. *Sci. China Ser. C Life Sci.* **2005**, *48*, 948–954.
17. Wu, Z.; Zhang, D.; Cai, Y.; Wang, X.; Zhang, L.; Chen, Y. Water quality assessment based on the water quality index method in Lake Poyang: The largest freshwater lake in China. *Sci. Rep.* **2017**, *7*, 17999. [CrossRef] [PubMed]
18. Boesch, D.F.; Brinsfield, R.B.; Magnien, R.E. Chesapeake Bay eutrophication: Scientific understanding, ecosystem restoration, and challenges for agriculture. *J. Environ. Qual.* **2001**, *30*, 303–320. [CrossRef] [PubMed]
19. Singh, A.; Imtiyaz, M.; Isaac, R.K.; Denis, D.M. Comparison of soil and water assessment tool (SWAT) and multilayer perceptron (MLP) artificial neural network for predicting sediment yield in the Nagwa agricultural watershed in Jharkhand, India. *Agric. Water Manag.* **2012**, *104*, 113–120. [CrossRef]
20. Li, Z.; Luo, C.; Xi, Q.; Li, H.; Pan, J.; Zhou, Q.; Xiong, Z. Assessment of the AnnAGNPS model in simulating runoff and nutrients in a typical small watershed in the Taihu Lake basin, China. *Catena* **2015**, *133*, 349–361. [CrossRef]
21. Redhead, J.W.; May, L.; Oliver, T.H.; Hamel, P.; Sharp, R.; Bullock, J.M. National scale evaluation of the InVEST nutrient retention model in the United Kingdom. *Sci. Total Environ.* **2018**, *610*, 666–677. [CrossRef] [PubMed]
22. Arnold, J.G.; Moriasi, D.N.; Gassman, P.W.; Abbaspour, K.C.; White, M.J.; Srinivasan, R.; Santhi, C.; Harmel, R.D.; Van Griensven, A.; Van Liew, M.W. SWAT: Model use, calibration, and validation. *Trans. ASABE* **2012**, *55*, 1491–1508. [CrossRef]
23. Sharp, R.; Tallis, H.T.; Ricketts, T.; Guerry, A.D.; Wood, S.A.; Chaplin-Kramer, R.; Nelson, E.; Ennaanay, D.; Wolny, S.; Olwero, N. *InVEST User's Guide*; The Natural Capital Project: Stanford, CA, USA, 2014.
24. Bingner, R.L.; Theurer, F.D. AGNPS 98: A Suite of Water Quality Models for Watershed Use. In Proceedings of the Federal Inter-Agency Sedimentation Conference 7, Reno, NV, USA, 25–29 March 2001; pp. 1–8.
25. Hoang, L.; van Griensven, A.; Mynett, A. Enhancing the SWAT model for simulating denitrification in riparian zones at the river basin scale. *Environ. Modell. Softw.* **2017**, *93*, 163–179. [CrossRef]
26. Shrestha, S.; Bhatta, B.; Shrestha, M.; Shrestha, P.K. Integrated assessment of the climate and landuse change impact on hydrology and water quality in the Songkhram River Basin, Thailand. *Sci. Total Environ.* **2018**, *643*, 1610–1622. [CrossRef] [PubMed]
27. Huang, Q.; Huang, Q. Tangbai River Water Pollution and Prevention Measures. *Yangtze River* **1999**, *30*, 30. (In Chinese)
28. STHJT. Monthly Report on Water Quality of Han River. Available online: http://sthjj.xiangyang.gov.cn/hjxx/tjsj/hjszyb/ (accessed on 1 May 2021). (In Chinese)
29. Yao, C. Pollution of Tangbai River Basin Endangers Hubei and Henan Provinces. Available online: http://www.xfmj.gov.cn/cms/html/jianyanxiance/20100304/65.html (accessed on 1 February 2021). (In Chinese)
30. State Council; Ministry of Natural Resources; National Bureau of Statistics. *Technical Regulations of the Third National Land Survey*; State Council; Ministry of Natural Resources; National Bureau of Statistics: Beijing, China, 2018. (In Chinese)

31. Cai, X.; Xu, Z.; Su, B.; Yu, W. Distributed simulation for regional evapotranspiration and verification by using remote sensing. *Trans. CSAE* **2009**, *25*, 154–160. (In Chinese)
32. Cai, L. Investigation on Agricultural Non-point Source Pollution in Henan Province. Master's Thesis, Henan University, Kaifeng, China, 2017. (In Chinese).
33. Lan, S. Investigation and Research on Agricultural Non-point Source Pollution in Nanyang City. *Agro-Environ. Dev.* **2009**, *26*, 59–61. (In Chinese)
34. Nanyang Agricultural Bureau. Guiding Opinions on Scientific Fertilization of Crops in Nanyang City in 2019. Available online: http://nyj.nanyang.gov.cn/ (accessed on 1 May 2021). (In Chinese)
35. Shi, J.; Zhou, S.; Zhao, J.; Dong, B. Livestock and Poultry Manure Excrements and Their Environmental Effect Evaluation in Nanyang. *J. Domest. Anim. Ecol.* **2014**, *35*, 76–81. (In Chinese)
36. Liu, X.; Kuang, M.; Dai, Z.; Wang, M.; Xie, B. Water pollution analysis and control measures of Tangbai river. *J. Chongqing Univ. Arts Sci.* **2014**, *33*, 5. (In Chinese)
37. Zhang, S.; Lin, L.; Wang, Z.; Pan, X.; Liu, M.; Dong, L.; Tao, J. Spatio-temporal Variation of Water Quality in the Middle-lower Hanjiang River. *J. Yangtze River Sci. Res. Inst.* **2020**, *38*, 7. (In Chinese)
38. Hass, J.E. Optimal taxing for the abatement of water pollution. *Water Resour. Res.* **1970**, *6*, 353–365. [CrossRef]
39. Branca, G.; Lipper, L.; Neves, B.; Lopa, D.; Mwanyoka, I. Payments for watershed services supporting sustainable agricultural development in Tanzania. *J. Environ. Dev.* **2011**, *20*, 278–302. [CrossRef]
40. Dunn, C.P.; Stearns, F.; Guntenspergen, G.R.; Sharpe, D.M. Ecological benefits of the conservation reserve program. *Conserv. Biol.* **1993**, *7*, 132–139. [CrossRef]
41. Delang, C.O.; Yuan, Z. *China's Grain for Green Program*; Springer: Cham, Switzerland, 2016.
42. Moriasi, D.N.; Gitau, M.W.; Pai, N.; Daggupati, P. Hydrologic and Water Quality Models: Performance Measures and Evaluation Criteria. *Trans. ASABE* **2015**, *58*, 1763–1785. [CrossRef]
43. Liu, R.; Xu, F.; Zhang, P.; Yu, W.; Men, C. Identifying non-point source critical source areas based on multi-factors at a basin scale with SWAT. *J. Hydrol.* **2016**, *533*, 379–388. [CrossRef]
44. Hoang, L.; Van Griensven, A.; van der Keur, P.; Refsgaard, J.C.; Troldborg, L.; Nilsson, B.; Mynett, A. Comparison and Evaluation of Model Structures for the Simulation of Pollution Fluxes in a Tile-Drained River Basin. *J. Environ. Qual.* **2014**, *43*, 86–99. [CrossRef]
45. Karki, R.; Srivastava, P.; Bosch, D.D.; Kalin, L.; Lamba, J.; Strickland, T.C. Multi-Variable Sensitivity Analysis, Calibration, and Validation of a Field-Scale SWAT Model: Building Stakeholder Trust in Hydrologic and Water Quality Modeling. *Trans. ASABE* **2020**, *63*, 523–539. [CrossRef]
46. Helfand, G.E.; House, B.W. Regulating nonpoint source pollution under heterogeneous conditions. *Am. J. Agric. Econ.* **1995**, *77*, 1024–1032. [CrossRef]
47. Richardson, C.W.; King, K.W. Erosion and nutrient losses from zero tillage on a clay soil. *J. Agric. Eng. Res.* **1995**, *61*, 81–86. [CrossRef]
48. Neitsch, S.L.; Arnold, J.G.; Kiniry, J.R.; Williams, J.R. *Soil and Water Assessment Tool Theoretical Documentation Version 2009*; Texas Water Resources Institute: Forney, TX, USA, 2011.
49. Van Kessel, C.; Clough, T.; van Groenigen, J.W. Dissolved organic nitrogen: An overlooked pathway of nitrogen loss from agricultural systems? *J. Environ. Qual.* **2009**, *38*, 393–401. [CrossRef] [PubMed]
50. Yuan, Y.; Chiang, L. Sensitivity analysis of SWAT nitrogen simulations with and without in-stream processes. *Arch. Agron. Soil Sci.* **2015**, *61*, 969–987. [CrossRef]
51. Chaubey, I.; Migliaccio, K.W.; Green, C.H.; Arnold, J.G.; Srinivasan, R. Phosphorus modeling in soil and water assessment tool (SWAT) model. In *Modeling Phosphorus in the Environment*; CRC Press: Boca Raton, FL, USA, 2006; pp. 163–187.
52. Davis, M. Nitrogen leaching losses from forests in New Zealand. *N. Z. J. For. Sci.* **2014**, *44*, 2. [CrossRef]
53. Zhang, Y.; Li, P.; Liu, X.; Xiao, L.; Shi, P.; Zhao, B. Effects of farmland conversion on the stoichiometry of carbon, nitrogen, and phosphorus in soil aggregates on the Loess Plateau of China. *Geoderma* **2019**, *351*, 188–196. [CrossRef]
54. Huo, D.; Gan, N.; Geng, R.; Cao, Q.; Song, L.; Yu, G.; Li, R. Cyanobacterial blooms in China: Diversity, distribution, and cyanotoxins. *Harmful Algae* **2021**, *109*, 102106. [CrossRef] [PubMed]
55. Malik, Z.; Malik, M.A.; Zong, Y.; Lu, S. Physical properties of unproductive soils of Northern China. *Int. Agrophys.* **2014**, *28*, 4. [CrossRef]
56. Liu, X.; Zhang, D.; Li, H.; Qi, X.; Gao, Y.; Zhang, Y.; Han, Y.; Jiang, Y.; Li, H. Soil nematode community and crop productivity in response to 5-year biochar and manure addition to yellow cinnamon soil. *BMC Ecol.* **2020**, *20*, 39. [CrossRef] [PubMed]
57. Wei, J.; Jin, Y.; Gao, H.; Chang, J.; Zhang, L. Effects of fertilization practices on infiltration in Shajiang black soils. *Zhongguo Shengtai Nongye Xuebao/Chin. J. Eco-Agric.* **2014**, *22*, 965–971.
58. Nanyang Municipal Bureau of Statistics. *Statistical Bulletin on National Economic and Social Development of Fujian Province in 2021*; Fujian International Investment Promotion Center: Xiamen, China, 2019.
59. Kosoy, N.; Martinez-Tuna, M.; Muradian, R.; Martinez-Alier, J. Payments for environmental services in watersheds: Insights from a comparative study of three cases in Central America. *Ecol. Econ.* **2007**, *61*, 446–455. [CrossRef]
60. Gaines, S.E. The polluter-pays principle: From economic equity to environmental ethos. *Tex. Int'l LJ* **1991**, *26*, 463.

61. He, S.; Su, Y.; Wang, L.; Gallagher, L.; Cheng, H. Taking an ecosystem services approach for a new national park system in China. *Resour. Conserv. Recycl.* **2018**, *137*, 136–144. [CrossRef]
62. Shang, W.; Gong, Y.; Wang, Z.; Stewardson, M.J. Eco-compensation in China: Theory, practices and suggestions for the future. *J. Environ. Manag.* **2018**, *210*, 162–170. [CrossRef]
63. Gersonius, B.; Ashley, R.; Pathirana, A.; Zevenbergen, C. Climate change uncertainty: Building flexibility into water and flood risk infrastructure. *Clim. Change* **2013**, *116*, 411–423. [CrossRef]
64. Li, W.; Chen, J.; Zhang, Z. Forest quality-based assessment of the Returning Farmland to Forest Program at the community level in SW China. *For. Ecol. Manag.* **2020**, *461*, 117938. [CrossRef]
65. Wu, Z.; Guo, X.; Lv, C.; Wang, H.; Di, D. Study on the quantification method of water pollution ecological compensation standard based on emergy theory. *Ecol. Indic.* **2018**, *92*, 189–194. [CrossRef]
66. Gu, Q.; Zhang, Y.; Ma, L.; Li, J.; Wang, K.; Zheng, K.; Zhang, X.; Sheng, L. Assessment of reservoir water quality using multivariate statistical techniques: A case study of Qiandao Lake, China. *Sustainability* **2016**, *8*, 243. [CrossRef]

 water 

Article

Comparison of the Calibrated Objective Functions for Low Flow Simulation in a Semi-Arid Catchment

Xue Yang [1,*], Chengxi Yu [1], Xiaoli Li [2], Jungang Luo [1], Jiancang Xie [1] and Bin Zhou [3,4]

1 State Key Laboratory of Eco-Hydraulics in Northwest Arid Region, Xi'an University of Technology, Xi'an 710048, China
2 Shandong Xinhui Construction Group Co., Ltd., Dongying 257091, China
3 State Key Laboratory of Environmental Criteria and Risk Assessment, Chinese Research Academy of Environmental Sciences, Beijing 100012, China
4 Tianjin Academy of Eco-Environmental Sciences, Tianjin 300191, China
* Correspondence: xue.yang@xaut.edu.cn

Abstract: Low flow simulation by hydrological models is a common solution in water research and application. However, knowledge about the influence of the objective functions is limited in relatively arid regions. This study aims to increase insight into the difference between the calibrated objective functions by evaluating eight objectives in three different classes (single objectives: KGE(log(Q)) and KGE(1/Q); multi objectives: KGE(Q)+KGE(log(Q)), KGE(Q)+KGE(1/Q), KGE(Qsort)+KGE(log(Qsort)) and KGE(Qsort)+KGE(1/Qsort); Split objectives: split KGE(Q) and split (KGE(Q)+KGE(1/Q))) in Bahe, a semi-arid basin in China. The calibrated model is Xin An Jiang, and the evaluation is repeated under varied climates. The results show a clear difference between objective functions for low flows, and the mean of KGE and logarithmic transformed-based KGE in time series (KGE(Q)+KGE(log(Q))) presents the best compromise between the estimation for low flows and general simulation. In addition, the applications of the inverse transformed-based KGE (KGE(1/Q)) and the Flow Duration Curve-based series (Qsort) in objectives are not suggested.

Keywords: low flow simulation; objective functions; hydrological calibration; semi-arid basin

Citation: Yang, X.; Yu, C.; Li, X.; Luo, J.; Xie, J.; Zhou, B. Comparison of the Calibrated Objective Functions for Low Flow Simulation in a Semi-Arid Catchment. *Water* **2022**, *14*, 2591. https://doi.org/10.3390/w14172591

Academic Editor: Leonardo V. Noto

Received: 22 July 2022
Accepted: 19 August 2022
Published: 23 August 2022

1. Introduction

Low flow research plays a significant role in water management, such as aquatic ecosystems, irrigation, water supply, and hydroelectricity [1,2]. Applying hydrological models to low flow analysis is essential, especially for basins lacking discharge data [3]. However, hydrological models have simplified the water cycling processes and include some model parameters that cannot be directly measured [4]. Therefore, before applying the model in the interested regions, model calibration is essential to optimize the model parameters [5]. Due to the changing climate, growing scientific efforts to assess hydrological changes for future scenarios have been made. Aiming to reduce the uncertainty of future predictions, generating well-calibrated models is imperative [6].

Model calibration is the process of identifying a suitable model parameter set to minimize the difference between the simulated and observed values, represented by the objective function [7]. Thus, an excellent objective function is always the backbone of a satisfactory scientific outcome. To understand the influence of objective functions and improve the model simulation, considerable research has been carried out in recent decades (e.g., [8–11]). The most critical improvement is replacing the single objective with multi-objective (e.g., [12,13]), making the multi-objective calibration widely used in water resource applications, especially for hydrological simulations [14]. Efstratiadis and Koutsoyiannis [9] reviewed different case studies about multi-objective applications in hydrology and found that the multi-objective approach improved the identifiability of parameters in complex parameterization.

Even though a significant number of studies have applied various multi-objective functions in hydrological model calibration, studies focusing on low flow analysis are limited. Shafii and De Smedt [15] calibrated the WetSpa model by combining the normal and log-transformed Nash–Sutcliffe Efficiency (NSE) as the objective function and found that it is possible to find a compromise with equal attention to both high-flows and low-flows. Kim [16] also applied the normal and log-transformed NSE in the objective function to emphasize high and low flow in a hydrograph and concluded that it worked better. Garcia et al. [3] conducted a comprehensive evaluation, particularly on low flow simulations with different objective functions in hundreds of French basins but applied the inverse transformation to make the low flow sensitive to the objective functions. Their result suggested that the combination between normal and inverse transformed Kling Gupta Efficiency (KGE) is recommended. Apart from the transformed format of objective functions based on time series, studies including the hydrological signatures in the objective function are increasing. According to the comparison between the time series-based and Flow Duration Curve (FDC)-based transformed format of objective functions, Garcia et al. [3] found using FDC based transformation is worse than the time series-based objective function for low flow indices simulation. At the same time, Lombardi et al. [17] deduced that including the match of the FDC statistic in the calibration outperformed the time domain calibration on an excellent reproduction of the low-to-average flow quantiles, based on 52 Italian catchments. Consistent with Lombardi et al. [17], Chilkoti et al. [14] found the inclusion of FDC-based signatures in objective functions could improve the performance for low flow simulation, according to the calibration of a SWAT model in a small snow-fed catchment. From the above studies, there are consistent answers to the question of whether or not taking the FDC-based signatures could help low flow simulation. On the other hand, the above studies were conducted in humid regions and little attention has been paid to relatively arid areas.

To enhance the knowledge about the influence of the calibrated objective functions in relatively arid regions, this study proposes a comprehensive evaluation by considering eight different objective functions in a semi-arid Chinese basin. The evaluated objective functions consist of varied formats, transformations, and bases, and are compared from three aspects: the hydrograph simulation, FDC simulation, and the low flow indices. To additionally explore more about the climatic influence on the objective functions, different climatic conditions are also considered in the evaluation.

2. Study Area and Data

2.1. Study Area

The study was conducted in the Bahe basin of China, which is in the northern part of the Qinling Mountains. The Ma Du Wang (MDW) hydrologic station was selected, located downstream in the Bahe basin; the watershed station before the Bahe River flows into the Weihe River. There is no large reservoir in the catchment. The catchment area is about 1760 km^2 (see Figure 1), the average elevation is 1170 m, and the land use is dominated by agriculture and forest. The average annual precipitation in the Bahe region is about 720 mm, and nearly 60% of precipitation occurs between July and October. Precipitation is the primary source of runoff, and the summer runoff accounts for more than 40% of annual runoff. According to the Köppen–Geiger climate classification, the watershed controlled by the MDW station belongs to Dwa and Dwb classes: monsoon-influenced hot/warm summer, semi-arid continental climate.

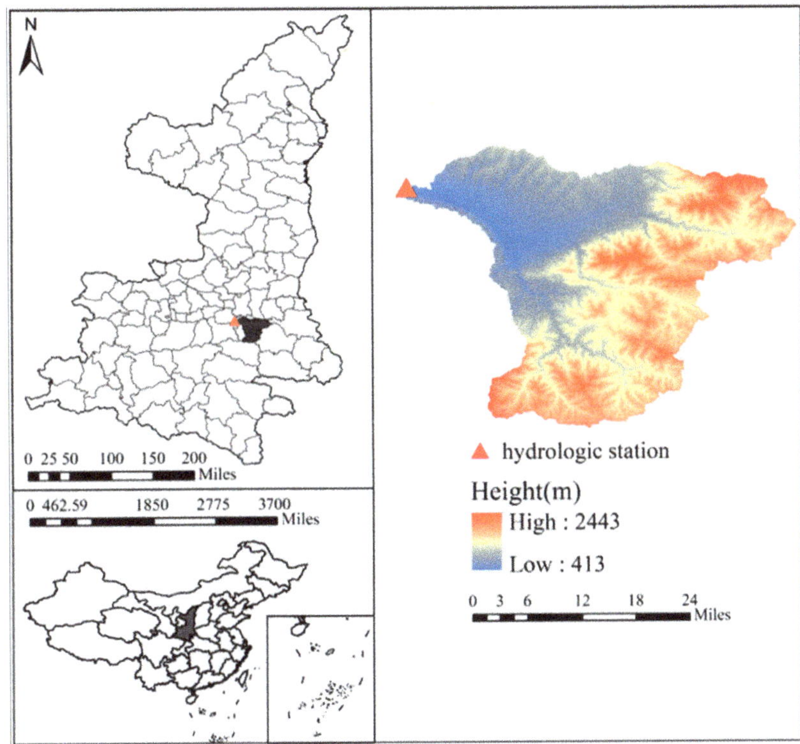

Figure 1. The location and Digital Elevation Model (DEM) information of the study area.

2.2. Data

The meteorological data in this study come from the National Meteorological Information Centre (NMIC) and applies the same site station information as He et al. [18]. In addition, the spatial interpolation method for areal mean precipitation and ET calculation is by the Simple Kriging, which is also the same as He et al. [18]. The runoff data are at the daily time scale in this study, obtained from the Yellow River Conservancy Commission (YRCC).

3. Methods

For this comparative analysis, a conceptual hydrological model, Xin An Jiang (XAJ), is calibrated with different objective functions under varied climates. More detailed information is shown in the following.

3.1. Hydrological Model and Model Optimization

In this study, the XAJ model, a conceptual rainfall-runoff model at a daily time step, is selected. The XAJ model was developed for relatively humid regions in China by Zhao et al. [19,20], which has become a widely used model in runoff simulation, water resources assessment, and climate change assessments (e.g., [21]). In this study area, this model has been validated [22], and the model structure applied here is the same as Lin et al. [23]. For a detailed model description, please check there.

To optimize the hydrological model parameter set, an effective global optimization algorithm, the shuffled complex evolution (SCE-UA) algorithm was used in this study. This algorithm is mainly based on the concept of information-sharing and natural biological evolution [24,25]. It has been widely used in hydrological model calibration (e.g., [26,27]).

3.2. Calibration Objective Functions

Summarizing currently used objective functions, three different classes of objectives were evaluated; Table 1 gives more detailed information. About the criteria, both NSE and KGE are widely used in hydrology, while KGE is free of the influence of unhelpful interactions among components [28]. Therefore, KGE has been analyzed and recommended by many studies (e.g., [29–31]) and is applied here.

The calculation of KGE follows Equation (1):

$$\mathrm{KGE} = 1 - \sqrt{(\mathrm{r}-1)^2 + (\alpha-1)^2 + (\beta-1)^2} \tag{1}$$

With

$$\begin{cases} r = \frac{1}{N}\sum_{i=1}^{N} \frac{(\mathrm{Q_o}-\mu_\mathrm{o})(\mathrm{Q_s}-\mu_\mathrm{s})}{\sigma_o\sigma_s} \\ \alpha = \frac{\mu_\mathrm{s}}{\mu_\mathrm{o}} \\ \beta = \frac{\sigma_s}{\sigma_o} \end{cases} \tag{2}$$

As shown in Table 1, it includes the single objectives, multi objectives, and split objectives. The single objective class, which includes OBJ1 and OBJ2, applies two different transformation approaches to the discharge in the KGE calculation. These transformations are considered to emphasize the low flow goodness of fit: one is the logarithmic transformed discharge [5,32,33] and another is the inverse transformed discharge [34]. For the multi objective class, which is from OBJ3 and OBJ6, follows the format from Garcia et al. [3], who combined the normal KGE with the inverse transformation-based KGE in the objective by the same weights for both the time-based and FDC-based series. Additionally, this study includes the logarithm transformation-based partners to explore the influence from the transformation selection. The last class considers the recommendation from Fowler et al. [35], who proposed the split KGE as the objective function and found it could significantly improve the model performance. To validate the improvement of this strategy, this split KGE is set as OBJ7. OBJ8 is proposed in this study, which applies this strategy to the suggested objective function (OBJ4) by Garcia et al. [3]. Regarding the above description, some connections or similarities exist between the evaluated objective functions, which is helpful to explore the characteristics by pair comparison.

Table 1. The information of evaluated objective functions in this study.

Classes	Criteria	Name	Description	Reference
Single objective	$\mathrm{KGE_{(log(Q))}}$	OBJ1	KGE calculated on logarithmic transformed discharges	Oudin et al. [33]
	$\mathrm{KGE_{(1/Q)}}$	OBJ2	KGE calculated on inverse transformed discharges	Pushpalatha et al. [34]
Muti objective	$\mathrm{KGE_{(Q)}}+\mathrm{KGE_{(log(Q))}}$	OBJ3	Sum of KGE calculated on discharges and logarithmic transformed discharges	Proposed in this study
	$\mathrm{KGE_{(Q)}}+\mathrm{KGE_{(1/Q)}}$	OBJ4	Sum of KGE calculated on discharges and inverse transformed discharges	Garcia et al. [3]
	$\mathrm{KGE_{(Qsort)}}+\mathrm{KGE_{(log(Qsort))}}$	OBJ5	Sum of KGE calculated on the FDC and logarithmic transformed of the FDC	Proposed in this study
	$\mathrm{KGE_{(Qsort)}}+\mathrm{KGE_{(1/Qsort)}}$	OBJ6	Sum of KGE calculated on the FDC and logarithmic transformed of the FDC	Garcia et al. [3]
Split objective	split $\mathrm{KGE_{(Q)}}$	OBJ7	Averaged KGE calculated on discharges in each year	Fowler et al. [35]
	split $(\mathrm{KGE_{(Q)}}+\mathrm{KGE_{(1/Q)}})$	OBJ8	Averaged sum of KGE calculated on discharges and inverse transformed discharges in each year	Proposed in this study

3.3. Model Performance Assessment

Climatic Robustness Assessment

The Differential Split-Sample Test (DSST) is applied to test the objective functions, where two independent periods are in different conditions [36]. According to the statistical climate analysis, the climate will be drier in the future in this region [18]. Considering this, the model is calibrated with a relatively wet climate and validated in a dry climate.

Figure 2 displays the precipitation information from 1998 to 2019 in the MDW station. It is easy to see that 2000–2002 is the only continuous period that every annual precipitation is lower than the average value; the annual precipitation is around 540 mm. To provide more valuable information for the application, as considered above, the period 2000–2002 is thus set as the validation period. Correspondingly, a relatively humid period is considered to be the calibration period. Through the plot, 2003–2005 shows the highest 3-year mean precipitation value (about 681 mm per year), making it as the calibration period. In order to increase the climatic robustness, the period 2007–2009 is also set as a calibration period, since its annual precipitation (about 661 mm per year) is higher than the average value in each year. In summary, two relatively humid periods (2003–2005 and 2007–2009) are applied for the calibration evaluation, and a relatively arid period (2000–2002) is used for the evaluation in this study.

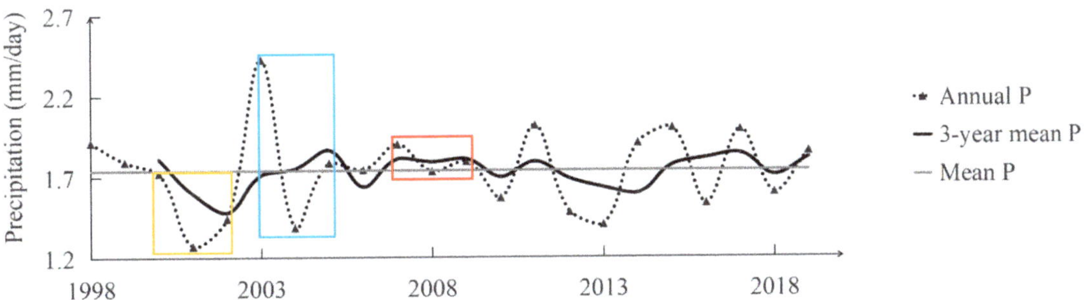

Figure 2. The precipitation (P) information in Ma Du Wang station, the blue and red window marks the calibration periods, and the yellow window marks the validation period.

3.4. Assessment Criteria

Paying more attention to the low flow simulation does not mean reducing the general performance. Therefore, the evaluation criteria used to compare the objective functions in this study are based on the general and the low flow simulation. Table 2 shows the applied assessment criteria correspondingly, including many low-flow indices used in hydrology [2,37]. For instance, the logarithmic transformed criteria have been widely used in studies, which shows overall goodness of fitting but emphasizes low flow [4,5,33]. Another class of low flow indices measure the low flow severity at different time steps, which is more concerned by water management agencies; for example, the mean annual 3-day minimum discharges. Moreover, the usage of FDC statistics increases, since it could provide valuable information in the frequency domain [11,38]. In this class, the LFD, Q95, and Q75 are applied in the study.

Table 2. The applied criteria of performance evaluation in this study.

Criteria	Description
KGE	Kling-Gupta Efficiency (see Equation (1))
KGE_{log}	KGE calculated on logarithmic transformed flow
MAM3	Mean Annual Minimum 3-day mean flow at 3-year return period
MAM10	Mean Annual Minimum 10-day mean flow at 3-year return period
MAM30	Mean Annual Minimum 30-day mean flow at 3-year return period
LFD	The duration of low flow smaller than 30% of the time
Q95	Flow exceeded 95% of the time
Q75	Flow exceeded 75% of the time

4. Results

4.1. Objective Functions Evaluation

4.1.1. Hydrograph Simulation

The time series of flow observation presents the temporal change in the water cycle in a basin, which is the base information for hydrological statistical analysis, such as the trend, the seasonality, etc. Therefore, assessing the performance of time series simulation is a vital evaluation aspect for objective functions.

To compare the objective function influence on the time series simulation, Figure 3 displays the probability density function (PDF) of the percent bias (Pbias) information for the period of 2003–2005. The Pbias here divides the model residual by the observed flow value, which measures the general information for relative simulation errors. From the left subplot, the logarithmic transformed objective functions are better than the inverse transformed objective functions, regardless of whether the objective is single or multi. Taking the performance classification from Moriasi et al. [39], the days that achieved good simulation ($|Pbias| < 15\%$) account for 45% and 44% during the calibration period by the OBJ1 and OBJ3, followed by OBJ4 with 40%, which is much higher than OBJ2. The result of acceptable performance ($|Pbias| < 25\%$) also supports the above finding, where 61%, 60%, and 57% days are achieved by the OBJ1, OBJ3, and OBJ4, respectively. When comparing the single and multi-objectives, the above results indicate that the difference between single objectives (OBJ1 and OBJ2) is much more significant than the multi objectives (OBJ3 and OBJ4). Moving to the middle subplot, which shows the result from the multi objectives, the general probability of achieving smaller Pbias for objectives based on the time series seems higher than that based on the FDC. For instance, for OBJ5 and OBJ6, the days showing a good performance account for 45% and 26%, respectively, and the values change to 62% and 48% for acceptable performance. The right subplot shows the result from all three different kinds of objectives, and OBJ4 presents a better result than others, which means the split objective functions did not improve the simulation for the hydrological time series, while between two split objective functions, OBJ8 provides a slightly better simulation performance, which makes 2% and 1% days achieve a good and acceptable performance than OBJ7, correspondingly.

Figure 4 shows the same information as Figure 3, but for the calibration result during 2007–2009. Even though the general characteristics here are in line with Figure 3, minor differences exist. Although a clear distinction appears between single objectives (OBJ1 and OBJ2), the difference between multi objectives (see the middle subplot) is smaller than the period 2003–2005. Statistically, the days in good performance accounts for 45%, 44%, 36%, and 35%, and the values change to 64%, 58%, 53%, and 50% for acceptable performance for the OBJ3, OBJ4, OBJ5, and OBJ6, respectively.

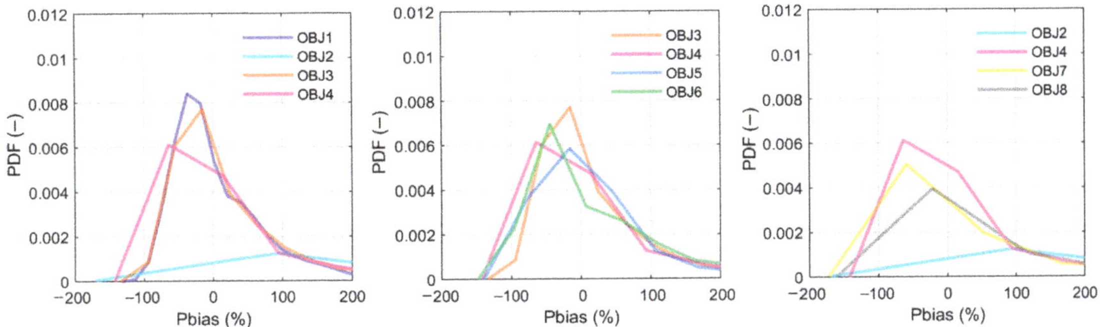

Figure 3. The probability density function (PDF) comparison for the objective functions evaluating by the percent bias (Pbias) during the calibration period 2003–2005.

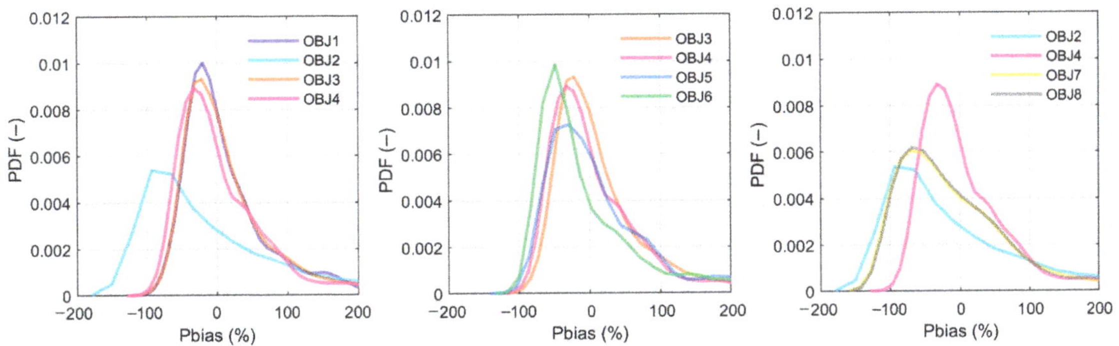

Figure 4. The probability density function (PDF) comparison for the objective functions evaluating by the percent bias (Pbias) during the calibration period 2007–2009.

The goodness of hydrograph fitting is also an essential measure for flow time series simulation, which has been evaluated frequently by KGE in recent years. Since this study focuses more on the low flow, the logarithmic transformed KGE results (KGE_{log}) are also included. Table 3 presents the calculated values of KGE and KGE_{log} during two calibration periods with all eight objective functions. In the table, the highest values for each period among objective functions are highlighted in bold, and '/' is used when the value is lower than 0.

Table 3. The calibrated KGE and KGElog values during two calibration periods.

Evaluation Criteria	KGE		KGElog	
Calibration Period	**2003–2005**	**2007–2009**	**2003–2005**	**2007–2009**
OBJ1	0.85	0.63	**0.78**	**0.84**
OBJ2	0.60	0.25	/	/
OBJ3	0.90	**0.78**	0.77	0.83
OBJ4	**0.92**	**0.78**	0.70	0.79
OBJ5	0.89	0.62	0.74	0.69
OBJ6	0.90	0.55	0.68	0.61
OBJ7	0.85	0.68	/	/
OBJ8	0.74	0.69	/	/

Note: '/' is used when the value is lower than 0.

When looking at the KGE values, almost all objective functions show an acceptable performance in both calibration periods but with considerable differences. For example, during 2003–2005, the highest KGE is 0.92 from OBJ4, and the lowest value is 0.6 from OBJ2, while the difference between multi objectives is slight, which is 0.03 according to the result from 2003–2005. Comparing three different objectives classes, multi objectives show relatively higher KGE values, followed by split objectives and single ones. Focusing on the low flow simulation assessment by KGE_{log}, three objectives produce values lower than 0, which means unacceptable. At the same time, all the multi objective functions provide good results, whose KGE_{log} values are higher than 0.61. The highest KGE_{log} value appears for OBJ1; this is mainly because the evaluation criterion is the same as the objective function and the KGE_{log} values for OBJ3 are very close to OBJ1.

Considering the balance between general and low flow simulation through two periods, OBJ3 and OBJ4 yield relatively better results, followed by OBJ1. Taking the averaged KGE and KGE_{log} values for the two periods as the example, the result is 0.821 and 0.797 for OBJ3 and OBJ4, respectively, followed by 0.773 for OBJ1. Among the multi objectives, regardless of whether it is time series-based or FDC-based, the logarithmic transformed objectives tend to yield higher averaged measurements than the inverse transformed objectives. The averaged KGE and KGE_{log} value of the two periods for OBJ5 is 0.736, which is 0.685 for OBJ6.

4.1.2. Flow Duration Curves

Unlike the time series evaluation, FDC statistics could provide valuable frequency domain information. Figure 5 presents the FDC assessment result overall for the eight objective functions during 2003–2005, and each subplot contains two zoomed subplots to more clearly present the results for high and relatively low flow simulations.

According to the left subplots, the simulated FDC from OBJ2 is far from all other curves, including the observation one. With the two zoomed subplots, OBJ2 presents substantial overestimation-to-observation for the highest 10% flow and heavy underestimation-to-observation for the lowest 50% flow. While the curves from OBJ1 and OBJ3 seem closer to the observation through two zoomed subplots, especially the low flow one. Compared with the left subplot, these multi objectives evaluated in the middle subplot produce more similar FDC simulations, especially for the high flows. While according to the zoomed low flow subplot, OBJ5 presents the closest FDC simulation to the observation, followed by OBJ3, and OBJ6 stays furthest. All simulated curves show a visible difference from the right subplot, more significant between each curve than from the left subplot. Among these objective functions, OBJ4 provides the closest simulation; the split objective functions work similarly to OBJ2.

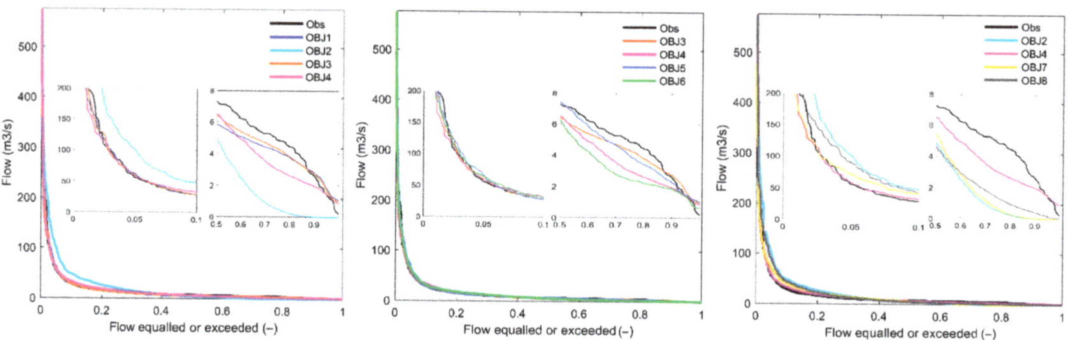

Figure 5. The result of the observed and simulated FDCs by all objective functions during the calibration period 2003–2005. The zoomed plots in each subplot show the result for the highest 10% flow (**left**) and the lowest 50% flow (**right**) simulations.

Figure 6 presents the results in the same way as in Figure 5 but is based on the calibration during 2007–2009. The general characteristics presented here totally agree with findings from Figure 5, regardless of the scale difference. For instance, the curve simulated by the OBJ2 stays visibly far from the observation curve, and OBJ5 and OBJ3 yield the closest simulation curve to the observation. However, the curves from the multi objectives keep close to each other.

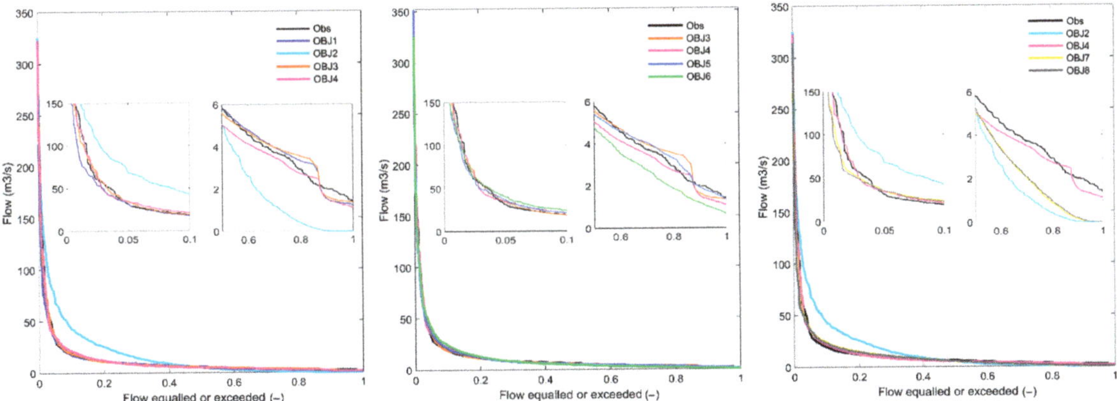

Figure 6. The result of the observed and simulated FDCs by all objective functions during the calibration period 2007–2009. The zoomed plots in each subplot show the result for the highest 10% flow (**left**) and the lowest 50% flow (**right**) simulations.

4.1.3. Low Flow Indices

Since this study emphasizes low flow simulation, different low flow indices are thus applied in Figure 7, where the line shows the observed value and the bar presents the simulated value.

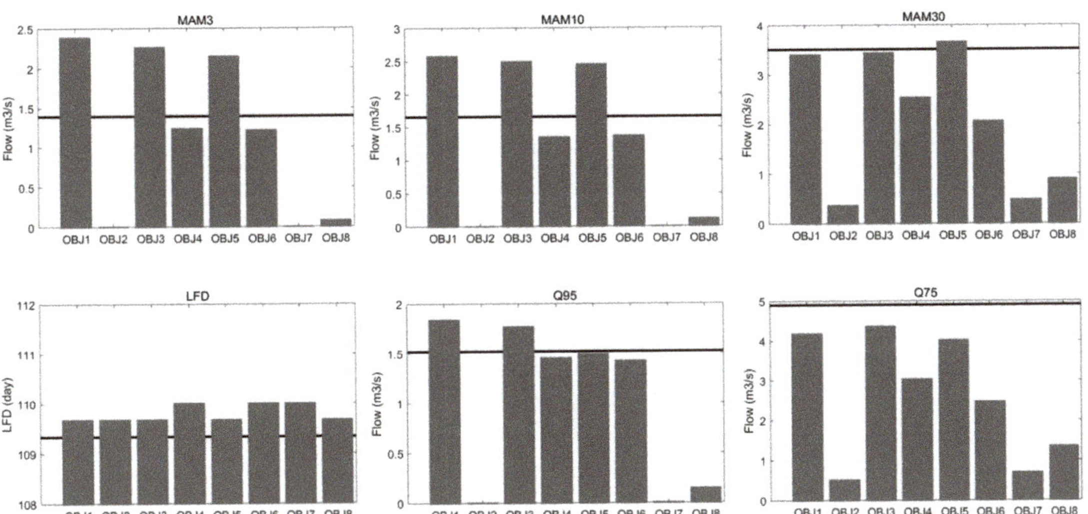

Figure 7. The observed (the line) and simulated (the bars) low flow indices by all objective functions during the calibration period 2003–2005.

Through all subplots, the objective functions provide similarly good simulations to the observed LFD. Apart from the simulation of LFD, the objectives OBJ2, OBJ7, and OBJ8 vastly underestimate the other observed low flow indices, which are not comparable with other evaluated partners. Between the rest of the objectives, the inverse transformed objectives (OBJ4 and OBJ6) estimate the indices visibly lower than the logarithmic transformed ones (OBJ1, OBJ3, and OBJ5). For example, the average estimation for MAM3 is about 1.3 m^3/s from the inverse transformed objectives, which is about 2.2 m^3/s from the logarithmic transformed objectives.

Assessing the performance aspect, the inverse transformed objectives better estimate the indices sensitive to the extreme low flows (e.g., MAM3, MAM10, and Q95). According to the subplot for MAM10, the averaged estimation error is about 0.3 m^3/s from the inverse transformed objectives, which climbs to 0.8 m^3/s from the logarithmic partners. Conversely, the logarithmic transformed objectives provide a better estimation for the indices less sensitive to the extreme low flows (e.g., MAM30 and Q75). Observing the subplot for MAM30, the averaged estimation error is about 1.4 m^3/s from the inverse transformed objectives, which is about 13 times for the logarithmic partners.

In Figure 8, the information is summarized similarly to Figure 7 but applies the data calibrated during 2007–2009. There are some of the same characteristics observed here as in Figure 7, such as the similar simulation from all objectives for LFD; unacceptable estimation by OBJ2, OBJ7, and OBJ8, and the higher estimation by logarithmic transformed objectives than the inverse transformed partners. However, the performance preference shows some differences here from Figure 7. First, OBJ6 produces a minor estimation error for the MAM3 and MAM10, while the OBJ5 yields almost the exact same estimation as the observed Q95. This result cannot support the finding in Figure 7 that the inverse transformed objectives produce a better estimation for the indices sensitive to the extreme low flows. Second, OBJ4 and OBJ5 provide the most similar estimation to the observed MAM30 and Q75, respectively. This result is not consistent with the finding that logarithmic transformed objectives provide a better estimation for the indices less sensitive to the extreme low flows.

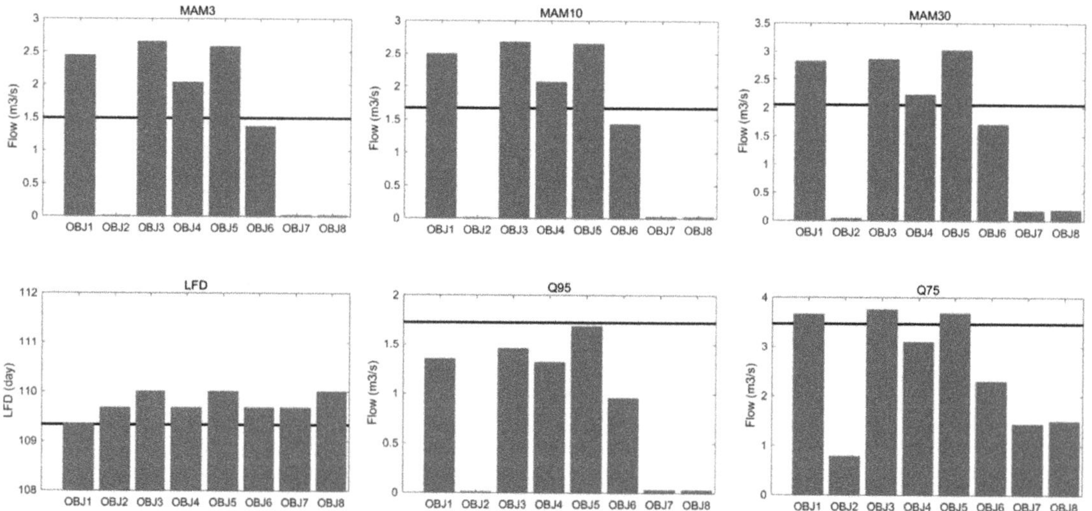

Figure 8. The observed (the line) and simulated (the bars) low flow indices by all objective functions during the calibration period 2007–2009.

4.2. Climatic Robustness Assessment

As mentioned above, the DSST method is applied to assess the climatic robustness of the objectives. To enhance the finding reliability and applicability, the climatic robustness evaluation validates the calibration result achieved in two different wet climate periods in a relatively dry climate.

4.2.1. Hydrograph Simulation

Figure 9 displays the observed and simulated hydrographs during the validation period, with the evaluated objectives, except for OBJ2, OBJ7, and OBJ8 due to the bad calibration.

At first sight, even though the objectives are different, all simulations follow the observation temporal change pattern, and no apparent time jags appear in both subplots. From the upper subplot, OBJ5 shows a relatively better estimation for high flows, especially the peaks, followed by OBJ4, and other objective measures are comparable. In the lower subplot, OBJ4 tends to overestimate the high flows, except for the peak flow. The rest objectives present similar simulations for most time steps, except OBJ5 for some high flows. Evaluating the simulation performance between two periods, the estimated hydrographs based on the period 2003–2005 are generally closer to the observation than based on the period 2007–2009.

Due to the serious overlaps between hydrograph simulations, the information about the evaluated statistics (KGE and KGE_{log}) is presented in Table 4 to provide more valuable information for hydrograph simulation evaluation.

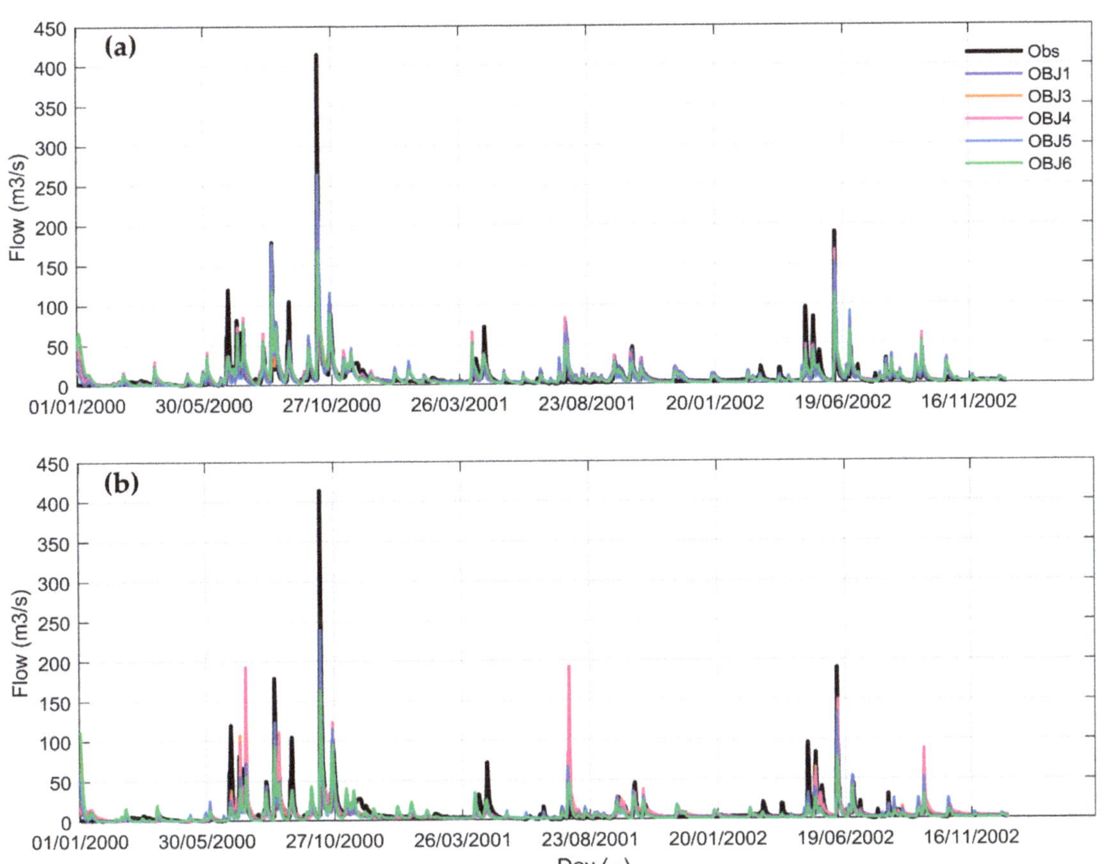

Figure 9. The hydrograph plot during the validation period based on the calibration period (**a**) 2003–2005 (**b**) 2007–2009.

Table 4. The validated KGE and KGElog values yield by different calibrated models.

Calibration Period	2003–2005		2007–2009	
Evaluation Criteria	**KGE**	**KGElog**	**KGE**	**KGElog**
OBJ1	0.61	0.67	0.42	0.69
OBJ3	0.68	**0.70**	0.58	0.68
OBJ4	0.68	0.62	**0.61**	**0.71**
OBJ5	**0.79**	0.67	0.58	0.63
OBJ6	0.61	0.64	0.49	0.67

Most of the objectives produce acceptable validation results based on both calibration periods. According to the values shown in the table, all the KGElog values are higher than 0.62, and most of the KGE values are higher than 0.58. As shown in bold text, all the highest values for both criteria appear in the multi objective group, and all the values are higher than 0.7, except the KGE value during 2007–2009. Between the evaluations based on two different calibration periods, all the KGE values based on the calibrated model during 2003–2005 are higher than 2007–2009, but the KGElog values are comparable. For example, when applying the OBJ1, the KGE value based on 2003–2005 is 0.19 higher than in 2007–2009, but the difference between the two KGElog values is only 0.02. Focusing on the low flow simulation through both validation results, if taking the averaged KGElog value as the measure, OBJ3 presents the best performance, with an averaged KGElog value of 0.69.

4.2.2. Flow Duration Curves

As mentioned above, FDC statistics could provide additional information to the time series simulation. Thus, the validation evaluation also includes the FDC assessment result. Figure 10 presents the corresponding result and the left and right penal subplot show the result based on the calibrated model during 2003–2005 and 2007–2009, respectively.

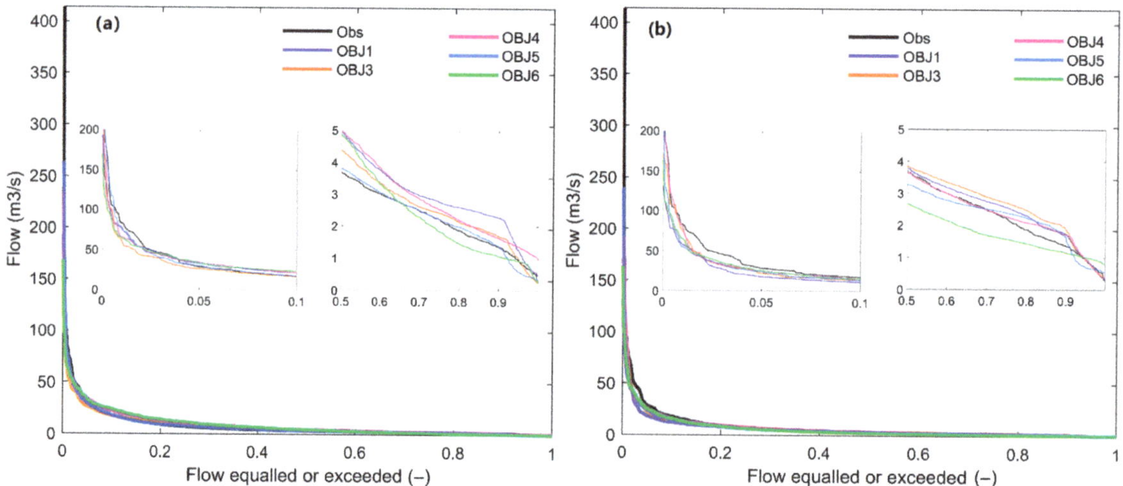

Figure 10. The observed and simulated FDCs for the validation period based on the calibration in (**a**) 2003–2005 (**b**) 2007–2009. The zoomed plots in each subplot show the result for the highest 10% flow (**left**) and the lowest 50% flow (**right**) simulations.

Overall, the simulated FDCs from all the objectives are comparable and not far from the observed FDC, consistent in both periods. Through the simulation for the highest 10% flow and the lowest 50% flow, the results of which are shown in the left and right zoomed subplots, respectively, the difference between objectives is large for the low flow simulations. Among the low flow simulation objectives, the FDC from OBJ5 seems closer to the observed curve in subplot (a), which changes to OBJ4 in subplot (b). In contrast, the simulated FDC from OBJ1 and OBJ6 in corresponding subplots (a) and (b) present a clear distance from the observation. In addition, the simulated FDCs tend to be higher than the observation in subplot (a), while they spread mixed on both sides of the observation in subplot (b).

4.2.3. Low Flow Indices

To further explore the objective influence on low flow simulation, Figure 11 displays the observed and simulated low flow indices during the validation period by applying the calibrated model based on different periods. Through the validated simulations, there is no apparent conflict result shown between the different calibration periods, therefore, the generated result over both periods is described below.

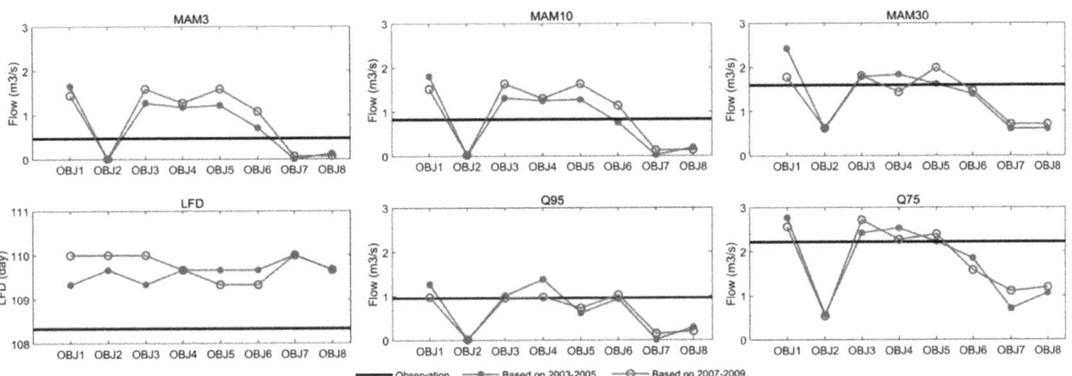

Figure 11. The observed and simulated low flow indices by all objective functions during the validation period.

Consistent with the results shown in the calibration period. First, the OBJ2, OBJ7, and OBJ8 provide significantly different and worse estimations than other objectives for all evaluated low flow indices except LFD. Second, all the objectives based on both calibration periods appear similar to the simulation for LFD, which is about 1.2 days averagely longer than the observation. Third, all the left logarithmic transformed objectives (e.g., the OBJ1, OBJ3, and OBJ5) provide a relatively higher estimation than the inverse transformed partners (e.g., the OBJ4 and OBJ6), except for the Q95 simulation here.

Going through the simulation over upper subplots, the inverse transformed objectives appear closer to the observation, even though the observations between the three indices are clear. For instance, the estimated MAM30 from OBJ6 is only about 0.1 lower than the observation, while for the quartile indices, the simulations from the left objectives (e.g., the OBJ1, OBJ3, OBJ4, OBJ5, and OBJ6) are comparable, especially for Q95, whose range between those objectives is smaller than 0.5. Another interesting point is that for the logarithmic transformed objectives, the difference between the single objective and multi objectives is relatively smaller for extreme low flow indices. For example, the difference between the simulations from the three objectives is about 0.5 for the evaluation of MAM3, which increases to about 1 when assessing the MAM30.

5. Discussion

The hydrological models have been popularly applied in water research and application, while the objective functions that are suitable for calibrating the hydrological models for low flow simulation are unclear, especially in relatively arid regions. Therefore, a comprehensive evaluation of different kinds of objective functions in relatively dry areas will provide valuable information.

5.1. Objective Functions Evaluation

In the current study, eight objective functions were selected for model calibration, which belong to different classes. According to the above results, the logarithmic and inverse transformation formats shows a pronounced difference. This may be due to the high sensitiveness of the inverse transformation to extreme low values, as analyzed by Pushpalatha et al. [34]. In the model error term part, the inverse transformation gave more emphasis on low flows than the logarithm transformation. Another explanation may be that, at relatively arid regions, very low values appear frequently in observed flow, which enhances the weight on low flows. Second, our results suggest applying logarithm transformation rather than the inverse transformation in low flow studies, which differs from Pushpalatha et al. [34]. One possible reason may be that Pushpalatha et al. [34] applied NSE, and NSE is regarded to give more emphasis on high flow than KGE, which, thus, balances the low flow weight to some extent. Furthermore, despite the FDC simulation, the FDC-based multi objectives did not exhibit a better performance than the time series-based partners, which concurs with Garcia et al. [3]. As confirmed by the above results and analysis, OBJ3 is suggested for low flow simulation studies in relatively arid regions. It fails to agree with the finding by Garcia et al. [3], who recommended the OBJ4 as the sufficient calibration objective for low flow simulation based on the evaluation in a humid region. Therefore, the climate and geographic conditions may be the main factor for the disagreement between the two studies.

Comparing the single and the combined multi objective group from the performance over different aspects, the general performance applying combined multi objectives seems better. This result confirms the notion that the combined objectives could achieve an overall better benefit (e.g., [40]). The single objective is difficult to simulate all the hydrograph shape characteristics simultaneously (e.g., [41,42]). Furthermore, whether based on the time series or FDC, the performance difference between multi objectives appears smaller than between the single objectives. That means the simulation uncertainty is relatively smaller when applying the multi objectives, which is in line with the knowledge that multi objective calibration could mitigate the uncertainty issues (e.g., [43]). Regarding the performance from spit objectives, which is proposed and recommended by Fowler et al. [35], it seems the worst among all groups, especially for the simulation of low flow indices.

5.2. Climatic Robustness Assessment

To additionally explore their climatic robustness, the evaluation of their performance was based on two different calibration periods and a validation period, whose climatic conditions varied. Assessing from two calibration periods, the general observed characteristics of the results are consistent, although the performance values showed some differences. Furthermore, even though the average precipitation changed more than 20%, shown as the validation and calibration periods in this study, no noticeable changes were detected for the general observed characteristics. This is unlike the finding by Garcia et al. [3], who asserts that the robustness depends on the climate variability rather than the objective function. This difference of opinion may be related to the different magnitude. For example, the climate difference between the calibration and validation period in the current study is not big enough to explore the climatic influence. At the same time, a minor but interesting characteristic from the validation results is that the general performance difference between single and multi-objectives seems more significant based on the calibration with more considerable climate variability.

6. Conclusions

The accuracy of low flow simulation yield from the hydrological models presents an apparent effect on water management. Research on the suitableness evaluation of the calibration objectives is of importance. Aiming to enhance insight into the objective influence on low flow simulation in relatively arid regions, which prior to our research was very limited, this study evaluated eight different kinds of objective functions with varied climate conditions. The analysis was performed using the observation at Ma Du Wang station in the Bahe basin, China, located in a semi-arid and semi-humid continental climate region. The main conclusions from the study are summarized in the following points:

- The influence of the included transformation formats in objective functions on low flow simulation is pronounced, and logarithmic transformation is recommended.
- Among the three classes of objective functions, the combined multi-class is highly recommended, and the mean of $KGE_{(Q)}$ and $KGE_{(\log(Q))}$ remains a first choice. In contrast, the class of split objectives is regarded as the last choice as it demonstrated the worst performance.
- Replacing the objective function from the time series based on the FDC could not improve the simulation performance.

Although this study evaluated the performance of different objectives under varied climates and achieved additional valuable knowledge, the current study has some limitations. First, including more hydrological models could help obtain more solid conclusions and deepen the understanding of model influence. In addition, assessing the performance under an increased number of varied climate conditions could broaden the knowledge about the climatic influence, which is essential for research concerning the changing climate.

Author Contributions: Conceptualization, X.Y. and B.Z.; methodology, X.Y.; software, X.Y.; validation, C.Y. and X.L.; formal analysis, X.L.; investigation, C.Y.; resources, J.L.; data curation, J.X.; writing—original draft preparation, X.Y.; writing—review and editing, X.Y. and C.Y.; visualization, X.L.; supervision, B.Z.; project administration, J.L.; funding acquisition, J.X. All authors have read and agreed to the published version of the manuscript.

Funding: This research was funded by the Education Department of Shaanxi Provincial Government (Project No. 21JT032) and Xi'an University of Technology (Project No. 256082016). The APC was funded by Xi'an University of Technology (Project No. 256082016).

Data Availability Statement: Data available upon request.

Conflicts of Interest: The authors declare that they have no known competing financial interests or personal relationships that could have influenced the work reported in this paper.

References

1. Engeland, K.; Hisdal, H. A comparison of low flow estimates in ungauged catchments using regional regression and the HBV-model. *Water Resour. Manag.* **2009**, *23*, 2567–2586. [CrossRef]
2. Lang Delus, C. Les étiages: Définitions hydrologique, statistique et seuils réglementaires. *Cybergeo Eur. J. Geogr.* **2011**. [CrossRef]
3. Garcia, F.; Folton, N.; Oudin, L. Which objective function to calibrate rainfall-runoff models for low-flow index simulations? *Hydrol. Sci. J.* **2017**, *62*, 1149–1166. [CrossRef]
4. Zhang, R.; Liu, J.; Gao, H.; Mao, G. Can multi-objective calibration of streamflow guarantee better hydrological model accuracy? *J. Hydroinform.* **2018**, *20*, 687–698. [CrossRef]
5. Kim, H.S.; Lee, S. Assessment of the adequacy of the regional relationships between catchment attributes and catchment response dynamics, calibrated by a multi-objective approach. *Hydrol. Process.* **2014**, *28*, 4023–4041. [CrossRef]
6. Peel, M.C.; Blöschl, G. Hydrologic modelling in a changing world. *Prog. Phys. Geogr.* **2011**, *35*, 249–261. [CrossRef]
7. Misgana, M.K. Model Performance Sensitivity to Objective Function during Automated Calibrations. *J. Hydrol. Eng.* **2012**, *17*, 756–767.
8. Reed, P.M.; Hadka, D.; Herman, J.D.; Kasprzyk, J.R.; Kollat, J.B. Evolutionary multiobjective optimization in water resources: The past, present, and future. *Adv. Water Resour.* **2013**, *51*, 438–456. [CrossRef]
9. Efstratiadis, A.; Koutsoyiannis, D. One decade of multi-objective calibration approaches in hydrological modelling: A review. Hydrol. *Sci. J. J. Sci. Hydrol.* **2010**, *55*, 58–78. [CrossRef]
10. Pfannerstill, M.; Guse, B.; Fohrer, N. Smart low flow signature metrics for an improved overall performance evaluation of hydrological models. *J. Hydrol.* **2014**, *510*, 447–458. [CrossRef]
11. Asadzadeh, M.; Leon, L.; McCrimmon, C.; Yang, W.; Liu, Y.; Wong, I.; Fong, P.; Bowen, G. Watershed derived nutrients for Lake Ontario inflows: Model calibration considering typical land operations in Southern Ontario. *J. Gt. Lakes Res.* **2015**, *41*, 1037–1051. [CrossRef]

12. Coello, C.A.; Aguirre, A.H.; Zitzler, E. Evolutionary multi-objective optimization. *Eur. J. Oper. Res.* **2007**, *181*, 1617–1619. [CrossRef]
13. Tian, F.; Hu, H.; Sun, Y.; Li, H.; Lu, H. Searching for an Optimized Single-objective Function Matching Multiple Objectives with Automatic Calibration of Hydrological Models. *Chin. Geogr. Sci.* **2019**, *29*, 934–948. [CrossRef]
14. Chilkoti, V.; Bolisetti, T.; Balachandar, R. Multi-objective autocalibration of SWAT model for improved low flow performance for a small snowfed catchment. *Hydrol. Sci. J.* **2018**, *63*, 1482–1501. [CrossRef]
15. Shafii, M.; De Smedt, F. Multi-objective calibration of a distributed hydrological model (WetSpa) using a genetic algorithm. Hydrol. *Earth Syst. Sci.* **2009**, *13*, 2137–2149. [CrossRef]
16. Kim, H.S. Adequacy of a Multi-objective Regional Calibration Method Incorporating a Sequential Regionalisation. *Water Resour. Manag.* **2014**, *28*, 5507–5526. [CrossRef]
17. Lombardi, L.; Toth, E.; Castellarin, A.; Montanari, A.; Bratha, A. Calibration of a rainfall-runoff model at regional scale by optimising river discharge statistics: Performance analysis for the average/low flow regime. *Phys. Chem. Earth* **2012**, *42*, 77–84. [CrossRef]
18. He, Y.; Qiu, H.; Song, J.; Zhao, Y.; Zhang, L.; Hu, S.; Hu, Y. Quantitative contribution of climate change and human activities to runoff changes in the Bahe River watershed of the Qinling Mountains, China. *Sustain. Cities Soc.* **2019**, *51*, 101729. [CrossRef]
19. Zhao, R.J.; Zhang, Y.L.; Fang, L.R.; Liu, X.R.; Zhang, Q.S. The Xinanjiang Model. In *Hydrological Forecasting*; IAHS Press: Wallingford, UK, 1980; pp. 351–356.
20. Zhao, R.J. The Xinanjiang model applied in China. *J. Hydrol.* **1992**, *135*, 371–381.
21. Yang, X.; Magnusson, J.; Huang, S.; Beldring, S.; Xu, C.Y. Dependence of regionalization methods on the complexity of hydrological models in multiple climatic regions. *J. Hydrol.* **2020**, *582*, 124357. [CrossRef]
22. An, D.; Li, Z.J.; Kan, G.Y.; Li, Q.L. Comparison between the Application of Data-driven Model and Conceptual Model. *Water Power* **2013**, *39*, 9–12.
23. Lin, K.; Liu, P.; He, Y.; Guo, S. Multi-site evaluation to reduce parameter uncertainty in a conceptual hydrological modeling within the GLUE framework. *J. Hydroinform.* **2014**, *16*, 60–73. [CrossRef]
24. Duan, Q.; Sorooshian, S.; Gupta, V. Effective and efficient global optimization for conceptual rainfall-runoff models. *Water Resour. Res.* **1992**, *28*, 1015–1031. [CrossRef]
25. Duan, Q.; Sorooshian, S.; Gupta, V.K. Optimal use of the SCE-UA global optimization method for calibrating watershed models. *J. Hydrol.* **1994**, *158*, 265–284. [CrossRef]
26. Jeon, J.H.; Park, C.G.; Engel, B. Comparison of performance between genetic algorithm and SCE-UA for calibration of SCS-CN surface runoff simulation. *Water* **2014**, *6*, 3433–3456. [CrossRef]
27. Zeng, Q.; Chen, H.; Xu, C.Y.; Jie, M.X.; Chen, J.; Guo, S.L.; Liu, J. The effect of rain gauge density and distribution on runoff simulation using a lumped hydrological modelling approach. *J. Hydrol.* **2018**, *563*, 106–122. [CrossRef]
28. Gupta, H.V.; Kling, H.; Yilmaz, K.K.; Martinez, G.F. Decomposition of the mean squared error and NSE performance criteria: Implications for improving hydrological modelling. *J. Hydrol.* **2009**, *377*, 80–91. [CrossRef]
29. Lobligeois, F.; Andréassian, V.; Perrin, C.; Tabary, P.; Loumagne, C. When does higher spatial resolution rainfall information improve streamflow simulation? An evaluation using 3620 flood events. *Hydrol. Earth Syst. Sci.* **2014**, *18*, 575–594. [CrossRef]
30. Magand, C. Influence de la Représentation des Processus Nivaux sur L'hydrologie de la Durance et sa Réponse Auchangement Climatique. Ph.D. Thesis, Université Pierre et Marie Curie, Paris, France, 2014.
31. Osuch, M.; Romanowicz, R.J.; Booij, M.J. The influence of parametric uncertainty on the relationships between HBV model parameters and climatic characteristics. *Hydrol. Sci. J.* **2015**, *60*, 1299–1316. [CrossRef]
32. Legates, D.R.; McCabe, G.J. Evaluating the Use of "Goodness-of-Fit" Measures in Hydrologic and Hydroclimatic Model Validation. *Water Resour. Res.* **1999**, *35*, 233–241. [CrossRef]
33. Oudin, L.; Andréassian, V.; Mathevet, T.; Perrin, C.; Michel, C. Dynamic averaging of rainfall-runoff model simulations from comple-mentary model parameterizations. *Water Resour. Res.* **2006**, *42*, W07410.1–W07410.10. [CrossRef]
34. Pushpalatha, R.; Perrin, C.; Moine, N.L.; Andréassian, V. A review of efficiency criteria suitable for evaluating low-flow simulations. *J. Hydrol.* **2012**, *420*, 171–182. [CrossRef]
35. Fowler, K.; Peel, M.; Western, A.; Zhang, L. Improved Rainfall-Runoff Calibration for Drying Climate: Choice of Objective Function. *Water Resour. Res.* **2018**, *54*, 3392–3408. [CrossRef]
36. Klemeš, V. Operational testing of hydrological simulation models. *Hydrol. Sci. J.* **1986**, *31*, 13–24. [CrossRef]
37. Laaha, G.; Blöschl, G. Seasonality indices for regionalizing low flows. *Hydrol. Process.* **2006**, *20*, 3851–3878. [CrossRef]
38. Price, K.; Purucker, S.T.; Kraemer, S.R.; Babendreier, J.E. Tradeoffs among watershed model calibration targets for parameter estimation. *Water Resour. Res.* **2012**, *48*, W10542.1–W10542.16. [CrossRef]
39. Moriasi, D.N.; Arnold, J.G.; Van Liew, M.W.; Bingner, R.L.; Harmer, R.D.; Veith, T.L. Model Evaluation guidelines for systematic quantification of accuracy in watershed simulations. *Trans. ASABE* **2007**, *50*, 885–900. [CrossRef]
40. Matrosov, E.S.; Huskova, I.; Kasprzyk, J.R.; Harou, J.J.; Lambert, C.; Reed, P.M. Many-objective optimization and visual analytics reveal key trade-offs for London's water supply. *J. Hydrol.* **2015**, *531*, 1040–1053. [CrossRef]
41. Jie, M.X.; Chen, H.; Xu, C.Y.; Zeng, Q.; Tao, X. A comparative study of different objective functions to improve the flood forecasting accuracy. *Hydrol. Res.* **2016**, *47*, 718–735. [CrossRef]

42. Kamali, B.; Mousavi, S.J.; Abbaspour, K.C. Automatic calibration of HEC-HMS using single-objective and multiobjective PSO algorithms. *Hydrol. Process.* **2013**, *27*, 4028–4042. [CrossRef]
43. Her, Y.; Seong, C. Responses of hydrological model equifinality, uncertainty, and performance to multi-objective parameter calibration. *J. Hydroinform.* **2018**, *20*, 864–885. [CrossRef]

Article

Evaluating Monthly Flow Prediction Based on SWAT and Support Vector Regression Coupled with Discrete Wavelet Transform

Lifeng Yuan [1,2] and Kenneth J. Forshay [1,*]

1 U.S. Environmental Protection Agency, Center for Environmental Solutions and Emergency Response, Robert S. Kerr Environmental Research Center, Ada, OK 74820, USA
2 U.S. Environmental Protection Agency, Center for Environmental Solutions and Emergency Response, Homeland Security Materials Management Division, Durham, NC 27711, USA
* Correspondence: forshay.ken@epa.gov

Abstract: Reliable and accurate streamflow prediction plays a critical role in watershed water resources planning and management. We developed a new hybrid SWAT-WSVR model based on 12 hydrological sites in the Illinois River watershed (IRW), U.S., that integrated the Soil and Water Assessment Tool (SWAT) model with a Support Vector Regression (SVR) calibration method coupled with discrete wavelet transforms (DWT) to better support modeling watersheds with limited data availability. Wavelet components of the simulated streamflow from the SWAT-Calibration Uncertainty Procedure (SWAT-CUP) and precipitation time series were used as inputs to SVR to build a hybrid SWAT-WSVR. We examined the performance and potential of the SWAT-WSVR model and compared it with observations, SWAT-CUP, and SWAT-SVR using statistical metrics, Taylor diagrams, and hydrography. The results showed that the average of *RMSE*-observation's standard deviation ratio (*RSR*), Nash–Sutcliffe efficiency (*NSE*), percent bias (*PBIAS*), and root mean square error (*RMSE*) from SWAT-WSVR is 0.02, 1.00, −0.15, and 0.27 $m^3 s^{-1}$ in calibration and 0.14, 0.98, −1.88, and 2.91 $m^3 s^{-1}$ in validation on 12 sites, respectively. Compared with the other two models, the proposed SWAT-WSVR model possessed lower discrepancy and higher accuracy. The rank of the overall performance of the three SWAT-based models during the whole study period was SWAT-WSVR > SWAT-SVR > SWAT-CUP. The developed SWAT-WSVR model supplies an additional calibration approach that can improve the accuracy of the SWAT streamflow simulation of watersheds with limited data.

Keywords: SWAT; support vector regression; streamflow prediction; wavelet transform; Illinois River watershed

Citation: Yuan, L.; Forshay, K.J. Evaluating Monthly Flow Prediction Based on SWAT and Support Vector Regression Coupled with Discrete Wavelet Transform. *Water* **2022**, *14*, 2649. https://doi.org/10.3390/w14172649

Academic Editors: Dengfeng Liu, Hui Liu and Xianmeng Meng

Received: 30 June 2022
Accepted: 4 August 2022
Published: 27 August 2022

1. Introduction

A precise and reliable monthly streamflow prediction model is helpful in the planning, management, development, and protection of water resources, such as future flood and drought forecasting, reservoirs, and/or agricultural water management [1–3]. However, many hydrological elements (e.g., precipitation, runoff, sediment, flood, and streamflow) have highly complex, nonlinear, non-stationary, and uncertainty features, which is a challenge for conventional hydrological methods when analyzing and predicting the complex patterns and inherent variabilities of rainfall–runoff relationships [4–8]. Hence, accurate rainfall–runoff prediction became a difficult task in stochastic hydrology; subsequently, new theories and methods, such as machine learning methods [9,10], have been introduced to improve rainfall–runoff forecasting.

Support vector machine (SVM) is a machine learning method, which can be considered a data-driven black-box model [11]. SVM focuses on determining a kernel function and searching for an optimum separating hyperplane based on the kernel function selected.

This separating hyperplane determines an optimal parameter combination fitting of the observations. Meanwhile, the search process avoids overfitting and make SVM present better generalization characteristics [12]. Compared to an artificial neural network (ANN), SVM performs better [13] in some hydrological applications because it applies a structural risk minimization principle to obtain a global optimum solution rather than a solution based on the empirical risk minimization as applied in ANN [4,13–15]. SVM can solve the nonlinear problem in a low dimension input space by projecting to a higher dimension feature space where an original nonlinear problem is converted into a linear problem [16,17]. SVM has been widely applied in recent decades to hydrological prediction worldwide. For instance, Shabri and Suhartono (2012) [18] used the least-squares SVM (LSSVM) method to predict streamflow in Peninsular Malaysia to compare performance of different models' including LSSVM, autoregressive integrated moving average (ARIMA), ANN, and regular SVM. Kalteh (2013) [4] compared the prediction accuracy of monthly flow discharge from an artificial neural network (ANN) method and the support vector regression (SVR) model coupled with a wavelet transform in two stations in northern Iran. Chiogna et al. (2018) [19] combined the Soil and Water Assessment Tool (SWAT) model and SVR method to predict hydropeaking for the Upper Adige River watershed in northeast Italy and applied a wavelet method to analyze the price of energy. However, these studies only used a few hydrological stations to test their methods and evaluate the model performance. Nourani et al. (2015) [20] proposed a two-stage SVM method with spatial statistics to simulate monthly river suspended sediment load for 15 sites within the Ajichay River in northwest Iran. Yuan and Forshay (2021) [21] developed a seasonal SWAT model coupled with SVR for 13 hydrological stations using a spatial calibration method to improve the accuracy of monthly streamflow prediction in the Illinois River watershed. From these efforts, it is clear that the SVM method has powerful predictive ability depending on calibration datasets and can accurately capture nonlinear relationships between the input and output variables. However, the SVM prediction method neglects the detailed characteristics and processes of a watershed system [22] and simplifies the complexity of the rainfall–runoff relationship [23].

Wavelet analysis is a mathematical function with an auto-adaptive time-frequency window (i.e., the width of time and frequency may change) and is suited to analyze and calculate stationary or non-stationary time series signals [4]. In the high-frequency period of signals, the size of a frequency window becomes larger while the size of a time window becomes smaller and otherwise occurs in the signals low-frequency period. Wavelet analysis has an excellent capability to reduce data noise, analyze variabilities, periodicities, and trends of hydrological time series and is suitable for handling the non-stationary flow signals [24,25]. To date, the wavelet analysis method has been widely applied in hydrological prediction. For example, Nourani et al. (2014) [7] reviewed applications of hybrid wavelet-Artificial Intelligence (AI) models in hydrology and presented remarkable progress when integrating wavelet analysis and AI models to improve the prediction accuracy of hydrologic models in the recent couple of decades. Noraini and Norhaiza (2017) [13] compared different wavelet denoising techniques and decomposition levels, and input streamflow time series after wavelet denoising into SVR based on a radial basis function (RBF) kernel to improve 1-month-ahead streamflow prediction in the Segamat River basin in southern Malaysia. Sun et al. (2019) [26] integrated multiple methods, such as the autoregressive model, autoregressive moving average (ARMA) model, ANN model, and linear regression (LR) model, with wavelet transform and compared their differences in performance while predicting daily streamflow in the Heihe River basin of northern China, where they found that the wavelet-based method can effectively improve streamflow simulations compared to other models. Nalley et al. (2020) [27] used several wavelet-transform-based methods to improve the performance of extending streamflow records for areas in Canada with limited data available. From the works mentioned above, it was inferred that wavelet transform typically worked as a data pre-processing tool before employing conventional hydrological analysis methods or data-driven models and could raise the accuracy of the flow prediction. The main concerns regarding wavelet applications in hydrological forecasting are the proper selection of the mother wavelet function and the determination of decomposition levels corresponding to the specific hydrological time series [7,28]. Integrating these machine learning methods with physically based hydrologic models could better describe streamflow by incorporating the constraints of the physical world that include advanced mathematical techniques to discover the complex hidden or obscured signals that are difficult or impossible to model in a physically based modeling system.

Many physically based hydrologic models have been developed and applied to predict streamflow [29]. Among these models, the Soil and Water Assessment Tool (SWAT) is a conceptual, physically based, watershed-scale hydrologic model that has been extensively applied worldwide [30]. SWAT takes a large number of physical processes of hydrology into account. Hence, it requires extensive data and parameter inputs. Often, data are unavailable in certain regions due to time or economic cost or limitations of measurement technologies, especially in some developing countries. Therefore, the unknown values of many parameters in SWAT can only be determined via the procedure of calibration [3]. Calibration is a time-consuming and complicated process since it involves parameterization, optimal algorithm determination, and extensive iterative computation to find optimal value ranges and parameter combinations [11,31,32].

Moreover, the issue of parameter non-uniqueness is that different parameter sets might produce very similar simulated signals with the observed flow time series, which makes effective calibration harder to achieve [31]. To obtain appropriate parameter combinations, it requires researchers to have a profound understanding of hydrological parameters and processes and familiarity of local hydrological conditions and physical features in a study area. To raise the accuracy of hydrological models prediction, especially for a region with limited data available, several efforts have evaluated the performance and potential of SWAT coupling with the SVR methods in streamflow prediction [11,15,33,34], yet few efforts [19,35] have attempted to couple a distributed physically based model and a machine learning method to improve the rainfall–runoff simulation. In some cases, we do not achieve a desirable result of flow prediction with acceptable accuracy, even after conducting comprehensive model calibration. Here, we attempt to improve the accuracy of a commonly used physically based model (SWAT) by integrating discrete wavelet transform functions and support vector machines in a system with complex non-linear rainfall–runoff relationships to support modeling efforts in watersheds with limited data.

The object of this study aims to show how the SVM method with wavelet transforms can be used to improve the monthly flow prediction of the calibrated SWAT hydrological model in the Illinois River watershed (IRW), U.S. This work studied and compared the performance of calibrated SWAT (SWAT-CUP), regular SWAT-SVR without applying wavelet transform, and SWAT-WSVR coupled with wavelet transform for monthly flow prediction for 12 hydrological stations in the IRW. This study is helpful to improve streamflow prediction at a month scale in some areas with sparse data.

2. Methodology

To improve monthly flow prediction, we developed the hybrid model SWAT-WSVR based on SWAT and SVR with discrete wavelet transforms. First, we developed SWAT models for twelve sites after SWAT-CUP (SWAT Calibration and Uncertainty Program) calibration progresses. Then, the flow at month t simulated by SWAT-CUP calibration and corresponding precipitation (i.e., applying the Thiessen polygons divide method to allocate the NCDC meteorological stations to the USGS hydrological sites) of each site served as SVR input variables to predict flow on month t. It was a regular SWAT-SVR model. Next, the simulated monthly flow from SWAT-CUP was decomposed using the discrete wavelet transform (DWT) to obtain wavelet coefficients and approximation at different scale resolutions, which were served as inputs of SWAT-WSVR with precipitation data. The results from SWAT-CUP worked as a benchmark compared with the results from

SWAT-SVR and SWAT-WSVR. To estimate the performance of different models, we used a metric such as the Nash–Sutcliffe efficiency (*NSE*), Percent Bias (*PBIAS*), Root Mean Square Error (*RMSE*), and *RMSE*-observations standard deviation ratio (*RSR*) to estimate the model results quantitatively. Finally, we combined calibrated (or training) and validated (or testing) time series data and re-estimated the entire model performance based on Pearson's correlation coefficient (r), *RMSE*, and normalized standard deviation (*NSD*) and plotted the Taylor diagram and hydrography to compare their performance difference graphically. Figure 1 showed the construction flowchart of three hydrology models.

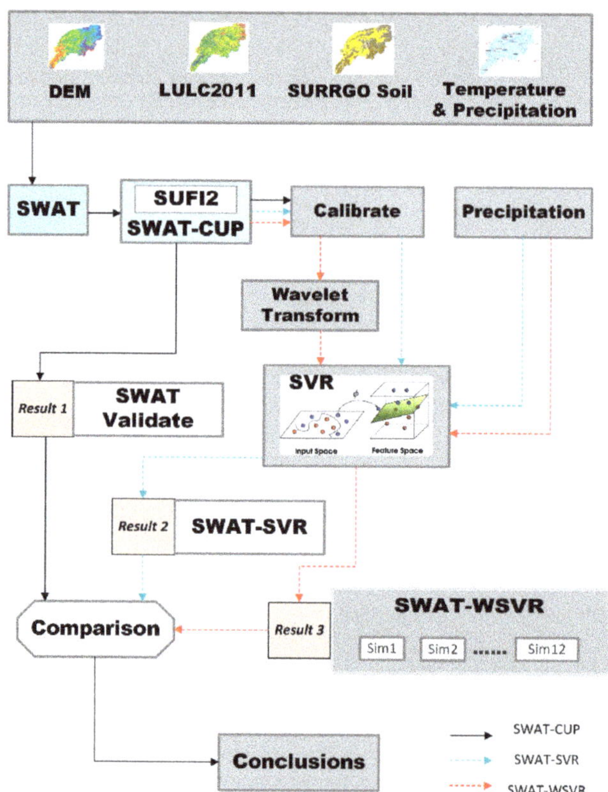

Figure 1. The developing flowchart of different models: SWAT-CUP, SWAT-SVR, and SWAT-WSVR.

2.1. Watershed Description and Data Source

The IRW (35°31′–36°9′ N, 94°12′–95°2′ W) covers about a 4200 km^2 drainage area and crosses Arkansas and Oklahoma, U.S. The average basin slope is about 5.6%. The average annual temperature and precipitation are about 16 °C and 1198 mm, respectively. Monthly statistical discharge data from twelve U.S. Geological Survey (USGS) hydrological sites were downloaded from the official website [21]. These monthly statistics generated from sites are based on USGS-approved daily-mean data. The basic information of 12 hydrologic stations and selected descriptive statistics for monthly flow time series are listed in Table 1. Additionally, Figure 2 shows the spatial distribution of meteorological and hydrological stations, terrain, lakes, and rivers in the IRW.

Table 1. Watershed properties and selected descriptive statistics of USGS hydrological stations.

No.	USGS Station	Upstream Area (km^2)	Data Period (month.year)	Number of Data	Average Monthly Streamflow (m^3 s^{-1})	Flow Descriptive Statistics (m^3 s^{-1})			
						Max	Min	Median	Standard Deviation
1	07195800	36.8	1.1995–12.2013	228	0.41	2.90	0.05	0.25	0.44
2	07195855	155.0	1.1995–12.2013	228	1.27	9.53	0.11	0.74	1.43
3	07196000	300.7	1.1995–12.2013	228	3.01	22.26	0.42	1.78	3.25
4	07195500	1633.0	1.1995–12.2013	228	18.71	149.42	2.73	10.75	20.02
5	07195430	1490.5	1.1996–12.2013	216	17.68	144.61	1.89	10.58	19.29
6	07196090	2138.5	7.2010–12.2013	42	23.19	178.54	2.95	11.77	33.59
7	07196973	64.8	1.1995–12.2002	96	0.66	3.57	0.00	0.38	0.75
8	07196500	2462.5	1.1995–12.2013	228	27.76	190.80	2.99	17.09	30.17
9	07197000	808.7	1.1995–12.2013	228	9.27	69.73	0.33	4.93	11.44
10	07196900	105.2	1.1995–12.2013	228	1.31	10.35	0.00	0.59	1.76
11	07197360	233.8	1.1998–12.2013	192	2.41	15.18	0.10	1.46	2.88
12	07198000	4186.2	1.1995–12.2013	228	44.03	378.65	0.98	25.59	46.81

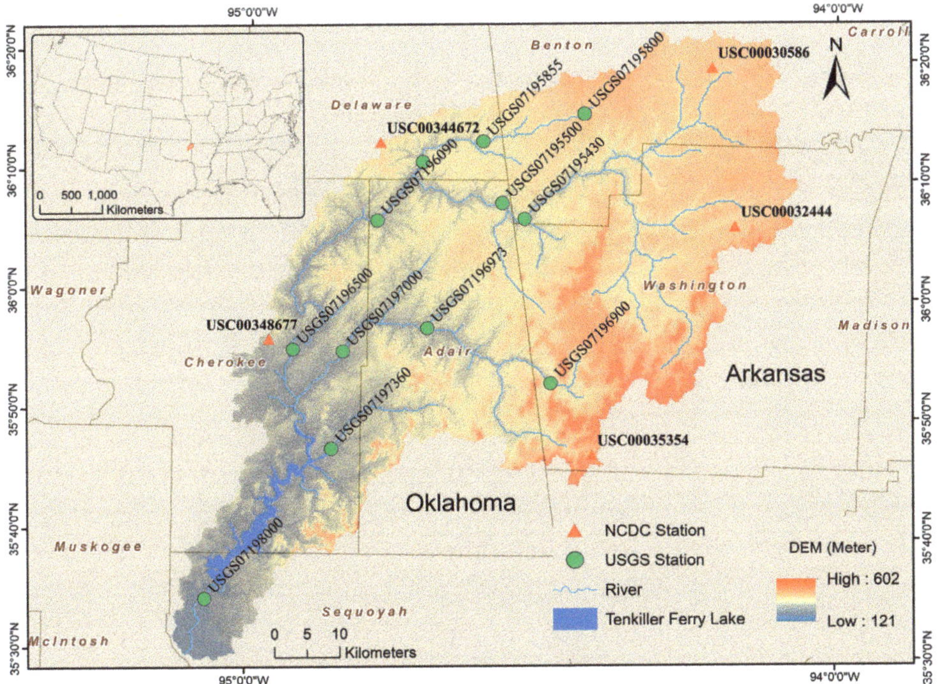

Figure 2. The geographic distribution of NCDC meteorological stations, USGS hydrological stations, lakes, and rivers in the IRW.

The data used to set up the SWAT model include the following: (1). Digital elevation model (DEM): 1 Arc-Second Global Database from Shuttle Radar Topography Mission (SRTM) were downloaded from the USGS website (https://earthexplorer.usgs.gov/, accessed on 28 January 2018). Its spatial resolution is about 30 m × 30 m. (2). Land use and land cover data: 2011 NLCD dataset (https://www.mrlc.gov/, accessed on 31 January 2018) were applied in this study, and spatial resolution is 100 m × 100 m. (3). Soil data: We downloaded soil data from the SSURGO database (https://websoilsurvey.nrcs.usda.gov/, accessed on 5 February 2018). (4). Climate data: Daily climate data came from the National Climatic Data Center (NCDC) (https://www.ncdc.noaa.gov/, accessed on 7 February 2018). Due to incomplete precipitation and temperature records from January 1990 to December

2013, we downloaded alternative Climate Forecast System Reanalysis (CFSR) data from the SWAT official website (https://globalweather.tamu.edu/, accessed on 31 January 2018), then filled missing NCDC data using climate data from the closest CFSR stations.

2.2. Hydrological Model

SWAT was developed by the U.S. Department of Agriculture Agricultural Research Service (USDA-ARS) and has been extensively used worldwide [30,36]. It is a conceptual, semi-distributed, and physically based hydrologic model used to simulate water cycles, crop growth, sediment yields, and agricultural chemical transport in a large river basin with varying soils, slopes, and land use management conditions [30]. A more detailed description of the SWAT model is available from online documentation [37].

ArcSWAT 2012.10_4.19 within ArcGIS 10.4.1 was selected to build the SWAT-based model in the study area. The entire IRW watershed was discretized as 86 subwatersheds with 1023 hydrologic response units (HRUs) using a threshold area of 3000 ha. Each HRU consisted of various land use/soil/slope attributes and was defined with a threshold of land use (10%), soil (10%), and slope (5%). The SCS curve number method [38] and the variable storage routing method [37] were applied to calculate the surface runoff and river flow, respectively. A five-year warm-up period (1990–1994) was set up to initialize the model input and stabilize the SWAT model. The SWAT simulation running period is from 1 January 1995 to 31 December 2013.

We used SWAT-CUP with Sequential Uncertainty Fitting (SUFI2) method to conduct sensitivity analysis, calibration, and validation procedures of the SWAT model [31]. The all-at-a-time approach was applied in the procedure of parameterization with 1000 SWAT-CUP simulations. SWAT-CUP was set up for all twelve stations and run at one time with two iterations. The nine sensitive parameters range were determined and their range and fitted values in calibration listed in Table 2. The results from SWAT-CUP were regarded as a benchmark to compare with results from SWAT-SVR and SWAT-WSVR. The optimal parameters combination and sensitivity of the SWAT model depends on precipitation input, interpolation of weather data, and the number of iterations. It has been investigated in previous publications [11,32,39,40] and will not be discussed further in this article.

Table 2. The sensitive parameters range and their fitted values in calibration.

No.	Parameter Name †	Parameter Description	Range	Fitted Value
1	R__CN2.mgt	SCS runoff curve number II	−0.25–0.25	−0.179
2	V__GWQMN.gw	Threshold depth of water in the shallow aquifer required for return flow to occur (mm H_2O)	0–2000	1764
3	V__GW_REVAP.gw	Groundwater "revap" coefficient	0.02–0.2	0.135
4	V__REVAPMN.gw	Threshold depth of water in the shallow aquifer for 'revap' to occur (mm)	0–500	121
5	V__EPCO.hru	Plant uptake compensation factor	0–1	0.154
6	V_ESCO.hru	Soil evaporation compensation factor	0–1	0.354
7	R__SOL_AWC (1).sol	Available water capacity of the 1st soil layer (mm H_2O mm soil^{-1})	0.08–0.2	0.177
8	A__OV_N.hru	Manning's "n" value for overland flow	0.01–30	26.941
9	R__HRU_SLP.hru	Average slope steepness (m m^{-1})	0–1	0.034

Note: "A__", "V__", and "R__" mean an absolute increase, a replacement, and a relative change to the initial parameter values, respectively.

2.3. Support Vector Machine

SVM is built on the principle of the statistical learning and structural risk minimization theory [41]. When SVM technology is applied in regression analysis, it is called SVR. The SVR function is expressed as below [41]:

$$f(x) = w \cdot \Phi(x) + b \tag{1}$$

where w is a weight vector, Φ is a nonlinear transfer function, and b is offset. An SVR function $f(x)$ can be expressed as the below formulation [17]:

$$\begin{gathered} \min \quad \frac{1}{2}||w||^2 + C\sum\nolimits_{n=1}^{n}(\xi_i + \xi_i^*) \\ y_i - (w \cdot \Phi(x_i) + b) \le \varepsilon + \xi_i \\ \text{subject to } (w \cdot \Phi(x_i) + b) - y_i \le \varepsilon + \xi_i^* \\ \xi_i,\ \xi_i^* \ge 0,\ i = 1, 2, \ldots, n \end{gathered} \tag{2}$$

where ξ_i and ξ_i^* are slack variables that estimate the deviation of training data falling out of the ε-insensitive zone. The C is a penalty factor that determines the tradeoff between the flatness of $\Phi(x_i)$ and the amount up to which the deviation ε can be tolerated [42].

In application, SVR includes four commonly used kernel functions such as the linear, polynomial, Gaussian radial basis (RBF), and sigmoid. In this paper, we selected the Gaussian RBF kernel function due to its computational efficiency, and its expression is described below [43]:

$$K(x_i, x_j) = \exp\left(-\gamma \|x_i - x_j\|^2\right) \tag{3}$$

The most critical three parameters in a SVR ε-regression application based on the RBF kernel include: the penalty error parameter C ($C > 0$), the Gaussian RBF kernel parameter γ, and the deviation of the error margin ε [14]. We applied the grid search and the *k*-fold cross-validation method to optimize these parameters. The parameters value range in the grid-searching was set up as: C (begin = 2^{-6}, end = 2^8, step = 1), γ (begin = 2^4, end = 2^{-8}, step = -1), and ε (begin = 2^{-8}, end = 2^{-1}, step = 0.5). The *k*-value in cross-validation was set to 5 for tuning the SVR. Before training the SVR, all input data were normalized to the value range [0, 1] by the formula $(x - x\_min)/(x\_max - x\_min)$. Additionally, for each site, we used the first 70% of data to train the model, then applied the remaining 30% subset for validation purposes. The SVR ε-regression model was used to develop both SWAT-SVR and SWAT-WSVR. R version 4.1.0 running on RStudio version 1.4.1717 and the 'e1071' package [44] were used for the development, training, and testing of the hybrid model [45].

2.4. Wavelet Transforms

The wavelet transform is a mathematical function that has an adjustable time-frequency window and can decompose time series into multiple resolution levels by controlling the scaling and shifting factors of a mother wavelet [46]. A mother wavelet needs to be determined before applying a wavelet analysis. The wavelet transform of time series data generates sets of wavelet coefficients for different scales and provides a time-scale localization of processes [7]. The wavelet transform has two forms: the continuous wavelet transform (CWT) and the discrete wavelet transform (DWT). In this paper, we applied DWT method to build the hybrid SWAT-WSVR model. The DWT discretizes the parameters of scales and positions before implementing the wavelet transform to decrease the redundancy. The DWT of signal $f(t)$ is defined as [47]:

$$W_f(j,k) = a_0^{-j/2} \int_{-\infty}^{\infty} \psi^*\left(a_0^{-j} t - kb_0\right) f(t) dt \tag{4}$$

where the dilation parameter a and temporal translation parameter b of the CWT are discretized as $a = a_0^j,\ b = kb_0 a_0^j,\ a_0 > 0 \text{ and } a_0 \ne 1,\ b_0 \in R$. In most cases, parameter $a_0 = 2$ and $b_0 = 1$. Then, a discrete wavelet can be expressed as [47]:

$$\psi_{j,k}(t) = a_0^{-j/2} \psi\left(a_0^{-j} t - kb_0\right) \quad j,k \in Z \tag{5}$$

The DWT obtains wavelet details (D) and approximations (A) of the original hydrological time series through high-pass and low-pass filters, respectively. Approximations at various resolution levels can be further decomposed by high-pass and low-pass filters (Figure 3). Commonly used DWT wavelets have 'Daubechies', 'Symlets', 'Coiflets', and 'Biorthogonal'. More details about wavelet transform can be found in Labat (2005) [48] and Mallat (2009) [49].

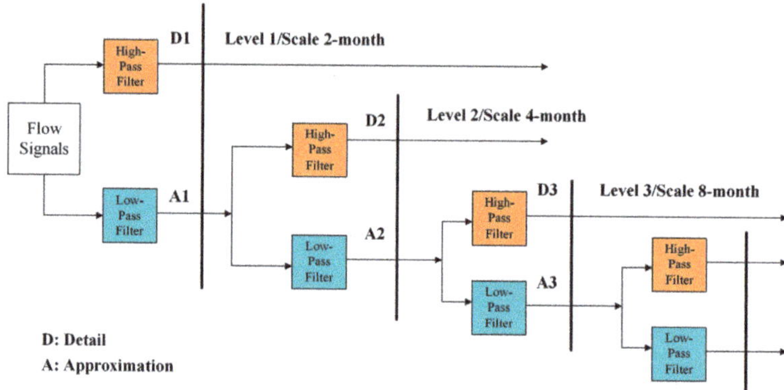

Figure 3. The discrete wavelet transforms (DWT) decomposition procedure of flow time series.

We examined Daubechies wavelet family with different wavelengths, such as extremal phase filter with length 1 ('haar' or 'd1'), filter with length 2 ('d2'), filter with length 4 ('d4'), filter with length 6 ('d6'), and Least Asymmetric filter with length 8 ('la8'), and found that 'haar' wavelet is suitable for this study. To date, there is not a standard method to determine the optimum DWT decomposition levels (D) for a specific time series. Some applied the equation as $D = int(Log(N))$ [50]; others used the formula as $D = Log(N/(2m-1))/Log(2)$ [27], where N is the length of monthly time series, and m is the number of vanishing moments of a Daubechies wavelet.

In this study, the maximum and minimum length of the monthly flow time series of 12 sites is 228 and 42 (Table 1), respectively. Regardless of applying either the formula mentioned above to calculate D, the maximum decomposition levels of monthly flow for all sites were between 1.62 and 7.83. Therefore, wavelet decomposition levels of 1 to 7 were tested to obtain the optimal resolution levels. The results indicated that the decomposition level of 3 and 2 attained the best model performance.

2.5. Model Performance Evaluation

We applied four statistics such as *NSE* (Nash–Sutcliffe efficiency), *PBIAS* (percent bias), *RMSE* (root mean square error), and *RSR* (*RMSE*-observation's standard deviation ratio) to evaluate the model performance in calibration and validation. Table 3 listed these statistical indicators, their mathematic expressions, and their value range. Preferred statistics combination is the lower *RSR*, *PBIAS*, and *RMSE* but the higher *NSE*, which present the better the model prediction performance. We used the 'hydroGOF' package [51] in R to calculate the mentioned statistical indicators.

To further compare the performance of SWAT-CUP, SWAT-SVR, and SWAT-WSVR on the entire time series (i.e., combined calibration and validation together), we plotted hydrography for each site and applied the Taylor diagram [52] to examine the relative importance of different statistics such as r, *RMSE*, and *NSD* between the observed and simulated flow for three models and twelve sites. The advantage of the Taylor diagram is that it can highlight the goodness-of-fit of multiple models and compare their difference from observed data at the same graph.

Table 3. Evaluation indicators of the model performance and their mathematic expressions.

Indicator Name	Calculation Equation †	Description
Pearson's Correlation Coefficient (*r*)	$r = \frac{n(\sum yy_i)-(\sum y)(\sum y_i)}{\sqrt{[n\sum y^2-(\sum y)^2][n\sum y'-(\sum y')^2]}}$	Range [−1, 1]
Nash–Sutcliffe efficiency (*NSE*)	$NSE = 1 - \frac{\sum_{i=1}^{n}(y_i-y_i')^2}{\sum_{i=1}^{n}(y_i-\overline{y})^2}$	Range (−∞, 1], and 1 is the optimal value
Percent Bias (*PBIAS*)	$PBIAS = 100 \times \frac{\sum_{i=1}^{n}(y_i'-y_i)}{\sum_{i=1}^{n}y_i}$	Range (−∞, +∞), and 0 is the optimal value
RMSE-observations standard deviation ratio (*RSR*)	$RSR = \frac{\sqrt{\sum_{i=1}^{n}(y_i-y_i')^2}}{\sqrt{\sum_{i=1}^{n}(y_i-\overline{y})^2}}$	Range [0, +∞), and 0 is the optimal value
Root Mean Square Error (*RMSE*)	$RMSE = \sqrt{\frac{\sum_{i=1}^{n}(y_i-y_i')^2}{n}}$	Range [0, +∞), and 0 is the optimal value

Note: y_i is the observed data series, y_i' is the simulated results series, the overbar represents the mean value of data series, and n is the sample number.

3. Results and Discussion

We developed a total of 72 models for the flow prediction at 12 sites by three methods: SWAT-CUP, SWAT-SVR, and SWAT-WSVR. Each method included 12 calibrated models and 12 validated models. The difference between SWAT-SVR and SWAT-WSVR is that model inputs of SWAT-SVR had only the flow outputted from SWAT-CUP (a calibrated SWAT model) and precipitation data. Instead, we replaced the simulated flow with its wavelet components at different resolution levels in SWAT-WSVR. Table 4 lists the statistical performance of the three above-mentioned models in calibration, validation, and the whole time series data combining calibration and validation time series data.

3.1. Flow Prediction by SWAT-CUP

Table 4 summarizes the average *RSR*, *NSE*, *PBIAS*, and *RMSE* for twelve sites from SWAT-CUP, and the corresponding values are 1.67, 0.22, 57.57, and 11.50 $m^3 s^{-1}$ in calibration and 0.84, 0.26, 35.98, and 11.04 $m^3 s^{-1}$ in validation, respectively. The *RSR*, *NSE*, and *RMSE* had approximately similar performances between calibration and validation, but the value of *PBIAS* in validation was lower than one in calibration, which indicated that the predicted discrepancy from validation was less than one from calibration. SWAT-CUP overestimated monthly flow in both calibration and validation. The low average *NSE* value ($\leq$0.26) indicated that SWAT-CUP has a poor goodness-of-fit between the observed and simulated flow for both calibration and validation. Additionally, the 07195430 site had the best performance among all sites in validation with lowest *PBIAS* (−6.7) and *RSR* (0.58) and the highest *NSE* value (0.66). The model performance of 07195500 and 07196500 were also acceptable in validation. After combined calibration and validation data together, the averages of *RSR*, *NSE*, *PBIAS*, and *RMSE* for 12 sites are 0.85, 0.13, 50.67, and 11.38 $m^3 s^{-1}$, respectively. Simulations from SWAT-CUP greatly overestimated the observed flow according to *PBIAS* (i.e., a positive mean of 50.67 for 12 sites) and presented a low fitting degree due to a low average value (0.13) of *NSE*. Overall, SWAT-CUP had a poor simulation performance for most of sites during both calibration and validation periods, with only few exceptions.

3.2. Flow Prediction by SWAT-SVR

Due to the unsatisfactory overall performance of SWAT-CUP, we developed the SWAT-SVR and SWAT-WSVR model integrating the simulated flow (or its wavelet components) and precipitation to improve the prediction accuracy. Table 5 listed the model structure and optimal parameter sets for SWAT-SVR and SWAT-WSVR. In 24 SVR models, the value of C is inconstant from 2.015625 to 255.015625, the value range of γ is from 1 to 14, and ε keeps a constant value of 0.00390625.

Table 4. Performance of flow calibration, validation, and combined data series on each site by SWAT-CUP, SWAT-SVR, and SWAT-WSVR.

	Station	SWAT-CUP				SWAT-SVR				SWAT-WSVR			
		RSR	*NSE*	*PBIAS*	*RMSE* (m^3 s^{-1})	*RSR*	*NSE*	*PBIAS*	*RMSE* (m^3 s^{-1})	*RSR*	*NSE*	*PBIAS*	*RMSE* (m^3 s^{-1})
Calibration	07195800	0.76	0.41	24.00	0.33	0.58	0.68	−10.80	0.25	0.08	0.99	−1.90	0.04
	07195855	0.95	0.09	66.50	1.32	0.63	0.60	−10.00	0.88	0.01	1.00	0.10	0.01
	07196000	0.96	0.07	67.90	3.18	0.57	0.68	0.50	1.88	0.01	1.00	0.10	0.02
	07195500	0.74	0.46	44.60	13.68	0.55	0.70	−5.50	10.23	0.01	1.00	0.10	0.23
	07195430	0.67	0.54	29.30	11.40	0.55	0.70	−5.30	9.30	0.01	1.00	0.20	0.22
	07196090	0.52	0.72	31.40	19.81	0.14	0.98	−3.30	5.18	0.02	1.00	−0.60	0.70
	07196973	10.70	−0.15	69.90	0.78	0.73	0.46	−6.50	0.53	0.02	1.00	0.20	0.01
	07196500	0.78	0.39	52.30	22.62	0.58	0.67	−10.20	16.71	0.02	1.00	0.10	0.48
	07197000	0.87	0.24	69.50	10.04	0.57	0.67	−11.80	6.64	0.01	1.00	0.10	0.09
	07196900	0.93	0.14	86.00	1.69	0.57	0.68	−13.60	1.04	0.01	1.00	0.00	0.02
	07197360	0.97	0.06	74.90	2.89	0.70	0.51	−21.30	2.07	0.00	1.00	0.00	0.01
	07198000	1.17	−0.39	74.50	50.21	0.78	0.39	−15.90	33.42	0.03	1.00	−0.20	1.41
	Mean	**1.67**	**0.22**	**57.57**	**11.50**	**0.58**	**0.64**	**−9.48**	**7.34**	**0.02**	**1.00**	**−0.15**	**0.27**
Validation	07195800	0.85	0.26	8.70	0.35	0.89	0.20	−20.40	0.36	0.11	0.99	−2.90	0.04
	07195855	0.89	0.19	22.80	1.35	0.81	0.33	−27.00	1.23	0.09	0.99	−1.10	0.14
	07196000	0.98	0.03	29.60	3.03	0.86	0.24	−16.00	2.68	0.14	0.98	−1.00	0.43
	07195500	0.59	0.65	15.40	13.59	0.58	0.65	−20.50	13.48	0.24	0.94	−4.60	5.49
	07195430	0.58	0.66	−6.70	13.88	0.58	0.68	−23.80	13.74	0.24	0.94	−4.80	5.77
	07196090	0.71	0.45	38.30	14.32	1.15	−0.43	−41.70	23.16	0.12	0.98	0.70	2.47
	07196973	1.03	−0.10	62.30	0.83	0.87	0.22	−15.70	0.69	0.06	1.00	0.70	0.05
	07196500	0.66	0.56	19.70	21.55	0.63	0.60	−25.50	20.62	0.13	0.98	−2.20	4.28
	07197000	0.88	0.22	64.80	9.84	0.72	0.47	−6.30	8.07	0.10	0.99	−0.40	1.17
	07196900	1.07	−0.17	87.90	1.72	0.96	0.06	3.60	1.54	0.05	0.99	0.60	0.08
	07197360	0.88	0.21	55.80	2.34	0.66	0.58	−25.10	1.76	0.07	0.99	−1.40	0.18
	07198000	0.90	0.18	33.20	49.62	0.84	0.29	−29.80	48.22	0.27	0.93	−6.10	14.81
	Mean	**0.84**	**0.26**	**35.98**	**11.04**	**0.80**	**0.32**	**−20.68**	**11.30**	**0.14**	**0.98**	**−1.88**	**2.91**
The whole series data †	07195800	0.79	0.37	19.50	0.33	0.68	0.53	−13.60	0.29	0.09	0.99	−2.20	0.04
	07195855	0.93	0.13	52.30	1.33	0.70	0.51	−15.50	1.00	0.06	1.00	−0.30	0.08
	07196000	0.97	0.06	55.90	3.13	0.66	0.56	−4.60	2.15	0.07	0.99	−0.30	0.24
	07195500	0.68	0.53	35.00	13.65	0.56	0.68	−10.40	11.30	0.15	0.98	−1.50	3.00
	07195430	0.63	0.60	16.80	12.19	0.56	0.68	−11.70	10.81	0.16	0.97	−1.60	3.15
	07196090	0.55	0.69	33.50	18.41	0.39	0.84	−14.70	13.13	0.04	1.00	−0.30	1.45
	07196973	1.06	−0.14	67.80	0.79	0.78	0.38	−9.10	0.58	0.04	1.00	0.30	0.03
	07196500	0.74	0.45	41.80	22.31	0.60	0.64	−15.10	17.97	0.08	0.99	−0.60	2.36
	07197000	0.87	0.24	68.20	9.98	0.62	0.61	−10.30	7.90	0.06	1.00	−0.10	0.64
	07196900	0.97	0.06	86.50	1.70	0.69	0.53	−8.70	1.21	0.03	1.00	0.20	0.05
	07197360	0.95	0.10	69.50	2.73	0.69	0.52	−22.40	1.98	0.03	1.00	−0.40	0.10
	07198000	1.07	−1.50	61.20	50.03	0.81	0.35	−20.30	37.70	0.17	0.97	−2.10	8.17
	Mean	**0.85**	**0.13**	**50.67**	**11.38**	**0.65**	**0.57**	**−13.03**	**8.84**	**0.08**	**0.99**	**−0.74**	**1.61**

† Note: The data series combined calibration and validation time series.

Table 5. Model inputs and optimum parameters of SWAT-SVR and SWAT-WSVR.

Station	SWAT-SVR			SWAT-WSVR			
	Model Input †	*C*	γ	Model Input	Decomposition Levels	*C*	γ
07195800	Flow + Prec	36.015625	3	Prec + D1 + D2 + D3 + A3	3	5.015625	1
07195855	Flow + Prec	22.015625	1	Prec + D1 + D2 + D3 + A3	3	2.015625	3
07196000	Flow + Prec	255.015625	1	Prec + D1 + D2 + D3 + A3	3	255.015625	1
07195500	Flow + Prec	103.015625	1	Prec + D1 + D2 + D3 + A3	3	96.015625	1
07195430	Flow + Prec	255.015625	1	Prec + D1 + D2 + D3 + A3	3	87.015625	1
07196090	Flow + Prec	255.015625	5	Prec + D1 + D2 + A2	2	255.015625	1
07196973	Flow + Prec	2.015625	1	Prec + D1 + D2 + A2	2	5.015625	1
07196500	Flow + Prec	130.015625	1	Prec + D1 + D2 + D3 + A3	3	125.015625	1
07197000	Flow + Prec	57.015625	1	Prec + D1 + D2 + D3 + A3	3	242.015625	1
07196900	Flow + Prec	4.015625	14	Prec + D1 + D2 + D3 + A3	3	15.015625	1
07197360	Flow + Prec	13.015625	1	Prec + D1 + D2 + D3 + A3	3	35.015625	1

† Note: Flow comes from SWAT-CUP simulated discharge output. Ds and As are wavelet components from the simulated flow of SWAT-CUP.

The average *RSR*, *NSE*, *PBIAS*, and *RMSE* of twelve sites from SWAT-SVR are 0.58, 0.64, −9.48, and 7.34 $m^3 s^{-1}$ in calibration and 0.80, 0.32, −20.68, and 11.30 $m^3 s^{-1}$ in validation (Table 4), respectively. Compared with SWAT-CUP, the performance of the SWAT-SVR model had the lower *RSR* and the absolute value of *PBIAS* and the higher *NSE* in calibration and validation, particularly for the calibrated simulations, as SVR has a strong learning ability for training data. The average *RMSE* in SWAT-SVR calibration is lower: only 7.34 $m^3 s^{-1}$ compared with the average one of 11.50 $m^3 s^{-1}$ in the SWAT-CUP calibration. The results showed that the SWAT-SVR calibration on all sites had lower deviation and higher *NSE* value in comparison with SWAT-CUP, but still underestimated monthly flows on most sites. SWAT-SVR was generally superior to SWAT-CUP on all sites. However, only a few sites (e.g., 07195500, 07195430) had lower *RSR*, *PBIAS*, and higher *NSE* values, which indicated that SWAT-SVR could greatly improve the model performance in calibration but did not possess good generalization capability, which means it failed to keep this prediction ability with high accuracy while it was applied in validation. From the perspective of the whole data series, the average *RSR*, *NSE*, *PBIAS*, and *RMSE* are 0.65, 0.57, −13.03, and 8.84 $m^3 s^{-1}$, respectively. Clearly, compared with SWAT-CUP, the performance of the SWAT-SVR model were improved but limited, although SWAM-SVR generally had a low *RSR*, absolute value of *PBIAS*, and higher *NSE* value for calibration, validation, and both periods.

3.3. SWAT-WSVR Development and Evaluation

3.3.1. Development of SWAT-WSVR

We conducted DWT of the simulated flow from SWAT-CUP at twelve hydrological sites using a 'haar' wavelet filter to obtain the flow series structure, trend, and temporal characteristics. The wavelet decomposition was implemented at three resolution levels: 2 months, 4 months, and 8 months. Figure 4 showed an example of the flow DWT at the 07195430 site, including temporal features of the flow 2-month mode (D1), 4-month mode (D2), 8-month mode (D3), and approximate mode (A3). From Figure 4a, we can observe a large deviation between the original observed (blue line) and SWAT-CUP simulated (orange dash-line) flow signals. In Figure 4b, D1, D2, and D3 modes indicated high-frequency details, and A3 mode revealed a low-frequency trend and the slowest flow changing of the simulated flow series at the 07195430 site. These wavelet components would be used to build the SWAT-WSVR model afterward.

To reduce the computational load and find the most related wavelet coefficients to participate in the construction of SWAT-WSVR, we analyzed Pearson's correlation coefficient matrices between the observed monthly flow, monthly precipitation, and its sub-time series D1, D2, D3, and A3 from wavelet decompositions (Figure 5). In Figure 5, the upper 'Pie' graphs are a corresponding display of the lower 'numeric' r in the diagonal direction where the flow, precipitation, and wavelet components are shown. Flow time series of 07196090 and 07196973 were decomposed into two resolution levels since their data lengths are shorter: 42 and 96, respectively.

The correlation analysis indicated that the flow and precipitation have a strong positive correlation, and the average value of r is 0.53 on twelve sites. Most wavelet components D1, D2, D3, and A3 on the flow also showed positive correlations. The average of r of D1, D2, D3, and A3 (or A2) on flow is 0.35, 0.47, 0.3, and 0.32. The wavelet coefficient of D2 had the strongest correlation with the flow. We chose wavelet components with r greater than 0.2 to participate in SVR prediction so that we can keep the intrinsic nonlinear features in wavelet components as much as possible while avoiding large computational burden. The specific model structure of SWAT-WSVR for each site is listed in Table 5.

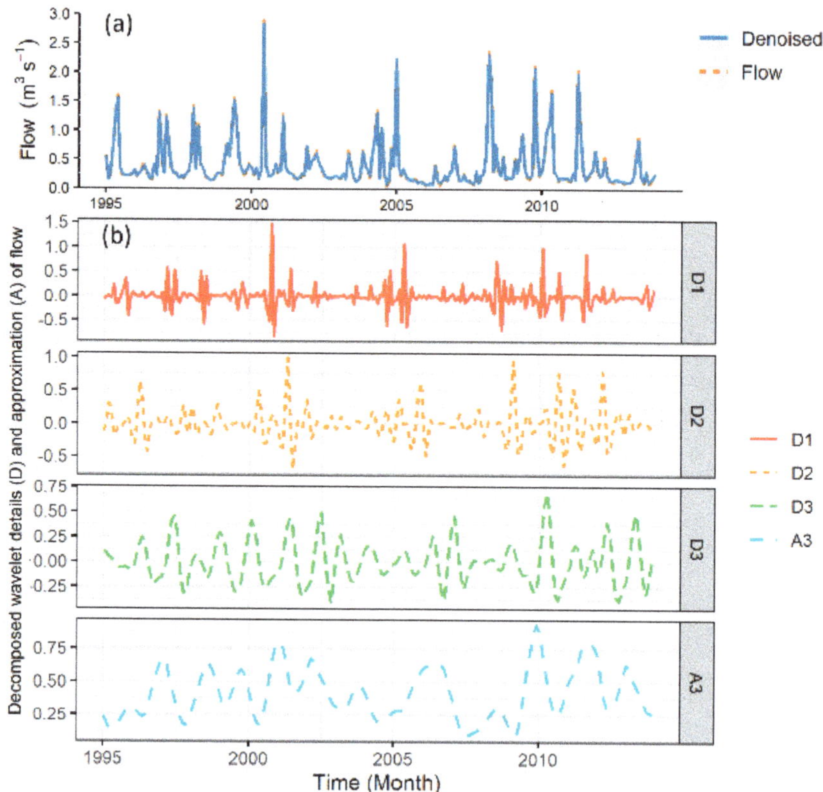

Figure 4. An example of the flow DWT at 07195430 site: (**a**) Observation and SWAT-CUP simulation, (**b**) wavelet decomposition.

3.3.2. Statistical Evaluation of SWAT-WSVR

Table 4 showed the statistical performance of SWAT-WSVR with the average *RSR*, *NSE*, *PBIAS*, and *RMSE* of 0.02, 1.00, −0.15, and 0.27 $m^3 s^{-1}$ in calibration; 0.14, 0.98, −1.88, and 2.91 $m^3 s^{-1}$ in validation; and 0.08, 0.99, −0.74, and 1.61 $m^3 s^{-1}$ in the whole data series. Compared with SWAT-SVR and SWAT-CUP, SWAT-WSVR had the lowest *RSR*, the absolute value of *PBIAS*, and *RMSE* but the highest *NSE* value in validation. This result clearly indicated that SWAT-WSVR could effectively decrease the discrepancy of the simulation and obtain the best prediction accuracy for validation in comparison with SWAT-SVR and SWAT-CUP. Based on the value of *PBIAS*, SWAT-WSVR slightly underestimated the monthly flow in calibration. SWAT-WSVR also presented the best performance on the whole data series, along with the lowest *RSR*, *PBIAS*, and *RMSE* but the highest *NSE* in comparison with SWAT-CUP and SWAT-SVR. By comparison, the SWAT-WSVR model outperformed the SWAT-CUP and SWAT-SVR model in calibration, validation, and both periods.

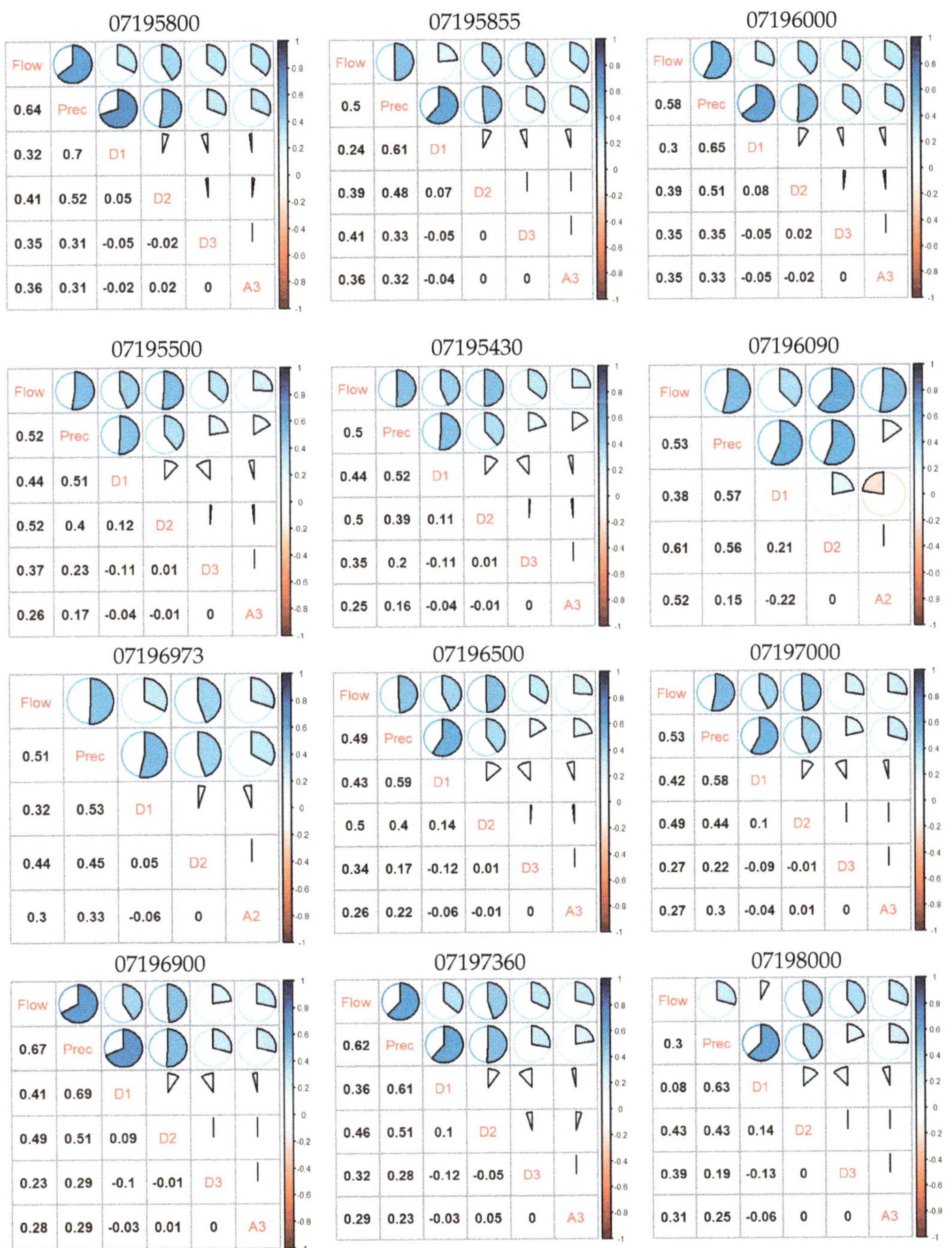

Figure 5. Correlation coefficient matrices of input variables in the SWAT-WSVR model (Note: 'Prec' denotes precipitation).

3.4. Taylor Diagram and Hydrographic Comparison between Different Models

To further compare the overall performance of three models, we combined the flow observations and the calibrated and validated simulations from SWAT-CUP, SWAT-SVR, and SWAT-WSVR; recalculated statistical indicators including r, $RMSE$, and NSD; and re-estimated the model performance on the whole time series (i.e., calibration and validation periods are considered together). The Taylor diagrams (Figure 6) depicted the overall performance of different models for each site by identifying the pattern correlations, variability, and $RMSE$ between observations and simulations. In Figure 6, both the x-axis and y-axis denote NSD; black dashed lines represent the r between observations and simulations; the normalized $RMSE$ of the simulation is proportional to the distance from the x-axis identified as "observation" (green contours); the NSD of the simulation is proportional to the radial distance from the origin point (black contours). The modeling results in Figure 6 demonstrated that the developed SWAT-WSVR had the best performance in comparison with the other two models since it had the lower $RMSE$ and NSD, but the higher r with observed flows in most cases. For example, SWAT-WSVR (redpoint in Taylor diagram) at all sites is closer to the reference point (observation) located on the x-axis than the other two models, which illustrated the SWAT-WSVR fitted best with the observed flow for most sites. The rank of the overall performance of three models from high to low follows SWAT-WSVR > SWAT-SVR > SWAT-CUP. The proposed SWAT-WSVR model that uses wavelet components as inputs of SVR presented a more satisfactory flow prediction than the SWAT-SVR with a single pattern input (i.e., the simulated flow from SWAT-CUP). Possible reasons include a periodical feature of sub-series represented by wavelet components is more obvious than those directly obtained from SWAT-CUP [53], and SVR captured the intrinsic nonlinear features between SWAT-CUP simulations and observed flows and built a mapping relationship at a higher dimension space successfully. This kind of fitting relationship is typically determined by using the trial-and-error method and a large number of iterations of many parameters sets in physically based hydrological models.

To investigate the entire and continuous performance of monthly flow prediction in the IRW, we plotted flow hydrography on the whole data series for each site (Figure 7). Here, we only labeled statistics of SWAT-WSVR for clarity, and other details related to estimate indicators can be found in Table 4. This figure reflected where the developed SWAT-WSVR model performed better than SWAT-CUP and SWAT-SVR methods. An oval region at the 07196090 site showed clear evidence that SWAT-WSVR agreed well with the observed flow, but SWAT-SVR missed the peak flow during this time window. Although SWAT-WSVR had desirable modeling performance, it missed few peak flows. For example, the SWAT-WSVR simulated flow (105.24 and 107.68 $m^3 s^{-1}$) was 27.2% and 27.9% lower than observed flow (144.61 and 149.42 $m^3 s^{-1}$) in April 2011 at 07195430 and 07195500 site, respectively. Overall, the other two models more or less capture the rising and recession of the observed monthly flow over time at all sites, but the SWAT-WSVR is more efficient at fitting with the observation and corrected errors compared to SWAT-CUP. This result is in line with others' conclusions that the application of wavelet transform in data-driven models can improve the accuracy of flow prediction [53,54]. The developed SWAT-WSVR model fit the observations well at all sites of the IRW based on the statistical results, Taylor diagram, and hydrography analysis.

In this study, we applied wavelet transforms on the simulated flow time series from SWAT-CUP rather than on the observed flow data to show that the developed SWAT-WSVR model can be applied in practice or future scenario prediction where observed data are impossible to access or has limited availability. An SVR coupled to a wavelet transform model approach based solely on observed monthly flow data could produce greater accuracy. This is because wavelet decompositions with less noise can represent the periodical and trend characteristics of flow series structure better than the original data series [25,53], and SVR has a strong learning capability to capture the corresponding relationship between wavelet decompositions and its original data series. However, the limitations of a purely mathematical or machine learning approach prior to constraint by a physically-based model could fail to capture the relationship between rainfall and runoff and exhibit less predictive or unrealistic behavior [28].

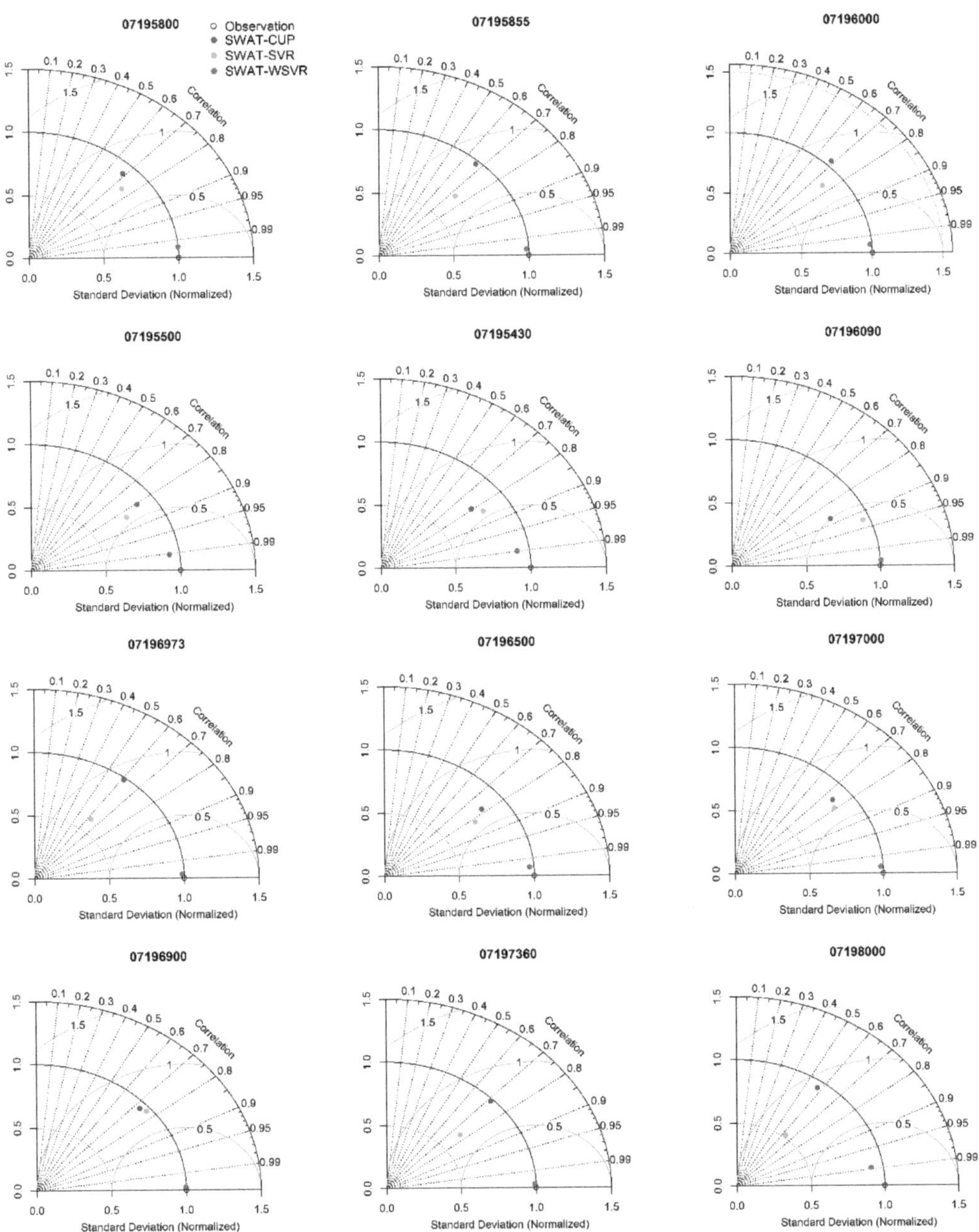

Figure 6. Taylor diagram of three models in comparison with the observation for each site.

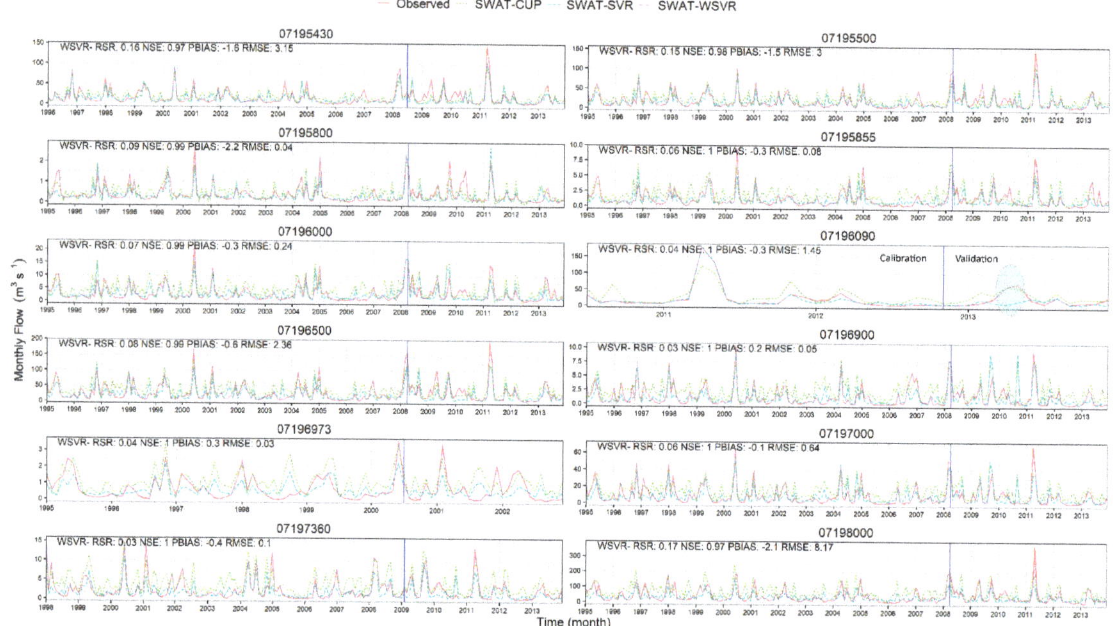

Figure 7. Comparison of SWAT-CUP, SWAT-SVR, SWAT-WSVR, and observed flow time series on each site. (Note: a vertical blue line separated the 70% training and 30% testing data. Only statistics from SWAT-WSVR were labeled for clarity).

Moreover, the current proposed SWAT-WSVR model can be regarded as a compromise method when we cannot attain a desirable prediction accuracy even after conducting extensive SWAT-CUP iterative computation, although we still encourage researchers to directly obtain a satisfactory prediction from SWAT-CUP if possible. In this work, the construction of SWAT-WSVR heavily depends on the procedure of SWAT-CUP parameter calibration. Alternatively, we can also build SWAT-WSVR based on wavelet decomposition of the initial output from SWAT without the procedure of calibration based on the methodology proposed if SWAT-CUP is not accessible.

4. Summary and Conclusions

This study developed the SWAT-WSVR monthly flow prediction model which was built on the basis of the SWAT and SVR with discrete wavelet transforms and investigated the performance and effectiveness of this model. Precipitation and wavelet components of flow outputted from SWAT-CUP were served as input variables into SWAT-WSVR. The methodology loosely integrated the physically based model and the data-driven model. The proposed SWAT-WSVR model had the best statistical performance with the lower *RSR*, the absolute value of *PBIAS*, *RMSE*, and higher *NSE* in comparison with regular SWAT-SVR and SWAT-CUP, which indicated that SWAT-WSVR possessed the lower discrepancy and higher goodness-of-fit between the simulated and observed flow. The rank of the overall performance of the three models on the entire study period was SWAT-WSVR > SWAT-SVR > SWAT-CUP. The SWAT-WSVR model can predict monthly flow more accurately than the other two models for all sites in the IRW.

The strength of the SWAT-WSVR is its ability to capture the intrinsic nonlinear and non-stationary features between rainfall and runoff while considering physical processes by integrating SWAT. It could be regarded as a compromise method when one cannot directly obtain a desirable accuracy from SWAT-CUP simulation or when one applies SWAT into a region with limited data available. In future work, more predictors (e.g., temperature, evaporation, relative humidity) will be considered as the model input variables to raise the forecasting accuracy of SWAT-WSVR further and increase its generalization ability.

Author Contributions: Conceptualization, L.Y. and K.J.F.; methodology, L.Y.; software, L.Y.; validation, L.Y.; formal analysis, L.Y.; investigation, L.Y.; resources, K.J.F.; data curation, K.J.F.; writing—original draft preparation, L.Y.; writing—review and editing, L.Y. and K.J.F.; visualization, L.Y.; supervision, K.J.F.; project administration, K.J.F.; funding acquisition, K.J.F. All authors have read and agreed to the published version of the manuscript.

Funding: This project is funded by the EPA Safe and Sustainable Water Resources National Program.

Institutional Review Board Statement: Not applicable.

Informed Consent Statement: Not applicable.

Data Availability Statement: https://doi.org/10.23719/1527817.

Acknowledgments: The simulation experiments and original draft were completed while Lifeng Yuan worked as an NRC Research Associate at Robert S. Kerr Environmental Research Center, Ada, OK, USA. The final draft was completed while Lifeng Yuan worked at Homeland Security and Materials Management Division, NC, USA. This work does not reflect the views of the U.S. EPA, and no official endorsement should be inferred. It has been reviewed by the Agency but does not necessarily reflect the Agency's views. No official endorsement should be inferred. EPA does not endorse the purchase or sale of any commercial products or services.

Conflicts of Interest: The authors declare no conflict of interest.

References

1. Alizadeh, M.J.; Kavianpour, M.R.; Kisi, O.; Nourani, V. A new approach for simulating and forecasting the rainfall-runoff process within the next two months. *J. Hydrol.* **2017**, *548*, 588–597. [CrossRef]
2. Huo, Z.; Feng, S.; Kang, S.; Huang, G.; Wang, F.; Guo, P. Integrated neural networks for monthly river flow estimation in arid inland basin of Northwest China. *J. Hydrol.* **2012**, *420–421*, 159–170. [CrossRef]
3. U.S. EPA. *A Review of Watershed and Water Quality Tools for Nutrient Fate and Transport*; EPA/600/R-19/232; Center for Environmental Solutions & Emergency Response | Groundwater Characterization & Remediation Division, Office of Research and Development (EPA): Ada, OK, USA, 2019.
4. Kalteh, A.M. Monthly River Flow Forecasting Using Artificial Neural Network and Support Vector Regression Models Coupled with Wavelet Transform. *Comput. Geosci.* **2013**, *54*, 1–8. [CrossRef]
5. Yuan, L.; Zhou, Q. Complexity of soil erosion and sediment yield system in a watershed. *J. Chongqing Inst. Technol. (Nat. Sci.)* **2008**, *22*, 112–116.
6. Zhang, Z.; Chen, X.; Xu, C.; Yuan, L.; Yong, B.; Yan, S. Evaluating the non-stationary relationship between precipitation and streamflow in nine major basins of China during the past 50 years. *J. Hydrol.* **2011**, *409*, 81–93. [CrossRef]
7. Nourani, V.; Hosseini Baghanam, A.; Adamowski, J.; Kisi, O. Applications of hybrid wavelet–Artificial Intelligence models in hydrology: A review. *J. Hydrol.* **2014**, *514*, 358–377. [CrossRef]
8. Yuan, L.; Yang, G.; Li, H.; Zhang, Z. Spatio-Temporal Variation Analysis of Precipitation during 1960–2008 in the Poyang Lake Basin, China. *Open J. Mod. Hydrol.* **2016**, *6*, 115–127. [CrossRef]
9. Kratzert, F.; Klotz, D.; Herrnegger, M.; Sampson, A.K.; Hochreiter, S.; Nearing, G.S. Toward improved predictions in ungauged basins: Exploiting the power of machine learning. *Water Resour. Res.* **2019**, *55*, 11344–11354. [CrossRef]
10. Hsu, K.l.; Gupta, H.V.; Sorooshian, S. Artificial neural network modeling of the rainfall-runoff process. *Water Resour. Res.* **1995**, *31*, 2517–2530. [CrossRef]
11. Yuan, L.; Forshay, K. Enhanced streamflow prediction with SWAT using support vector regression for spatial calibration: A case study in the Illinois River watershed, U.S. *PLoS ONE* **2021**, *16*, e0248489. [CrossRef]
12. Smola, A.J.; Schölkopf, B. A tutorial on support vector regression. *Stat. Comput.* **2004**, *14*, 199–222. [CrossRef]
13. Noraini, I.; Norhaiza, A. Comparative performance of support vector regressions for accurate streamflow predictions. *Malays. J. Fundam. Appl. Sci.* **2017**, 325–330.
14. Misra, D.; Oommen, T.; Agarwal, A.; Mishra, S.K.; Thompson, A.M. Application and Analysis of Support Vector Machine Based Simulation for Runoff and Sediment Yield. *Biosyst. Eng.* **2009**, *103*, 527–535. [CrossRef]
15. Zhang, X.; Srinivasan, R.; Van Liew, M. Approximating SWAT model using artificial neural network and support vector machine. *J. Am. Water Resour. Assoc.* **2009**, *45*, 460–474. [CrossRef]
16. Yuan, L.; Li, W.; Zhang, Q.; Zou, L. Debris flow hazard assessment based on support vector machine. In Proceedings of the IEEE International Symposium on Geoscience and Remote Sensing, Denver, CO, USA, 31 July–4 August 2006; pp. 4221–4224.

17. Raghavendra, N.S.; Deka, P.C. Support vector machine applications in the field of hydrology: A review. *Appl. Soft Comput.* **2014**, *19*, 372–386. [CrossRef]
18. Shabri, A.; Suhartono. Streamflow forecasting using least-squares support vector machines. *Hydrol. Sci. J.* **2012**, *57*, 1275–1293. [CrossRef]
19. Chiogna, G.; Marcolini, G.; Liu, W.; Perez Ciria, T.; Tuo, Y. Coupling hydrological modeling and support vector regression to model hydropeaking in alpine catchments. *Sci. Total Environ.* **2018**, *633*, 220–229. [CrossRef]
20. Nourani, V.; Alizadeh, F.; Roushangar, K. Evaluation of a two-stage SVM and spatial statistics methods for modeling monthly river suspended sediment load. *Water Resour. Manag.* **2015**, *30*, 393–407. [CrossRef]
21. USGS Water Data for the Nation. Available online: https://nwis.waterdata.usgs.gov (accessed on 18 January 2018).
22. Hrachowitz, M.; Savenije, H.H.G.; Blöschl, G.; McDonnell, J.J.; Sivapalan, M.; Pomeroy, J.W.; Arheimer, B.; Blume, T.; Clark, M.P.; Ehret, U.; et al. A decade of Predictions in Ungauged Basins (PUB)—A review. *Hydrol. Sci. J.* **2013**, *58*, 1198–1255. [CrossRef]
23. Devia, G.K.; Ganasri, B.P.; Dwarakish, G.S. A Review on Hydrological Models. *Aquat. Procedia* **2015**, *4*, 1001–1007. [CrossRef]
24. Liu, Z.; Zhou, P.; Chen, G.; Guo, L. Evaluating a coupled discrete wavelet transform and support vector regression for daily and monthly streamflow forecasting. *J. Hydrol.* **2014**, *519*, 2822–2831. [CrossRef]
25. Zhu, S.; Zhou, J.; Ye, L.; Meng, C. Streamflow estimation by support vector machine coupled with different methods of time series decomposition in the upper reaches of Yangtze River, China. *Environ. Earth Sci.* **2016**, *75*, 531. [CrossRef]
26. Sun, Y.; Niu, J.; Sivakumar, B. A comparative study of models for short-term streamflow forecasting with emphasis on wavelet-based approach. *Stoch. Environ. Res. Risk Assess.* **2019**, *33*, 1875–1891. [CrossRef]
27. Nalley, D.; Adamowski, J.; Khalil, B.; Biswas, A. A comparison of conventional and wavelet transform based methods for streamflow record extension. *J. Hydrol.* **2020**, *582*, 124503. [CrossRef]
28. Quilty, J.; Adamowski, J. Addressing the incorrect usage of wavelet-based hydrological and water resources forecasting models for real-world applications with best practices and a new forecasting framework. *J. Hydrol.* **2018**, *563*, 336–353. [CrossRef]
29. Yuan, L.; Sinshaw, T.; Forshay, K.J. Review of watershed-scale water quality and nonpoint source pollution models. *Geosciences* **2020**, *10*, 25. [CrossRef]
30. Arnold, J.G.; Moriasi, D.N.; Gassman, P.W.; Abbaspour, K.C.; White, M.J.; Srinivasan, R.; Santhi, C.; Harmel, R.; Van Griensven, A.; Van Liew, M.W. SWAT: Model use, calibration, and validation. *Trans. ASABE* **2012**, *55*, 1491–1508. [CrossRef]
31. Abbaspour, K.C. *SWAT-CUP: SWAT Calibration and Uncertainty Programs—A User Manual*; Eawag: Swiss Federal Institute of Aquatic Science and Technology: Zurich, Switzerland, 2015.
32. Yuan, L.; Forshay, K.J. Using SWAT to evaluate streamflow and lake sediment loading in the Xinjiang River basin with limited data. *Water* **2019**, *12*, 39. [CrossRef]
33. Jajarmizadeh, M.; Kakaei Lafdani, E.; Harun, S.; Ahmadi, A. Application of SVM and SWAT models for monthly streamflow prediction, a case study in South of Iran. *KSCE J. Civ. Eng.* **2014**, *19*, 345–357. [CrossRef]
34. Jimeno-Sáez, P.; Senent-Aparicio, J.; Pérez-Sánchez, J.; Pulido-Velazquez, D. A comparison of SWAT and ANN models for daily runoff simulation in different climatic zones of peninsular Spain. *Water* **2018**, *10*, 192. [CrossRef]
35. Noori, N.; Kalin, L. Coupling SWAT and ANN models for enhanced daily streamflow prediction. *J. Hydrol.* **2016**, *533*, 141–151. [CrossRef]
36. Gassman, P.W.; Balmer, C.; Siemers, M.; Srinivasan, R. The SWAT literature database: Overview of database structure and key SWAT literature trends. In Proceedings of the SWAT 2014 Conference, Pernambuco, Brazil, 28 July–1 August 2014. Available online: http://swat.tamu.edu/conferences/2014/ (accessed on 27 June 2018).
37. Neitsch, S.L.; Arnold, J.G.; Kiniry, J.R.; Williams, J.R. *Soil and Water Assessment Tool Theoretical Documentation Version 2009*; TR-406; Texas Water Resources Institute: College Station, TX, USA, 2009.
38. Soil Conservation Service. *National Engineering Handbook*; United States Department of Agriculture: Washington, DC, USA, 1972; Section 4 Hydrology.
39. Storm, D.E.; Busteed, P.R.; Mittelstet, A.R.; White, M.J. *Hydrologic Modeling of the Oklahoma/Arkansas Illinois River Basin Using SWAT 2005*; Oklahoma Department of Environmental Quality: Stillwater, OK, USA, 2010.
40. Mittelstet, A.R.; Storm, D.E.; White, M.J. Using SWAT to enhance watershed-based plans to meet numeric water quality standards. *Sustain. Water Qual. Ecol.* **2016**, *7*, 5–21. [CrossRef]
41. Vapnik, V. *The Nature of Statistical Learning Theory*; Springer: New York, NY, USA, 1995.
42. Danandeh Mehr, A.; Nourani, V.; Karimi Khosrowshahi, V.; Ghorbani, M.A. A hybrid support vector regression–firefly model for monthly rainfall forecasting. *Int. J. Environ. Sci. Technol.* **2018**, *16*, 335–346. [CrossRef]
43. Hsu, C.-W.; Chang, C.-C.; Lin, C.-J. *A Practical Guide to Support Vector Classification*; Department of Computer Science, National Taiwan University: Taipei, Taiwan, 2003.
44. Meyer, D.; Dimitriadou, E.; Hornik, K.; Weingessel, A.; Leisch, F. *e1071: Misc Functions of the Department of Statistics, Probability Theory Group (Formerly: E1071)*; Rstudio: Boston, MA, USA, 2019.
45. R Core Team. *R: A Language and Environment for Statistical Computing*; Rstudio: Boston, MA, USA, 2022.
46. Liu, Z.; Menzel, L. Identifying long-term variations in vegetation and climatic variables and their scale-dependent relationships: A case study in Southwest Germany. *Glob. Planet. Chang.* **2016**, *147*, 54–66. [CrossRef]
47. Makwana, J.J.; Tiwari, M.K. Intermittent Streamflow Forecasting and Extreme Event Modelling using Wavelet based Artificial Neural Networks. *Water Resour. Manag.* **2014**, *28*, 4857–4873. [CrossRef]

48. Labat, D. Recent advances in wavelet analyses: Part 1. A review of concepts. *J. Hydrol.* **2005**, *314*, 275–288. [CrossRef]
49. Mallat, S. *A Wavelet Tour of Signal Processing*; Elsevier: Amsterdam, The Netherlands, 2009.
50. Nourani, V.; Komasi, M.; Mano, A. A multivariate ANN-Wavelet approach for rainfall–runoff modeling. *Water Resour. Manag.* **2009**, *23*, 2877–2894. [CrossRef]
51. Zambrano-Bigiarini, M. *hydroGOF: Goodness-of-Fit Functions for Comparison of Simulated and Observed Hydrological Time Series*; Rstudio: Boston, MA, USA, 2017.
52. Taylor, K.E. Summarizing multiple aspects of model performance in a single diagram. *J. Geophys. Res. Atmos.* **2001**, *106*, 7183–7192. [CrossRef]
53. Kisi, O.; Cimen, M. A wavelet-support vector machine conjunction model for monthly streamflow forecasting. *J. Hydrol.* **2011**, *399*, 132–140. [CrossRef]
54. De Macedo Machado Freire, P.K.; Santos, C.A.G.; da Silva, G.B.L. Analysis of the use of discrete wavelet transforms coupled with ANN for short-term streamflow forecasting. *Appl. Soft Comput.* **2019**, *80*, 494–505. [CrossRef]

Article

Study on Urban Rainfall–Runoff Model under the Background of Inter-Basin Water Transfer

Jiashuai Yang [1], Chaowei Xu [1,*], Xinran Ni [2] and Xuantong Zhang [3]

1 College of Urban and Environmental Sciences, Peking University, No. 5 Yiheyuan Road, Haidian District, Beijing 100871, China
2 School of Architecture, Tsinghua University, No. 30 Shuangqing Road, Haidian District, Beijing 100084, China
3 China Academy of Building Research, Beijing 100013, China
* Correspondence: xucw@pku.edu.cn

Abstract: The imbalance of water supply and demand forces many cities to transfer water across basins, which changes the original "rainfall–runoff" relationship in urban basins. Long-term hydrological simulation of urban basins requires a tool that comprehensively considers the relationship of "rainfall–runoff" and the background of inter-basin water transfer. This paper combines the rainfall–runoff model, the GR3 model, with the background of inter-basin water transfer to simulate the hydrological process of Huangtaiqiao basin (321 km^2) in Jinan city, Shandong Province, China for 18 consecutive years with a 1 h time step. Twenty-one flood simulation results of different scales over 18 years were selected for statistical analysis. By comparing the simulation results of the GR3 model and the measured process, the results were verified by multiple evaluation indicators (the Nash–Sutcliffe efficiency coefficient, water relative error, the relative error of flood peak flow, and difference of peak arrival time) at different time scales. It was found that the simulation results of the GR3 model after inter-basin water transfer were considered to be in good agreement with the measured data. This study proves the long-term impact of inter-basin water transfer on rainfall–runoff processes in an urban basin, and the GR3-ibwt model can better simulate the hydrological processes of urban basins, providing a new perspective and method.

Keywords: basin water balance; GR3 model; hydrological model; long-term series; single flood process; urban hydrology simulation

Citation: Yang, J.; Xu, C.; Ni, X.; Zhang, X. Study on Urban Rainfall–Runoff Model under the Background of Inter-Basin Water Transfer. *Water* **2022**, *14*, 2660. https://doi.org/10.3390/w14172660

Academic Editor: Yurui Fan

Received: 5 July 2022
Accepted: 25 August 2022
Published: 28 August 2022

1. Introduction

The distribution of global water resources is not balanced, and the demand for water in many regions far exceeds the amount of available water resources, leading to a break in the balance between the supply and demand of water resources [1–4]. This phenomenon is particularly obvious in China, where the uneven distribution of water resources in time and space makes the problem of water shortage in China's regions increasingly serious [5–7]. Especially in urban areas, due to a large amount of industrial water, domestic water and other "urban water", the amount of available water resources, which is not rich, becomes more scarce [8–10], forcing cities to exploit groundwater in large quantities [11–14], causing many problems [15–17]. In view of the above problems, inter-basin water transfer is an effective and direct method to solve the problems [18–20] and has been widely used in water-shortage areas around the world [21–25]. The South-to-North Water Diversion Project is a strategic project in China. Since it was fully completed in 2014, southern water has become the main source of water for more than 140 million people in more than 40 large and medium-sized cities such as Beijing and Tianjin [22–25].

As areas with a highly concentrated population, cities alleviate many problems, such as water resource shortages, through inter-basin water transfer projects on the one hand, while on the other hand, they are inevitably affected by other impacts brought by inter-basin water transfer [26], mainly manifested as changes in water quality [27] and river hydrological factors [26,28–31]. However, changes in hydrological elements such as river runoff and river water levels are related to urban flood control and drainage [32]. Urban floods occur frequently around the world and are highly harmful and destructive, affecting urban economic development and seriously threatening the life and property safety of urban residents [25,33,34]. Therefore, urban flood simulation has always been a research hotspot [35,36]. In the current context of inter-basin water transfer, many river basins are connected in series, and it has become a common phenomenon for cities to implement inter-basin water transfer [22]. Therefore, the impact of inter-basin water transfer on urban flood simulation has also become common. Essenfelder studied the flow contribution of inter-basin water transfer by incorporating machine learning techniques into basin hydrological models [23]. Woo et al. used a SWAT model to study the impact of inter-basin water transfer on water quality in the basin [24]. Bui et al. also used a SWAT model to study the impact of inter-basin water transfer on Lake Urmia in Iran and provided data support for the management of inter-basin water transfer [25]. However, few studies consider both the hydrological model and inter-basin water transfer, and only a few experts and scholars consider inter-basin water transfer using other tools. Safavi et al. used an artificial neural network (ANN) and a fuzzy inference system (FIS) to establish a model to simulate the river runoff of inter-basin water transfer cities [32]. Wang et al. used the MIKE series of hydrodynamic models to analyze the impact of inter-basin water transfer on flood control in water-receiving areas [33]. However, the water balance of urban basins is changed by inter-basin water transfer. The increased water volume of outer basins in urban basins means the water balance is no longer the same as in the relationship between rainfall, infiltration, evaporation, interception, and runoff in natural basins [37,38]. Although the rainfall–runoff model is an important tool for basin hydrological simulation [39,40], it is rarely used to study urban flood problems under the background of inter-basin water transfer, because inter-basin water transfer causes runoff and runoff in urban basins to no longer mirror the "rainfall–runoff" relationship in natural basins. As long as the problem of water balance in the basin is dealt with, the rainfall–runoff model is still the preferred choice for the hydrological simulation of a basin [41,42].

In this paper, the GR3 model, which has been proven to have good applicability in the Yangtze River basin, Yellow River basin, Heilongjiang River basin, Huaihe River basin, and other basins in China [43,44], is selected as the rainfall–runoff model for research. Xu et al. concluded that the simulation accuracy of the GR3 model and the Xinanjiang model, used in seven representative basins in China, is at the same level [44], while the Xinanjiang model has been widely used in China for decades [45–47]. At the same time, the distributed model requires a great deal of data and is very complex, but the simulation at the watershed outlet is not always better than the lumped (conceptual) model. Therefore, we chose the lumped (conceptual) GR3 model, which has fewer parameters, simple calibration, and can reduce the uncertainty of model parameters [48–51]. In order to study the impact of inter-basin water transfer on urban hydrological simulation, this paper chooses two scales, a long-time series and a single flood, to conduct hydrological simulation considering inter-basin water transfer and not considering inter-basin water transfer. This paper focuses on how to integrate the inter-basin water transfer into the GR3 model, using the GR3 model and the GR3 model combined with the inter-basin water transfer (hereinafter referred to as GR3-ibwt) to carry out multi-year (18 years) continuous hydrological simulations. Twenty-one flood simulation results of different scales over 18 years were selected for statistical analysis. Comparing the performance of the GR3 model and the GR3-ibwt model, we verified the simulation results on two scales of long-term series and single-flood processes, forming a new processing method. Section 2 outlines the materials and methods, Section 3 presents the results, Section 4 discusses these results, and Section 5 provides conclusions.

2. Materials and Methods

2.1. Study Area and Data Sources

In this study, the basin controlled by the Huangtaiqiao Hydrological Station in Jinan City (hereinafter referred to as the Huangtaiqiao basin) was selected as the study area. Jinan is also known as "Spring City". There are many rivers and springs in the city, including The Yellow River, Xiaoqing River, Baotu Spring, Heihu Spring, and Pearl Spring, etc. [52], which are rich in water resources. Even so, Jinan still suffers from a severe water shortage. The average annual total water resources are 1.16 billion m^3, and the per capita water resources are 210 m^3, which is only 10% of China's national standard and far lower than the global average of 1000 m^3 [53,54]. The difference between the supply and demand of the available water resources is approximately 30% [54], which is a typical water-deficient city in northern China. The phenomenon of overexploitation of groundwater is serious, and the obstruction of spring water occurs from time to time [55]. Therefore, Jinan introduces approximately 600 million m^3 of Yellow River water every year to supplement the water resource gap [56]. Since the completion of the eastern route of the South-to-North Water Diversion Project in 2013, Jinan has had to divert a large amount of water from the Yangtze River every year to supplement its water source [55]. A large number of water diversions from other basins have fully guaranteed the available water resources in Jinan, and groundwater exploitation has been effectively replaced [55,56]. However, the original water balance in the Huangtaiqiao basin was broken, and the runoff process of the Huangtaiqiao Hydrological Station was no longer a simple "rainfall–runoff" relationship. At the same time, Jinan is a city with frequent urban floods [57]. Due to the mountains in the south and the Yellow River in the north, the terrain is high in the south and low in the north, and the urban section of the Yellow River is an "above-ground river". In addition, the annual precipitation is highly concentrated from June to September, which accounts for 70–80% of the annual precipitation. These factors cause frequent urban floods in Jinan [58]. On 18 July 2007, the "18 July" rainstorm in Jinan urban area had a maximum rainfall of 151 mm in one hour, which was the historical maximum since the meteorological records began in Jinan. The flood caused more than 30 deaths, more than 170 injuries, and direct economic losses of approximately 1.32 billion. RMB [59,60]. Jinan is not only a city with frequent floods [58], but also utilizes a large number of inter-basin water transfers [55,56]. Therefore, the Huangtaiqiao basin was selected as the study area to study the urban rainfall–runoff model under the background of inter-basin water transfer.

Jinan is located at the southeastern edge of North China Plain, with Mount Tai in the south and the Yellow River in the north. The terrain is high in the south and low in the north. The climate type is a temperate continental monsoon climate, and the annual average precipitation is 580–750 mm, accounting for 75% of the annual precipitation in the flood season [61]. Although the Yellow River passes through the northern part of the urban area of Jinan, Jinan does not belong to the Yellow River Basin but to the Xiaoqing River Basin, because the Jinan section of the Yellow River is an "overground river", and the flood control dam is more than 20 m above the urban area [62]. Huangtaiqiao River basin is located in the main urban area of Jinan city, covering an area of 321 km^2. Xiaoqing River is the only river flowing out of the basin and the final drainage channel of Jinan city. It originates from the northwest of Jinan city and eventually flows into the Bohai Sea. Huangtaiqiao Hydrology Station is the general control station of the Jinan urban area, located in the lower reaches of the Jinan urban section of Xiaoqing River, and the runoff of the urban river ultimately flows out through the Huangtaiqiao Hydrology Station [62]. Details of the study area are shown in Figure 1.

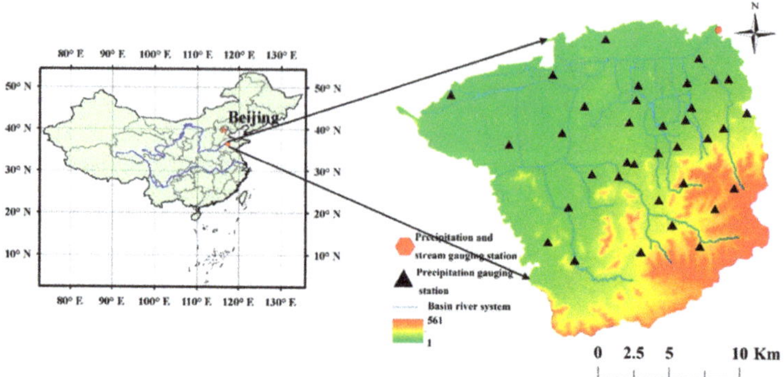

Figure 1. Map of the Huangtaiqiao basin.

Data used in this study include rainfall, evaporation, runoff, DEM, land use, and inter-basin water transfer. Hydrological data (rainfall, evaporation, and runoff) were obtained from The Hydrological Bureau of Jinan City, including hourly precipitation data from 37 rainfall stations in the study area for 2000–2017, daily evaporation data from the study area for 2000–2017, and hourly runoff data from the Huangtaiqiao hydrological station for 2000–2017. FABDEM, from the University of Bristol (https://data.bris.ac.uk/data/dataset/ accessed on 12 May 2022), is the world's first elevation model dataset that simultaneously removes trees and buildings [63]. The dataset is between 60° S and 80° N, and the resolution is 30 m. The land use data were obtained from 2015 Landsat remote sensing image data of Jinan city from China Geospatial Data Cloud (https://www.gscloud.cn/ accessed on 7 December 2021), with a resolution of 30 m. The data for the inter-basin water transfer are from the Hydrology Bureau of Jinan City, including data on the annual inter-basin water transfer amount and usage of inter-basin water transfer in Huangtaiqiao Basin from 2008 to 2017. The distribution of rainfall stations, hydrologic stations, and DEM in the study area is shown in Figure 1, and the land use distribution is shown in Figure 2. For the data on inter-basin water transfer, please refer to Supplementary material Table S1.

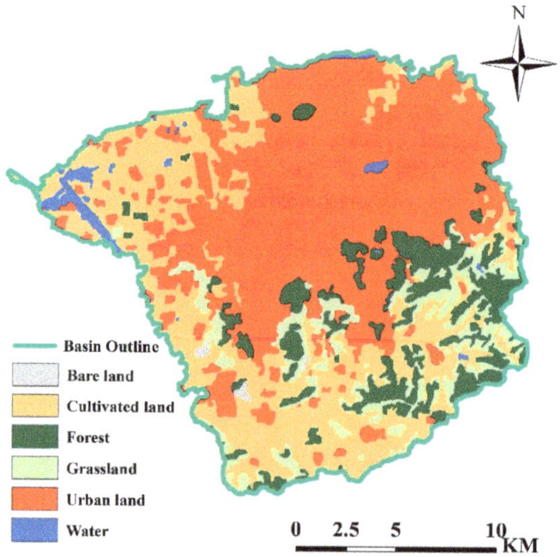

Figure 2. Land use of the Huangtaiqiao basin.

2.2. GR3

GR3 is a lumped conceptual rainfall–runoff model with three parameters. It is based on the storage, infiltration, migration, and evaporation of water in the soil, and uses empirical or semi-empirical mathematical expressions to describe the formation process of watershed runoff. The model is designed around the runoff generation tank and the runoff routing tank. The runoff generation tank reveals the storage process of soil water. The effective rainfall after deducting evaporation is calculated through the unit line of the slope after satisfying the soil water storage capacity. Part of it becomes direct runoff, and the other part enters the runoff routing tank, which is superimposed with the direct runoff after calculation. Then the outflow process is formed. The whole model is designed with three calculation units: Rainfall and evaporation calculation, runoff generation calculation, and runoff routing calculation, as shown in Figure 3.

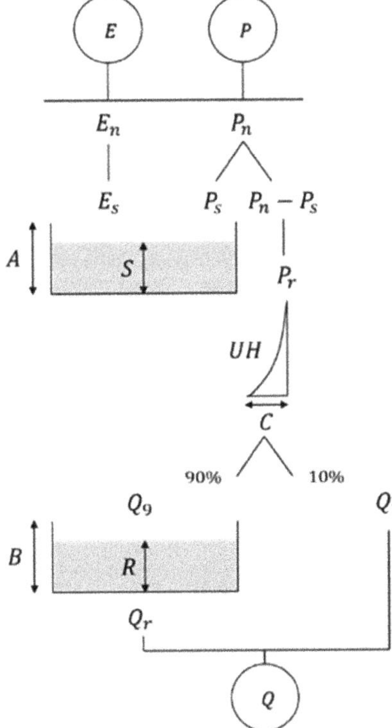

Figure 3. The framework of GR3 model.

The GR3 model contains three parameters to be optimized, namely A, B, and C (see Table 1 for details), which represent the maximum water depth of the runoff generation tank, the maximum water depth of the runoff routing tank, and the number of unit line periods, respectively. Automatic calibration can be achieved in the model. The 80% confidence interval in Table 1 is the statistical results obtained from large sample experiments in the United States, France, Australia, Ivory Coast, Brazil, and other places [64,65]. In addition, the GR3 model also includes six fixed parameters. These values are fixed after a large sample test [64,65].

Table 1. GR3 model parameters needed to be calibrated.

Parameters	Units	Physical Meaning	80% Confidence Interval
A	mm	Maximum water depth of the runoff generation tank	(100, 1500)
B	mm	Maximum water depth of the runoff routing tank	(20, 600)
C	The length of time period in model calculation	The number of unit line periods	(1.1, 2.9)

2.3. Processing of Inter-Basin Water Transfer and Integration into GR3 Model

Due to data limitations, data from 2008 to 2017 were used in this study. According to the Hydrology Bureau of Jinan city, in 2000, Jinan city began to make extensive use of water transferred across the basin. Due to the lack of complete data [66] for the annual inter-basin water transfer volume from 2000 to 2007, we chose to use the minimum inter-basin water transfer data from 2008 to 2017. At the same time, we learned from the Hydrology Bureau of Jinan that all the water transferred across the basin is used for urban water in Jinan, and the daily water transferred in each water transfer cycle (every year) is average, and the water transferred is used for domestic water, industrial water, irrigation water, and river water replenishment. According to the data on inter-basin water transfer provided by The Hydrological Bureau of Jinan city (see Table 2 for details), domestic water takes up the largest proportion in these water transfer directions, and the average proportion of multi-year (water transfer cycle) reaches 37%, that is, 37% of the water transferred across basin every year (water transfer cycle) is transported to every household for urban residents. The multi-year average proportions of other water transfer destinations are 13% for irrigation, 21% for industrial use, and 29% for river replenishment.

Table 2. The average proportion of multi-year use of inter-basin water transfer.

Use of Inter-Basin Water Transfer	Domestic Water	Irrigation Water	Industrial Water	Channel Filling Water
Multiyear average (%)	37	13	21	29

In addition to irrigation water being directly used for agricultural irrigation and water replenishment directly into rivers, domestic water and industrial water are transported to households and factories through water distribution networks. Some of the water enters the river from the sewer, and some of the water is treated by the sewage treatment plant and then enters the river. As a result, in a short time scale (1 h), there is a certain lag between the introduction of inter-basin water transfer and the use of inter-basin water transfer. According to the Jinan Statistical Yearbook, the Jinan Water Resources Bulletin, and related studies [67–69], the average daily water consumption in the study area tends to be stable for many years. Therefore, in the long time scale used here (18 years), the lag between the introduction of inter-basin water transfer and the use of inter-basin water transfer has little impact on the study and is therefore ignored in this study. Some of the urban water is reused through recycling [70,71]. According to the Jinan Water Resources Bulletin, the average comprehensive water consumption rate in the study area for years is 70%, that is, 30% of the water has been reused. This study did not integrate this part of the reused water into the rainfall and runoff model. Since more specific industry water consumption rates were not found, the study no longer distinguished between them. Land use distribution can roughly reflect the destination distribution of inter-basin water transfer [67–69,72]. It can be seen from Figures 1 and 2 that construction land, cultivated land, and river channels are widely distributed in the study area, so this study does not show special treatment for the destination distribution of inter-basin water transfer.

2.3.1. Downscaling of Inter-Basin Water Transfer

The section above points out that the daily water transfer volume is the average in the inter-basin water transfer cycle (every year). In order to unify the time scale of inter-basin water transfer data and hydrological data (precipitation, evaporation, and runoff), the annual water transfer volume is converted into an hourly water transfer volume through the following formula:

$$\text{In common year, } W_h = \frac{W_y}{8760} \tag{1}$$

$$\text{In leap year, } W_h = \frac{W_y}{8784} \tag{2}$$

In the formula, W_h is the hourly inter-basin water transfer volume (m^3) and W_y is the annual inter-basin water transfer volume (m^3). Considering that the inter-basin water transfer volume should be integrated into the GR3 model and combined with the multi-year average comprehensive water consumption rate in the basin, the hourly inter-basin water transfer volume (m^3) was converted into the hourly inter-basin water transfer depth (mm) by the following formula:

$$Q_T = \frac{W_h}{A_b} \times r_{mawc} \times 10^{-3} \tag{3}$$

In the formula, Q_T is the hourly inter-basin water transfer depth (mm), r_{mawc} is the multi-year average comprehensive water consumption rate (%), and A_b is the basin area (km^2). We allocate Q_T according to the proportion of water transfer purposes to obtain the hourly inter-basin water transfer depth for different purposes.

2.3.2. Integrating Inter-Basin Water Transfer into GR3

The hourly inter-basin water transfer depth for different purposes is calculated in Section 2.3.1. To integrate the inter-basin water transfer into the GR3 model, in addition to the data scale problem already solved in Section 2.3.1, two aspects should be considered, namely, the GR3 model structure and the use characteristics of water transfer for different purposes. From Section 2.2, it can be seen that the GR3 model includes three units: Rainfall and evaporation calculation, runoff generation calculation, and runoff routing calculation. While the inter-basin water transfer is integrated into the model, it is also necessary to analyze the use characteristics of water transfer for different purposes. As mentioned above, domestic water and industrial water are transported to each water-using unit through the water pipe network. After using the water, the water-using unit discharges it into the sewer, and then enters the river for routing. Therefore, the domestic and industrial water is integrated into the river routing part of the GR3 model and participate in the model calculation together with the amount of runoff generation (P_r). Irrigation water is transported to farmland for irrigation through the water pipe network. During irrigation, irrigation water will undergo evaporation and infiltration, followed by runoff generation and routing. These processes are similar to the "rain–runoff" process in the natural state. Therefore, irrigation water is integrated into the rainfall and evaporation part of the GR3 model and participates in the model calculation together with the precipitation (P). The river water replenishment directly flows into the river through the water pipe network, and then converges to the outlet of the basin through the river network. Therefore, the river water replenishment is integrated into the runoff routing part of the GR3 model and participates in the model calculation together with the amount of runoff generation (P_r). The specific process is shown in Figure 4.

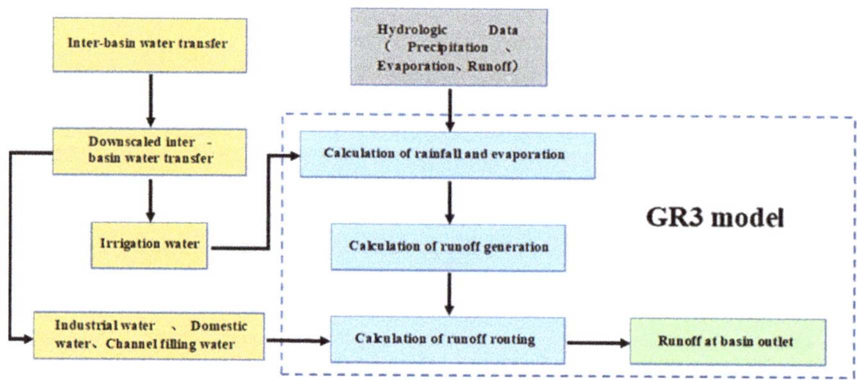

Figure 4. Flow chart of inter-basin water transfer processing and integration into GR3.

2.4. Model Setup and Data Processing

Continuous hydrological simulations for 18 years (2000–2017) were performed using the GR3 model and the GR3-ibwt model, respectively, with a time resolution of 1 h. Parameter calibration was performed using the minimum square error function (MSE). From the input side of the model, the time resolution of rainfall and runoff data in the hydrological data meets the simulation requirements; the evaporation data are daily evaporation, which needs to be processed to ensure the time resolution meets the simulation requirements. Evaporation changes more gradually over time than rainfall, so linear interpolation is performed on the daily evaporation data to meet the simulation requirements. From the perspective of the watershed distribution, the evaporation data obtained are the average data of the research area, so the evaporation data distribution is not processed; rainfall data come from all rainfall stations (including the Huangtaiqiao hydrological station). There are a large number (37) of rainfall stations in the study area, with uniform distribution, and the area of the study area is small (321 km^2). Therefore, the average value of measured data from all rainfall stations is calculated as the input data of the model. The data processing of inter-basin water transfer is detailed in Section 2.3 and will not be repeated here.

In this study, the hydrological simulation data volume is large, the time series is long (18 years), and the time resolution is high (1 h). Compared with the daily-scale (1d) simulation, more flood information can be captured. The period of 2000–2003 was used as the model warm-up period, 2004–2010 was the model calibration period, and 2011–2017 was the model validation period. According to the observed flood peak flow, the floods in the study area are divided into three levels: Big floods (peak flow > 130 m^3/s), medium floods (80 m^3/s < peak flow < 130 m^3/s), and small floods (peak flow < 80 m^3/s). The simulation results of 21 floods of different sizes in the long-term series simulation results were selected for statistical analysis for the single-flood process test. The selected single-flood process is shown in Table 3.

Table 3. The 21 typical flood processes (FPs).

Flood Process	Rainfall Depth (mm)	Peak Flow (m^3/s)	The Size of the Flood
20050817	37.63	33.60	
20070815	77.83	47.75	
20040511	43.84	48.30	
20080813	31.63	51.60	Small
20160712	39.82	53.00	

Table 3. *Cont.*

Flood Process	Rainfall Depth (mm)	Peak Flow (m³/s)	The Size of the Flood
20110818	82.23	54.90	
20160806	54.99	74.60	
20160720	76.11	82.96	
20090817	70.02	84.70	
20060804	71.00	92.50	
20080718	117.08	110.00	Medium
20170706	42.00	111.00	
20160801	62.95	116.17	
20040717	124.34	124.69	
20150730	76.30	133.62	
20140619	119.28	144.00	
20100819	248.83	161.00	
20130723	88.40	168.50	Big
20160816	96.79	169.00	
20070718	126.39	202.00	
20120708	179.33	209.67	

2.5. Model Evaluation Methods

In this paper, the performance of the model is evaluated on two scales, a long time series (18 years) and a single-flood process. The Nash–Sutcliffe efficiency coefficient (NSE) and the water relative error (RE) are selected as the long-term simulation evaluation indicators [73,74]; the Nash–Sutcliffe efficiency coefficient (NSE), the relative error of water volume (RE), the relative error of flood peak flow (PE), and difference in peak arrival time (DPAT) were selected as evaluation indicators for single-flood process simulation [75]. The calculation formula of each index is as follows:

$$NSE = 1 - \frac{\Sigma(Q_o - Q_s)^2}{\Sigma(Q_o - \overline{Q_o})^2} \tag{4}$$

$$RE = \frac{\Sigma(Q_s - Q_o)}{\Sigma Q_o} \tag{5}$$

$$PE = \frac{\max(Q_s) - \max(Q_o)}{\max(Q_o)} \tag{6}$$

$$DPAT = T_{\max(Q_s)} - T_{\max(Q_o)} \tag{7}$$

In the formula , Q_s is the simulated discharge (m^3/s), Q_o is the observed discharge (m^3/s), $\overline{Q_o}$ is the averaged observed discharge (m^3/s), $\max(Q_s)$ is the simulated peak discharge (m^3/s), $\max(Q_o)$ is the observed peak discharge (m^3/s), $T_{\max(Q_s)}$ is the arrival time of the simulated peak discharge (h), and $T_{\max(Q_o)}$ is the arrival time of the observed peak discharge (h).

The resulting range of the Nash–Sutcliffe efficiency coefficient (NSE) is $(-\infty, 1)$, whereby the closer it is to 1, the better the model simulation effect is. Generally, the model simulation effect is acceptable if it is above 0.5. The closer the relative error of water flow (RE) and flood peak flow (PE) are to 0, the better the simulation effect of the model will be. Generally, a range of $(-0.2, 0.2)$ indicates that the simulation effect of the model is acceptable [73–75].

3. Results

3.1. Long-Time Series Simulation Results

In this study, continuous hydrological simulations for 18 years (2000–2017) were performed using the GR3 model and the GR3-ibwt model, respectively, with a time resolution of 1 h. Among them, 2000–2003 is the model warm-up period, 2004–2010 is the model calibration period, and 2011–2017 is the model validation period. Excluding the warm-up period, the simulated runoff and the observed runoff at the outlet of the basin during the model calibration period and validation period are shown in Figure 5. It can be seen from the figure that compared with the observation process, the simulation results of the GR3 model are poor, and most of the simulated flood peaks are far away from the measured peaks. The simulated runoff in the drought period is close to 0, and the observed runoff is generally higher than the simulated runoff. The reason for this phenomenon is that the GR3 model is a rainfall–runoff model. Without considering the inter-basin water transfer, the model runoff is mainly related to the rainfall and evaporation of the basin [64,65]. When there is no rainfall, the runoff will not be generated. Due to the existence of inter-basin water transfer, the runoff can be observed at the outlet of the basin when no rainfall occurs. At the same time, long-term inter-basin water transfer has an impact on the nature of the underlying surface of the basin to a certain extent [76], which is mainly reflected in the change in soil moisture [45,77,78], which will affect the performance of the GR3 model and increase the error in runoff generation calculations. Compared with a single flood, the impact of cross-basin water transfer on the long-term series simulation is magnified. Therefore, this paper chooses to use long-term hydrological data for the simulation and selects the single-flood simulation results for statistical analysis. The purpose of this is to consider the long-term impact of inter-basin water transfer on the one hand, and on the other hand, the long-term series simulation will reduce the pseudo precision caused by direct single-flood simulation.

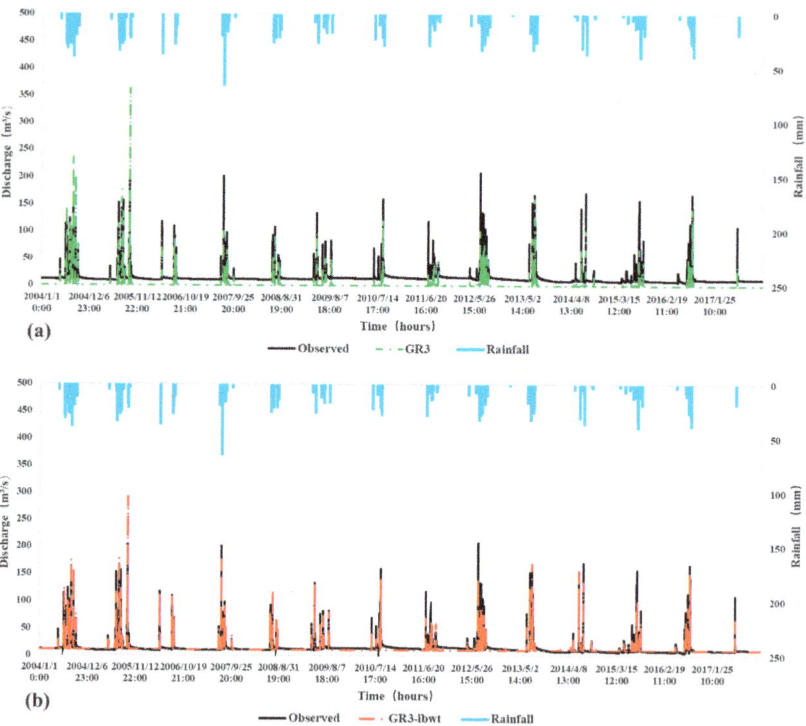

Figure 5. Time history of discharge at the outlet gauging station in long time series. (**a**): GR3 vs. observed; (**b**): GR3-ibwt vs. observed.

Compared with GR3, the simulated process of the GR3-ibwt model is closer to the observed process. From the flood peak simulation results, the GR3-ibwt model can simulate the flood peak flow well in most flood peaks, and only a few flood peaks have large simulation errors, which is greatly improved compared with the GR3 model simulation results. From the simulation results of the dry season, the simulation process of the GR3-ibwt model is almost identical to the observation process, which is a qualitative improvement compared with the simulation results of the GR3 model. It can be seen that when the inter-basin water transfer is considered, the study area achieved a water balance, and the urban basin with a high concentration of population is equivalent to a natural basin without human interference to a certain extent, and the simulation results of the rainfall–runoff model are generally better.

In order to further evaluate the performance of the model, the Nash–Sutcliffe efficiency coefficient (NSE) and the water relative error (RE) in the calibration and verification periods of the two hydrological simulations were calculated. The results are shown in Table 4, and the model parameters are shown in Table S2 of Supplementary Materials.

Table 4. The statistical analysis of the measured and simulated discharge in long time series.

Model	2004–2010 (Calibration)		2011–2017 (Validation)	
	NSE	RE	NSE	RE
GR3	−0.98	−0.86	−1.67	−0.90
GR3-ibwt	0.77	−0.06	0.70	−0.04

It can be seen that the Nash-Sutcliffe efficiency coefficient (NSE) of the GR3 model is far less than 0 in both the calibration period and the verification period, which is far from the lower limit of acceptability of 0.5, indicating that the simulation results are not credible. The relative error (RE) of water volume is below −0.85, which means that the simulated runoff is much smaller than the observed runoff, which is consistent with Figure 4. The Nash-Sutcliffe efficiency coefficients (NSE) of the GR3-ibwt model were 0.77 and 0.70 in the calibration period and validation period, respectively, and the model performance reached a good level [73–75]. Compared with the GR3 model, the Nash–Sutcliffe efficiency coefficients (NSE) increased by 178% and 142%, respectively. The water relative errors (RE) in the calibration period and validation period were −0.06 and −0.04, respectively, indicating that the simulated water volume of the GR3-ibwt model essentially reached a water balance with the actual observed water volume, and the water relative errors (RE) decreased by 94% and 96% compared with the GR3 model, respectively. Overall, the Nash–Sutcliffe efficiency coefficient (NSE) results of the GR3-ibwt model in the long-term series simulation essentially reached a good level, and the water balance was above 94%. Compared with the GR3 model, the results of the two indicators were significantly improved. From the perspective of a long time series, the hydrological simulation using the rainfall–runoff model without considering inter-basin water transfer is not in line with the actual situation, and the simulation results are greatly different from the measured process. Integrating the inter-basin water transfer into the rainfall–runoff model for hydrological simulation is closer to the actual situation, and the simulation results are relatively satisfactory.

3.2. Single Flood Simulation Results

In order to reduce the pseudo precision brought by the direct simulation of a single flood, 21 flood simulation results of different scales in the long-term series (see Table 3 for details) were selected for statistical analysis. According to the observed flood peak flow, the flood in the study area is divided into three levels: Big floods (peak flow > 130 m^3/s), medium floods (80 m^3/s < peak flow < 130 m^3/s), and small floods (peak flow < 80 m^3/s). Figures 6–8 show the simulated runoff of big, medium, and small floods and the observed runoff at the basin outlet, respectively.

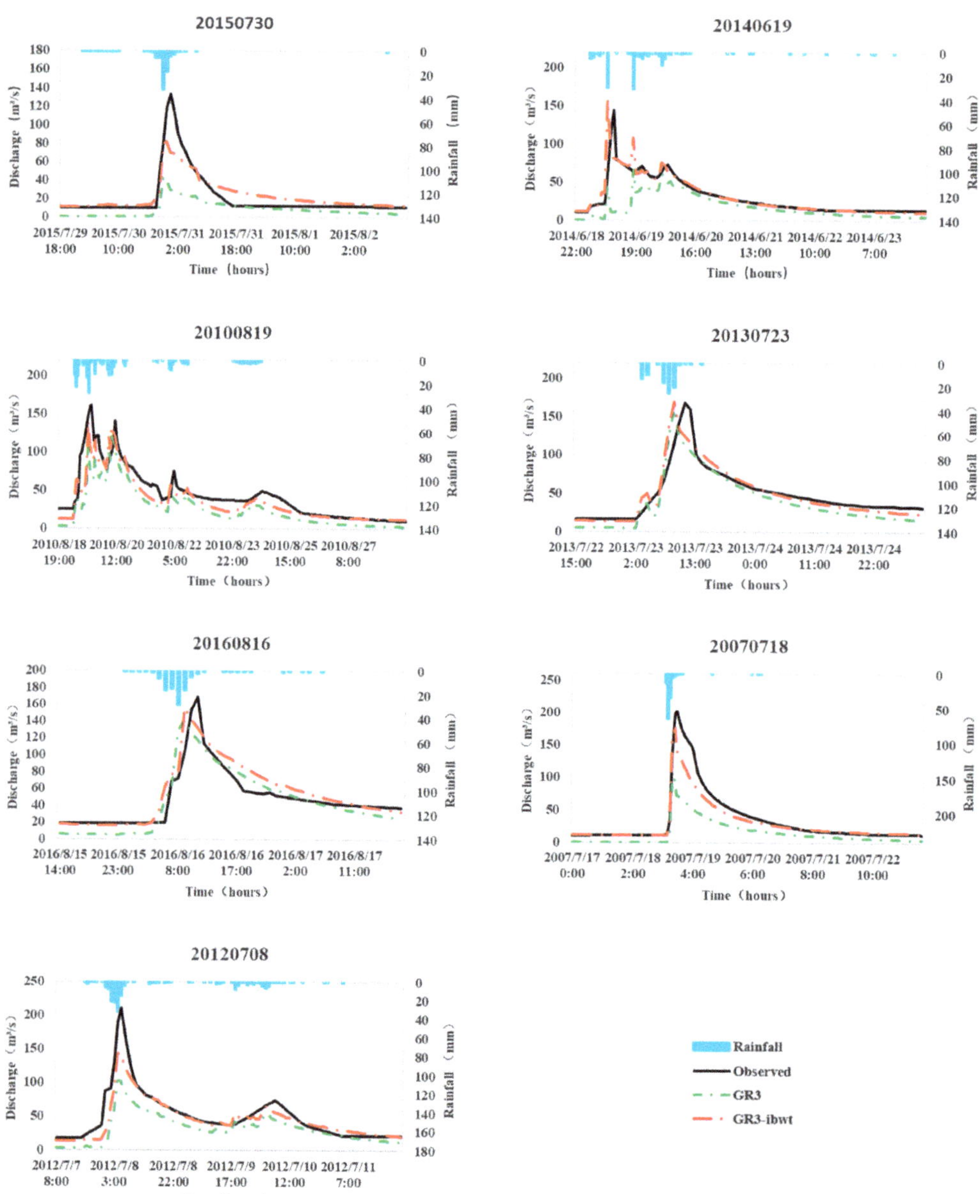

Figure 6. Time history of discharge at the outlet gauging station in big flood processes.

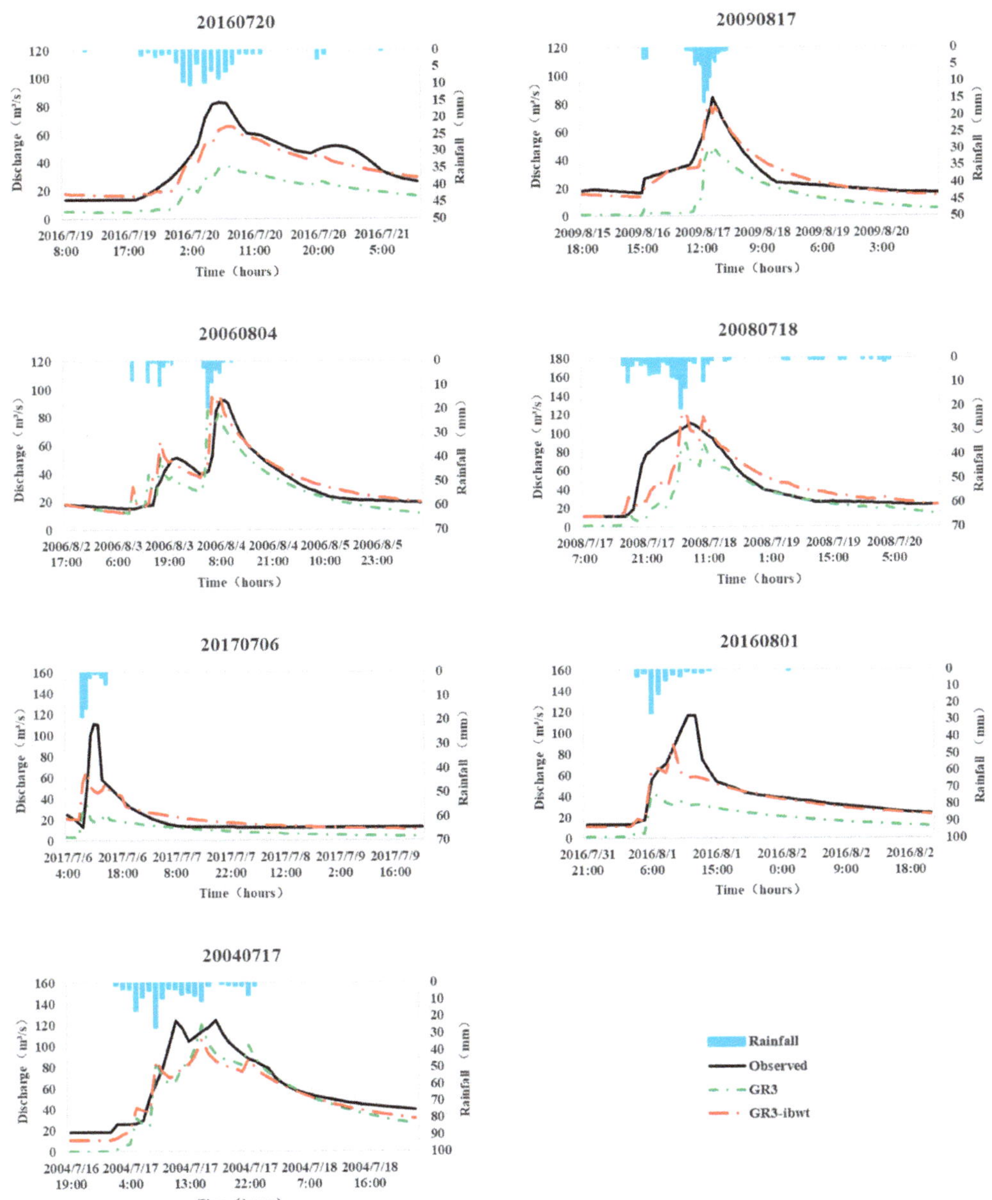

Figure 7. Time history of discharge at the outlet gauging station in medium flood processes.

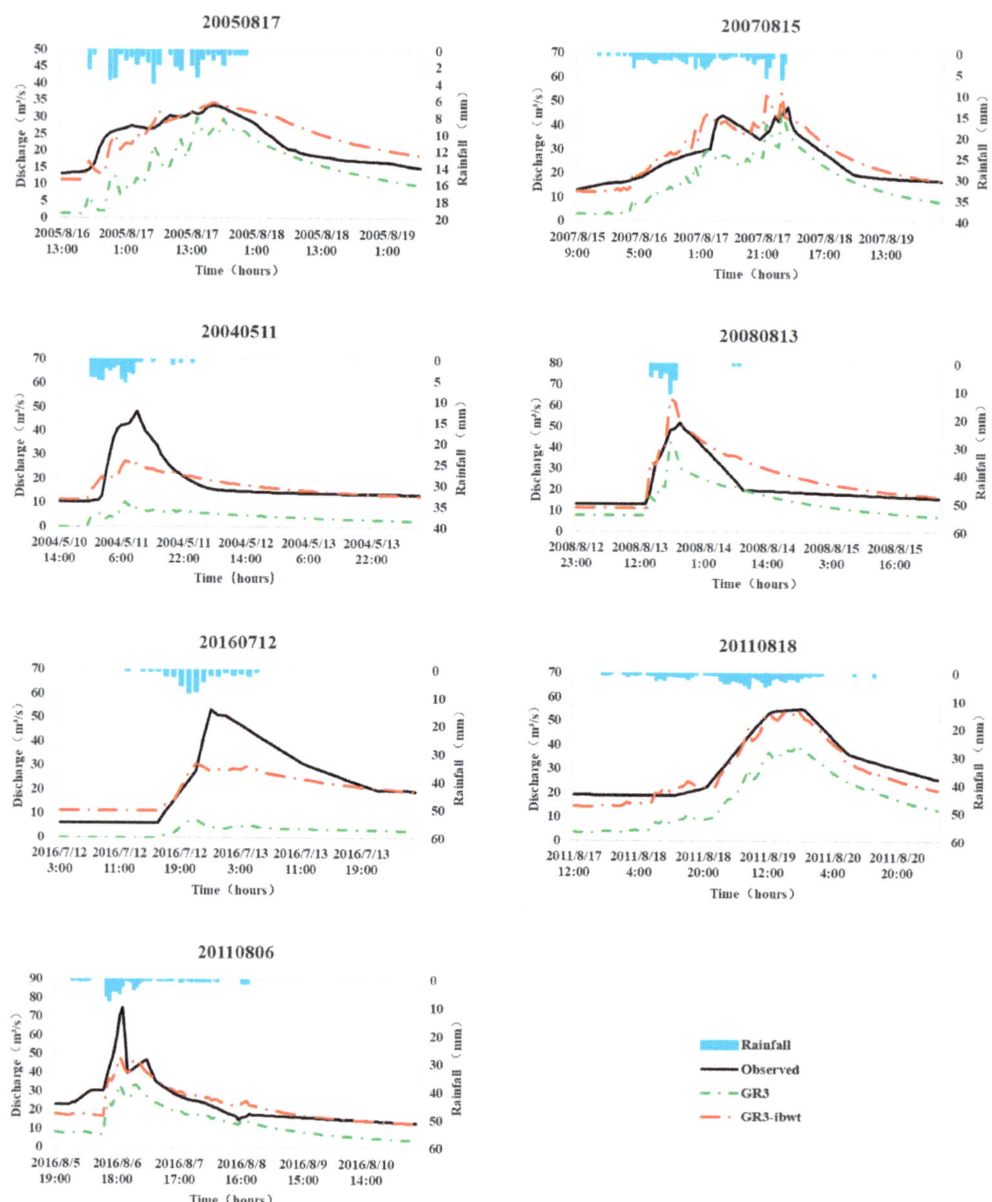

Figure 8. Time history of discharge at the outlet gauging station in small flood processes.

It can be seen from Figure 6 that the GR3 model has a general performance in the simulation of big floods, with good performance only in 20100819, 20130723, and 20160816, and poor performance in the other four floods. The GR3-ibwt model has a good overall performance in the simulation of big floods, which is slightly better than the GR3 model in 20100819, 20130723, and 20160816. The performance of the other four floods is significantly improved compared with the GR3 model.

As can be seen from Figure 7, the GR3 model has poor overall performance in the simulation of medium floods, with good performance only in the flood of 20060804 and poor performance in the other six floods. The GR3-ibwt model has a general performance in the simulation of medium floods, with good performance in 20090817, 20060804, and 20080718, while the other floods have a general performance, which is greatly improved compared with the GR3 model.

As can be seen from Figure 8, the GR3 model has poor overall performance in the simulation of small floods, and the performance of each flood is not satisfactory. The GR3-ibwt model has a general overall performance in the simulation of small floods, with good performance in the three floods of 20050817, 20070815, and 20110818. Although the performance of other floods is average, it is greatly improved compared with the GR3 model.

To summarize, the GR3 model and the GR3-ibwt model have similar performances in the single-flood scale from the basin outlet flow process diagram, that is, the simulation results of big floods are better than those of medium and small floods.

In order to further evaluate the model performance, the Nash–Sutcliffe efficiency coefficient (NSE), water relative error (RE), the relative error of flood peak flow (PE), and difference of peak arrival time (DPAT) of each flood simulation for the GR3 model and the GR3-ibwt model were calculated, respectively. The results are shown in Table 5. As can be seen from Table 5, (1) for the simulation of flood peak flow, the GR3-ibwt model is better than the GR3 model on the whole and closer to the observed flow. (2) For the Nash–Sutcliffe efficiency coefficient (NSE), the overall performance of the GR3-ibwt model is better than the GR3 model, with two floods above 0.9, 14 floods above 0.7, all floods above 0.5, and the performance of large floods is better than medium and small floods. (3) For the relative error of water volume (RE), the overall performance of the GR3-ibwt model is better than that of the GR3 model, and the relative error of water volume in all floods is within 0.2. (4) For the relative error of flood peak flow (PE), the overall performance of the GR3-ibwt model is better than that of the GR3 model. There are 12 flood peak flow relative errors within 0.2, and the maximum flood peak flow relative error is 0.42 (two floods). (5) For the difference in peak arrival time (DPAT), the overall performance of the GR3-ibwt model is better than the GR3 model, and the maximum difference in peak arrival time in all flood events does not exceed 2 h.

Statistical analysis was carried out on the four model evaluation indicators of all simulated flood events of the GR3 model and the GR3-ibwt model (See Figure 9 for details). As can be seen from Figure 9a, the Nash–Sutcliffe efficiency coefficient (NSE) of the GR3-ibwt model was greatly improved compared with the GR3 model, and the fluctuation range changed from [−1.01, 0.77] to [0.50, 0.92], and the median and mean increased from 0.07 and −0.04 to 0.78 and 0.75, respectively, and there was no abnormally low value. As can be seen from Figure 9b, the water relative error (RE) of the GR3-ibwt model was greatly reduced compared with the GR3 model, and the fluctuation range changed from [−0.88, −0.07] to [−0.18, 0.18]. Both the median and mean changed from −0.43 to −0.02. As can be seen from Figure 9c, the peak discharge relative error (PE) of the GR3-ibwt model is significantly lower than that of the GR3 model, and the fluctuation range changes from [−0.86, −0.01] to [−0.42, 0.22], while the median and mean changed from −0.43 and −0.38 to −0.13 and −0.13, respectively. As can be seen from Figure 9d, the difference in peak arrival time of the GR3-ibwt model is significantly lower than that of the GR3 model, and there are a large number of abnormal values in the different peak arrival times of the GR3 model. For the separate statistics of the model evaluation indicators of small, medium, and big floods, see Figures S1–S3 in the Supplementary Materials.

Table 5. The statistical analysis of the measured and simulated discharge in flood processes.

FP NO.	Peak Discharge (m^3/s)			NSE		RE		PE		DPAT (Hour)	
	O	H	C	H	C	H	C	H	C	H	C
20050817(S)	33.60	29.97	34.24	−1.01	0.50	−0.33	0.09	−0.11	0.02	−3	0
20070815(S)	47.75	47.33	53.92	0.07	0.76	−0.32	0.10	−0.01	0.13	−2	−2
20040511(S)	48.30	10.25	28.17	−1.92	0.54	−0.77	−0.07	−0.79	−0.42	−3	−1
20080813(S)	51.60	42.38	63.17	0.33	0.66	−0.34	0.18	−0.18	0.22	−2	−2
20160712(S)	53.00	7.58	30.53	−1.75	0.60	−0.88	−0.14	−0.86	−0.42	−2	−2
20110818(S)	54.90	40.36	55.15	−0.34	0.92	−0.45	−0.07	−0.26	0.00	−1	−1
20160806(S)	74.60	33.77	48.33	−0.17	0.74	−0.46	−0.03	−0.55	−0.35	−1	−1
20160720(M)	82.95	37.17	65.49	−0.37	0.84	−0.54	−0.11	−0.55	−0.21	1	1
20090817(M)	84.70	48.31	79.82	−0.39	0.92	−0.57	0.00	−0.43	−0.06	−1	−1
20060804(M)	92.50	88.11	95.55	0.73	0.87	−0.12	0.06	−0.05	0.03	−4	−1
20080718(M)	110.00	90.10	122.44	0.37	0.71	−0.33	0.01	−0.18	0.11	3	−1
20170706(M)	111.00	34.62	66.37	−0.02	0.58	−0.53	0.01	−0.69	−0.40	−2	−2
20160801(M)	116.17	42.51	88.80	−0.20	0.70	−0.54	−0.12	−0.63	−0.24	−5	−2
20040717(M)	124.69	120.18	106.26	0.72	0.78	−0.19	−0.16	−0.04	−0.15	−2	−2
20150730(B)	133.62	43.56	82.28	0.21	0.78	−0.59	0.09	−0.67	−0.38	−2	−1
20140619(B)	144.00	65.62	157.14	0.05	0.75	−0.43	0.01	−0.54	0.09	7	−2
20100819(B)	161.00	118.77	140.55	0.46	0.78	−0.40	−0.18	−0.26	−0.13	15	−2
20130723(B)	168.50	159.56	169.43	0.74	0.83	−0.20	−0.02	−0.05	0.01	−2	−2
20160816(B)	169.00	140.91	155.91	0.77	0.81	−0.07	0.12	−0.17	−0.08	−2	−2
20070718(B)	202.00	100.63	176.25	0.43	0.89	−0.59	−0.14	−0.50	−0.13	−1	−1
20120708(B)	209.67	102.20	142.85	0.44	0.82	−0.37	−0.10	−0.51	−0.32	−1	0

O: Observed data; H: Simulated by GR3; C: Simulated by GR3-ibwt; S: Small; M: Medium; B: Big.

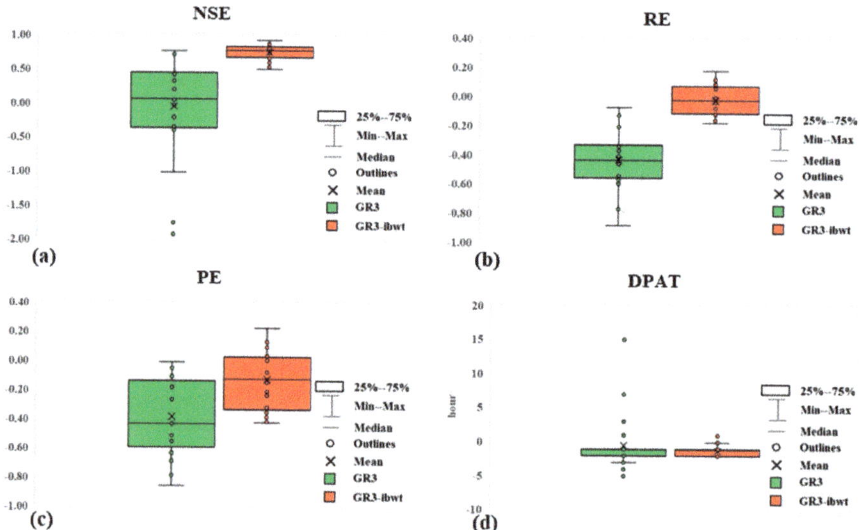

Figure 9. Box diagram of the measured and simulated discharge evaluation in flood processes. (**a**) is the box diagram of NSE; (**b**) is the box diagram of RE; (**c**) is the box diagram of PE; (**d**) is the box diagram of DPAT.

4. Discussion

In general, the GR3-ibwt model had better simulation results than the GR3 model in both long time series scale and single flood scale, and the results are relatively satisfactory. Specifically, for the simulation of the long time series, the Nash-Sutcliffe efficiency coefficient (NSE) of the GR3-ibwt model was increased by 178% and 142% in the calibration and validation periods, respectively, compared with the GR3 model. At the same time, the water relative error (RE) was reduced by 94% and 96% compared with the GR3 model during the calibration period and the validation period, respectively. Both model evaluation indicators prove that the GR3-ibwt model has good simulation performance on long time scales. The reason is that the GR3-ibwt model considers the inter-basin water transfer based on the GR3 model, and the rainfall–runoff simulation of the basin achieves a water balance and reduces the long-term impact of inter-basin water transfer on the hydrological effect of the basin.

For the simulation of the single-flood scale, the performance of the GR3-ibwt model is better than that of the GR3 model in four evaluation indicators: Nash–Sutcliffe efficiency coefficient (NSE), water relative error (RE), peak discharge relative error (PE), and difference of peak arrival time (DPAT). The simulation of big floods is better than that of medium and small floods. The reason is that in urban flood simulation, precipitation still plays a leading role [35,36], and big floods are less affected by other disturbances, which can be confirmed by the GR3 model's overall simulation of big floods also performing better than that of medium and small floods. Regarding the time lag of the flood peak, we suggest this is because the properties of the underlying surface (such as soil water requirement) of the basin change under the action of inter-basin water transfer over a long period of time, which makes it less sensitive to the flood peak. The results prove that this is more consistent with reality.

Hydrological simulation of urban basins is very common at present, but the phenomenon of inter-basin water transfer is rarely considered. However, the phenomenon of inter-basin water transfer is now very common, especially in urban basins where water resources are scarce. Therefore, the fundamental difference between this paper and other studies in this field is that the factor of inter-basin water transfer is taken into account in the hydrological simulation. We chose Jinan, the capital of Shandong Province in China, as the study area, which is an important city in East China. At the same time, the data we used are real data provided by local departments rather than scenario analysis, so we believe that this study reflects reality to a certain extent, both in terms of representativeness of the study area and data reliability. The biggest limitation of this study is the acquisition of measured data. On the one hand, high-precision hydrological data are required, and on the other hand, inter-basin water transfer data within the same period are also required, while at the same time, different uses of inter-basin water transfer are required. Once these data are obtained, they can be processed and combined with different types of hydrological models for hydrological simulations.

This study attempts to integrate the inter-basin water transfer into the rainfall–runoff model, and the results are relatively satisfactory. Subsequent research can focus on the following aspects: (1) Conducting a more comprehensive analysis of inter-basin water transfer so that it can be better integrated into the rainfall–runoff model, such as further refinement of the temporal and spatial distribution of inter-basin water transfer, and more comprehensive analysis of the use of inter-basin water transfer. Furthermore, researchers should (2) perform hydrological simulations under the background of inter-basin water transfer in other urban basins with data support and compare this with hydrological simulation results that do not consider inter-basin water transfer. Moreover, (3) the GR3-ibwt model could be applied to urban water resource management, urban water landscape design, basin ecological protection, and other aspects. Lastly, (4) by coupling the GR3-ibwt model with the hydraulic model and replacing the runoff routing part of the GR3-ibwt model with the hydraulic model for the simulation, one can obtain more flood information that has more physical meaning.

5. Conclusions

In this study, a new urban hydrological simulation method was proposed. Based on the GR3 rainfall–runoff model, the inter-basin water transfer is processed and integrated into the GR3 model, and then the hydrological simulation is conducted. The study area is located in the Huangtaiqiao basin of Jinan City, Shandong Province, China. The simulation object is the hydrological data of the Huangtaiqiao basin from 2000 to 2017, with a time resolution of 1 h, and 21 flood simulation results of different scales were selected for statistical analysis. By comparing the simulated results of the GR3-ibwt model with those of the GR3 model and measured data, their performances on the Nash–Sutcliffe efficiency coefficient (NSE), water relative error (RE), peak discharge relative error (PE), and difference peak arrival time (DPAT) were evaluated on the two scales of a long time series (18 years) and a single flood (21 flood events). The following conclusions are drawn:

(1) The GR3-ibwt model performs well in the hydrological simulation of an urban basin, and the long-term series and single-flood simulation results are satisfactory. (2) We demonstrate that inter-basin water transfer has a long-term impact on the rainfall–runoff process of urban basins and provide a new perspective and method for long-term hydrological simulation of urban basins. We also (3) revealed the potential application of the GR3-ibwt model in urban water resource management, urban water landscape design, basin ecological protection, and other aspects. Lastly, we (4) suggested the necessity of and made recommendations for further research.

Supplementary Materials: The following supporting information can be downloaded at: https://www.mdpi.com/article/10.3390/w14172660/s1, Table S1. The volume of inter-basin water transfer in Huangtaiqiao basin; Table S2. The parameters of GR3 and GR3-ibwt; Figure S1. Box diagram of the measured and simulated discharge evaluation in small flood processes; Figure S2. Box diagram of the measured and simulated discharge evaluation in medium flood processes; Figure S3. Box diagram of the measured and simulated discharge evaluation in big flood processes.

Author Contributions: Conceptualization, J.Y. and C.X.; methodology, J.Y.; software, J.Y.; validation, J.Y. and C.X.; formal analysis, J.Y.; investigation, J.Y., C.X., X.N. and X.Z.; resources, J.Y.; data curation, J.Y.; writing—original draft preparation, J.Y.; writing—review and editing, C.X. and X.N.; visualization, J.Y.; supervision, J.Y.; project administration, J.Y. All authors have read and agreed to the published version of the manuscript.

Funding: This paper was funded by the Major Science and Technology Program for Water Pollution Control and treatment, China (Grant No. 2014ZX07203-008).

Institutional Review Board Statement: Not applicable.

Informed Consent Statement: Not applicable.

Data Availability Statement: Not applicable.

Acknowledgments: We acknowledge the continued support and generosity of Xiaoliu Yang from the College of Urban and Environmental Sciences, Peking University. This work was financially supported by the Major Science and Technology Program for Water Pollution Control and treatment, China (Grant No. 2014ZX07203-008).

Conflicts of Interest: The authors declare no conflict of interest. The funders had no role in the design of the study; in the collection, analyses, or interpretation of data; in the writing of the manuscript, or in the decision to publish the results.

References

1. Schneider, R.L. Integrated, watershed-based management for sustainable water resources. *Front. Earth Sci. China* **2010**, *4*, 117–125.
2. Oki, T.; Kanae, S. Global hydrological cycles and world water resources. *Science* **2006**, *313*, 1068–1072. [PubMed]
3. Hashem, M.S.; Qi, X. Treated Wastewater Irrigation—A Review. *Water* **2021**, *13*, 1527.
4. Wang, J.; Hou, B.; Jiang, D.; Xiao, W.; Wu, Y.; Zhao, Y.; Zhou, Y.; Guo, C.; Wang, G. Optimal Allocation of Water Resources Based on Water Supply Security. *Water* **2016**, *8*, 237.
5. Dou, X. China's inter-basin water management in the context of regional water shortage. *Sustain. Water Resour. Manag.* **2018**, *4*, 519–526.
6. Li, T.; Qiu, S.; Mao, S.; Bao, R.; Deng, H. Evaluating Water Resource Accessibility in Southwest China. *Water* **2019**, *11*, 1708. [CrossRef]
7. Geng, Q.; Liu, H.; He, X.; Tian, Z. Integrating Blue and Green Water to Identify Matching Characteristics of Agricultural Water and Land Resources in China. *Water* **2022**, *14*, 685.

8. Di, D.; Wu, Z.; Guo, X.; Lv, C.; Wang, H. Value Stream Analysis and Emergy Evaluation of the Water Resource Eco-Economic System in the Yellow River Basin. *Water* **2019**, *11*, 710. [CrossRef]
9. Song, P.; Wang, C.; Zhang, W.; Liu, W.; Sun, J.; Wang, X.; Lei, X.; Wang, H. Urban Multi-Source Water Supply in China: Variation Tendency, Modeling Methods and Challenges. *Water* **2020**, *12*, 1199.
10. Hattingh, J.; Maree, G.A.; Ashton, P.J.; Leaner, J.J.; Rascher, J.; Turton, A.R. Introduction to ecosystem governance: Focusing on Africa. *Water Policy* **2007**, *9*, 5–10.
11. Su, D.; Zhang, Q.H.; Ngo, H.H.; Dzakpasu, M.; Guo, W.S.; Wang, X.C. Development of a water cycle management approach to Sponge City construction in Xi'an, China. *Sci. Total Environ.* **2019**, *685*, 490–496. [PubMed]
12. Sun, W.; Ren, J.M. Development and Utilization Status in the Basin of Shiyang River and Water Quality Evaluation. *Appl. Mech. Mater.* **2012**, *212*, 482–486. [CrossRef]
13. Ni, L.; Fan, M.; Qu, S.; Zheng, Q. Based on GMS management of shallow groundwater resource in Ningjin, China. *IOP Conf. Ser. Earth Environ. Science* **2019**, *237*, 32063. [CrossRef]
14. Li, F.; Yan, W.; Zhao, Y.; Jiang, R. The regulation and management of water resources in groundwater over-extraction area based on ET. *Theor. Appl. Climatol.* **2021**, *146*, 57–69.
15. Jia, X.; O'Connor, D.; Hou, D.; Jin, Y.; Li, G.; Zheng, C.; Ok, Y.S.; Tsang, D.C.W.; Luo, J. Groundwater depletion and contamination: Spatial distribution of groundwater resources sustainability in China. *Sci. Total Environ.* **2019**, *672*, 551–562.
16. Wang, H.; Wang, Y.; Jiao, X.; Qian, G. Risk management of land subsidence in Shanghai. *Desalin. Water Treat.* **2014**, *52*, 1122–1129.
17. Fan, Y.; Lu, W.; Miao, T.; Li, J.; Lin, J. Multiobjective optimization of the groundwater exploitation layout in coastal areas based on multiple surrogate models. *Environ. Sci. Pollut. R* **2020**, *27*, 19561–19576.
18. Bozorg-Haddad, O.; Abutalebi, M.; Chu, X.; Loáiciga, H.A. Assessment of potential of intraregional conflicts by developing a transferability index for inter-basin water transfers, and their impacts on the water resources. *Environ. Monit. Assess* **2019**, *192*, 40. [PubMed]
19. Zhang, C.; Wang, G.; Peng, Y.; Tang, G.; Liang, G. A Negotiation-Based Multi-Objective, Multi-Party Decision-Making Model for Inter-Basin Water Transfer Scheme Optimization. *Water Resour. Manag.* **2012**, *26*, 4029–4038.
20. Wang, Q.; Zhou, H.; Liang, G.; Xu, H. Optimal Operation of Bidirectional Inter-Basin Water Transfer-Supply System. *Water Resour. Manag.* **2015**, *29*, 3037–3054. [CrossRef]
21. Wang, Q.W.; Sun, R.R.; Guo, W.P. Study on Three-Dimensional Visual Simulation for Inter-Basin Water Transfer Project. *Appl. Mech. Mater.* **2013**, *256*, 2523–2527. [CrossRef]
22. Zhang, L.; Li, S.; Loáiciga, H.A.; Zhuang, Y.; Du, Y. Opportunities and challenges of interbasin water transfers: A literature review with bibliometric analysis. *Scientometrics* **2015**, *105*, 279–294.
23. Essenfelder, A.H.; Giupponi, C. A coupled hydrologic-machine learning modelling framework to support hydrologic modelling in river basins under Interbasin Water Transfer regimes. *Environ. Modell. Softw.* **2020**, *131*, 104779.
24. Woo, S.; Kim, S.; Lee, J.; Kim, S.; Kim, Y. Evaluating the impact of interbasin water transfer on water quality in the recipient river basin with SWAT. *Sci. Total Environ.* **2021**, *776*, 145984. [PubMed]
25. Tien Bui, D.; Talebpour Asl, D.; Ghanavati, E.; Al-Ansari, N.; Khezri, S.; Chapi, K.; Amini, A.; Thai Pham, B. Effects of Inter-Basin Water Transfer on Water Flow Condition of Destination Basin. *Sustainability* **2020**, *12*, 338. [CrossRef]
26. Cao, Y.; Chang, J.; Huang, Q.; Chen, X.; Chen, Y. Study of Discharge Model in South-to-North Water Diversion Middle Route Project Based on Radial Basis Function Neural Network. *MATEC Web Conf.* **2016**, *68*, 14011. [CrossRef]
27. Zhang, Y.; Li, G. Influence of south-to-north water diversion on major cones of depression in North China Plain. *Environ. Earth Sci.* **2014**, *71*, 3845–3853. [CrossRef]
28. Guo, Y.C.; Li, J.F.; Li, J.L. Agent Construction System Application and Improvement Discussion in the South-to-North Water Diversion Project. *Appl. Mech. Mater.* **2012**, *105*, 1096–1099. [CrossRef]
29. Arthington, A.H.; Pusey, B.J. Flow restoration and protection in Australian rivers. *River Res. Appl.* **2003**, *19*, 377–395.
30. Qin, G.; Liu, J.; Wang, T.; Xu, S.; Su, G. An Integrated Methodology to Analyze the Total Nitrogen Accumulation in a Drinking Water Reservoir Based on the SWAT Model Driven by CMADS: A Case Study of the Biliuhe Reservoir in Northeast China. *Water* **2018**, *10*, 1535.
31. Clark, W.A.; Wang, G.A. Conflicting Attitudes Toward Inter-basin Water Transfers in Bulgaria. *Water Int.* **2003**, *28*, 79–89. [CrossRef]
32. Safavi, H.R.; Golmohammadi, M.H.; Sandoval-Solis, S. Expert knowledge based modeling for integrated water resources planning and management in the Zayandehrud River Basin. *J. Hydrol.* **2015**, *528*, 773–789. [CrossRef]
33. Wang, K.; Wang, Z.; Liu, K.; Cheng, L.; Wang, L.; Ye, A. Impacts of the eastern route of the South-to-North Water Diversion Project emergency operation on flooding and drainage in water-receiving areas: An empirical case in China. *Nat. Hazard Earth Sys.* **2019**, *19*, 555–570. [CrossRef]
34. Du, J.; Qian, L.; Rui, H.; Zuo, T.; Zheng, D.; Xu, Y.; Xu, C.Y. Assessing the effects of urbanization on annual runoff and flood events using an integrated hydrological modeling system for Qinhuai River basin, China. *J Hydrol* **2012**, *464*, 127–139. [CrossRef]
35. Pour, S.H.; Wahab, A.K.A.; Shahid, S.; Asaduzzaman, M.; Dewan, A. Low impact development techniques to mitigate the impacts of climate-change-induced urban floods: Current trends, issues and challenges. *Sustain. Cities Soc.* **2020**, *62*, 102373. [CrossRef]
36. Yin, J.; Yu, D.; Yin, Z.; Liu, M.; He, Q. Evaluating the impact and risk of pluvial flash flood on intra-urban road network: A case study in the city center of Shanghai, China. *J. Hydrol.* **2016**, *537*, 138–145. [CrossRef]

37. Hammond, M.J.; Chen, A.S.; Djordjević, S.; Butler, D.; Mark, O. Urban flood impact assessment: A state-of-the-art review. *Urban Water J.* **2015**, *12*, 14–29. [CrossRef]
38. Hu, C.; Liu, C.; Yao, Y.; Wu, Q.; Ma, B.; Jian, S. Evaluation of the Impact of Rainfall Inputs on Urban Rainfall Models: A Systematic Review. *Water* **2020**, *12*, 2484. [CrossRef]
39. Bulti, D.T.; Abebe, B.G. A review of flood modeling methods for urban pluvial flood application. *Modeling Earth Syst. Environ.* **2020**, *6*, 1293–1302. [CrossRef]
40. Jillo, A.Y.; Demissie, S.S.; Viglione, A.; Asfaw, D.H.; Sivapalan, M. Characterization of regional variability of seasonal water balance within Omo-Ghibe River Basin, Ethiopia. *Hydrol. Sci. J.* **2017**, *62*, 1200–1215. [CrossRef]
41. Ismaiylov, G.K.; Fedorov, V.M. Year to year variations in water balance components in the Volga Basin and their interaction. *Water Resour.* **2008**, *35*, 247–263. [CrossRef]
42. Chen, H.; Xu, C.; Guo, S. Comparison and evaluation of multiple GCMs, statistical downscaling and hydrological models in the study of climate change impacts on runoff. *J. Hydrol.* **2012**, *434*, 36–45. [CrossRef]
43. Razavi, T.; Coulibaly, P. Streamflow Prediction in Ungauged Basins: Review of Regionalization Methods. *J. Hydrol. Eng.* **2013**, *18*, 958–975. [CrossRef]
44. Hromadka, T.V., II; Whitley, R.J. Approximating Rainfall-Runoff Modelling Response Using a Stochastic Integral Equation. *Hydrol. Process.* **1996**, *10*, 1003–1019. [CrossRef]
45. Lee, M.; Kang, N.; Joo, H.; Kim, H.S.; Kim, S.; Lee, J. Hydrological Modeling Approach Using Radar-Rainfall Ensemble and Multi-Runoff-Model Blending Technique. *Water* **2019**, *11*, 850. [CrossRef]
46. Chen, H.; Yang, X. A Three-parameter Hydrological Model and Its Application in China. *J. China Hydrol.* **2015**, *35*, 17–21.
47. Xu, S.; Yang, X. Comparison between GR3 Model and Xin'anjiang Model in Application for Watersheds in China. *J. China Hydrol.* **2015**, *35*, 7–13.
48. Hu, C.; Guo, S.; Xiong, L.; Peng, D. A modified Xinanjiang model and its application in northern China. *Hydrol. Res.* **2005**, *36*, 175–192. [CrossRef]
49. Ren-Jun, Z. The Xinanjiang model applied in China. *J. Hydrol.* **1992**, *135*, 371–381. [CrossRef]
50. Liu, J.; Chen, X.; Zhang, J.; Flury, M. Coupling the Xinanjiang model to a kinematic flow model based on digital drainage networks for flood forecasting. *Hydrol. Process.* **2009**, *23*, 1337–1348. [CrossRef]
51. Zhuo, L.; Han, D. Misrepresentation and amendment of soil moisture in conceptual hydrological modelling. *J. Hydrol.* **2016**, *535*, 637–651. [CrossRef]
52. Tran, Q.Q.; De Niel, J.; Willems, P. Spatially Distributed Conceptual Hydrological Model Building: A Generic Top-Down Approach Starting from Lumped Models. *Water Resour. Res.* **2018**, *54*, 8064–8085. [CrossRef]
53. Guo, W.; Wang, C.; Ma, T.; Zeng, X.; Yang, H. A distributed Grid-Xinanjiang model with integration of subgrid variability of soil storage capacity. *Water Sci. Eng.* **2016**, *9*, 97–105. [CrossRef]
54. Guo, W.; Wang, C.; Zeng, X.; Ma, T.; Yang, H. Subgrid Parameterization of the Soil Moisture Storage Capacity for a Distributed Rainfall-Runoff Model. *Water* **2015**, *7*, 2691–2706. [CrossRef]
55. Zhang, Z.; Liu, Y.; Zhang, F.; Zhang, L. Study on dynamic relationship of spring water in Jinan spring area based on gray relational analysis. *IOP Conf. Ser. Earth Environ. Sci.* **2018**, *128*, 12068. [CrossRef]
56. Wang, L.; Chen, C. Carrying Capacity Assessment of Water Environment in Jinan. *Enuivonmental Sci. Technol.* **2011**, *34*, 199–202.
57. Wu, Y.; Ma, Z.; Li, X.; Sun, L.; Sun, S.; Jia, R. Assessment of water resources carrying capacity based on fuzzy comprehensive evaluation–Case study of Jinan, China. *Water Supply* **2020**, *21*, 513–524.59. [CrossRef]
58. Li, Y.; Xiong, W.; Zhang, W.; Wang, C.; Wang, P. Life cycle assessment of water supply alternatives in water-receiving areas of the South-to-North Water Diversion Project in China. *Water Res.* **2016**, *89*, 9–19. [CrossRef] [PubMed]
59. Li, C.; Wang, Y.; Zhou, B. The Yellow River:A key of Eco-City Construction in Jian City in New Century. *Bull. Soil Water Conserv.* **2004**, *24*, 68–71, 78.
60. Cheng, T.; Xu, Z.; Hong, S.; Song, S.; Zhou, J.G. Flood Risk Zoning by Using 2D Hydrodynamic Modeling: A Case Study in Jinan City. *Math Probl. Eng.* **2017**, *2017*, 5659197. [CrossRef]
61. Zhao, Y.; Xia, J.; Xu, Z.; Zou, L.; Qiao, Y.; Li, P. Impact of Urban Expansion on Rain Island Effect in Jinan City, North China. *Remote Sens.* **2021**, *13*, 2989. [CrossRef]
62. Xu, J.; Bi, B.; Shen, Y. Analysis on Mesoscale Mechanism of Heavy Rainstorm in Jinan on 18 July 2007. *Plateau Meteorol.* **2010**, *29*, 1218–1229.
63. Chang, X.; Xu, Z.; Zhao, G.; Cheng, T.; Song, S. Spatial and temporal variations of precipitation during 1979–2015 in Jinan City, China. *J. Water Clim. Change* **2017**, *9*, 540–554. [CrossRef]
64. Wang, C.X.; Liu, L.Y. Empirical Research on the Impact to City Climate Caused by Urbanization–A Case of Jinan City. *Appl. Mech. Mater.* **2013**, *295*, 2669–2674. [CrossRef]
65. Xu, Q.Y.; Wang, W.P.; Deng, H.Y. Study on Ecological Effect of Urban Landscape Water in Jinan, Shandong Province. *Appl. Mech. Mater.* **2014**, *675*, 826–829. [CrossRef]
66. Hawker, L.; Uhe, P.; Paulo, L.; Sosa, J.; Savage, J.; Sampson, C.; Neal, J. A 30 m global map of elevation with forests and buildings removed. *Environ Res Lett* **2022**, *17*, 24016. [CrossRef]
67. Perrin, C.; Michel, C.; Andréassian, V. Improvement of a parsimonious model for streamflow simulation. *J. Hydrol.* **2003**, *279*, 275–289. [CrossRef]

68. Edijatno; Nascimento, N.D.; Yang, X.L.; Makhlouf, Z.; Michel, C. GR3J: A daily watershed model with three free parameters. *Hydrolog. Sci. J.* **1999**, *44*, 263–277. [CrossRef]
69. Borchani, H.; Chaouachi, M.; Ben Amor, N. Learning causal Bayesian networks from incomplete observational data and interventions. In Proceedings of the 9th European Conference on Symbolic and Quantitative Approaches to Reasoning with Uncertainty, Hammamet, Tunisia, 31 October–2 November 2007; Mellouli, K., Ed.; 2007; Volume 4724, p. 17.
70. Li, Y.; Ma, B.; Peng, X. Research on Hydraulic Factors of Water Conveyance Buried Culvert in Jinan Urban Section of Eastern Route of South-North Water Diversion. *China Water Wastewater* **2014**, *30*, 58–61.
71. Dong, N.; Tian, A.; Jiang, F. A Study on Strategy for Sustainable Use of Water Resources in Jinan City. *Shanghai Environ. Sci.* **2010**, *29*, 169–173.
72. Liu, Y.; Zhang, Z.; Zhang, F.; Liu, B. Coupling Correlation Measure and Prospect Forecast of Water Resources Environment and Economic Development in Jinan. *J. Yangtze River Sci. Res. Inst.* **2020**, *37*, 28–33.
73. Gu, J.; Liu, H.; Wang, S.; Zhang, M.; Liu, Y. An innovative anaerobic MBR-reverse osmosis-ion exchange process for energy-efficient reclamation of municipal wastewater to NEWater-like product water. *J. Clean Prod.* **2019**, *230*, 1287–1293. [CrossRef]
74. Villamar, C.; Vera-Puerto, I.; Rivera, D.; De la Hoz, F. Reuse and Recycling of Livestock and Municipal Wastewater in Chilean Agriculture: A Preliminary Assessment. *Water* **2018**, *10*, 817. [CrossRef]
75. Ren, Y.; Su, X.; He, Y.; Wang, X.; Ouyang, Z. Urban water resource utilization efficiency and its influencing factors in ecogeographic regions of China. *Acta Ecol. Sin.* **2020**, *40*, 6459–6471.
76. Knoben, W.; Freer, J.E.; Woods, R.A. Technical note: Inherent benchmark or not? Comparing Nash-Sutcliffe and Kling-Gupta efficiency scores. *Hydrol. Earth Syst. Sc.* **2019**, *23*, 4323–4331. [CrossRef]
77. Nash, J.E.; Sutcliffe, I.V. River flow forecasting through conceptual models. *J. Hydrol.* **1970**, *10*, 282–290. [CrossRef]
78. Wang, Y.; Yang, X. A Coupled Hydrologic-Hydraulic Model (XAJ-HiPIMS) for Flood Simulation. *Water* **2020**, *12*, 1288. [CrossRef]

Article

Identification of Time-Varying Parameters of Distributed Hydrological Model in Wei River Basin on Loess Plateau in the Changing Environment

Haizhe Wu [1,2], Dengfeng Liu [2,*], Ming Hao [1,*], Ruisha Li [1], Qian Yang [3], Guanghui Ming [4] and Hui Liu [5]

1 The Second Monitoring and Application Center, China Earthquake Administration, Xi'an 710054, China
2 State Key Laboratory of Eco-Hydraulics in Northwest Arid Region, School of Water Resources and Hydropower, Xi'an University of Technology, Xi'an 710048, China
3 Shaanxi Water Resources and Reservoir Dispatching Center, Xi'an 710004, China
4 Key Laboratory of Water Management and Water Security for Yellow River Basin (Ministry of Water Resources), Yellow River Engineering Consulting Co., Ltd., Zhengzhou 450003, China
5 China Institute of Water Resources and Hydropower Research, Beijing 100038, China
* Correspondence: liudf@xaut.edu.cn (D.L.); ha_mg@163.com (M.H.)

Abstract: In the watershed hydrological model, the parameters represent the characteristics of the watershed. Usually, the parameters are assumed to be constant in the stable environment. However, in the changing environment, the parameters may change and the constant parameters would not represent the change of the characteristics of the runoff generation and routing in the watershed. The identification of the time-varying characteristics of the watershed hydrological model parameters will help to improve the performance of the simulation and prediction of hydrological models in changing environments. Based on the measured data at the ground stations in the Wei River Basin on the Loess Plateau in China, the temporal and spatial evolution of the ecohydrological and meteorological factors was analyzed, and the SWAT model was used to identify the relationship between the model parameters and the factors, such as precipitation, potential evapotranspiration, NDVI and the other environmental characterization factors of the river basin. The results showed that the annual precipitation in the basin showed a decreasing trend, and the annual potential evapotranspiration, the annual average temperature, the annual runoff and the annual average NDVI all showed an increasing trend. The model parameters fluctuated with time during the study period. The change of the soil evaporation compensation coefficient (ESCO) was similar with the annual potential evapotranspiration, and the model parameters all showed a certain correlation with the potential evaporation of the basin, which indicates that the changes of the hydrological model parameters in the upper reach of the Wei River are closely related to the changes of the basin potential evapotranspiration. Potential evapotranspiration is a characterization factor for dynamic changes of the hydrological model parameters in the upper reach of the Wei River.

Keywords: hydrological simulation; time-varying parameters; potential evapotranspiration; Loess Plateau

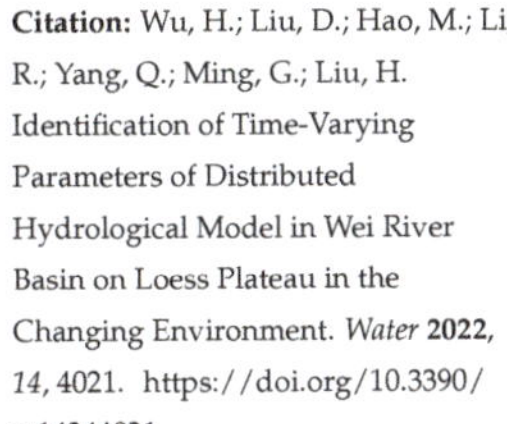

Citation: Wu, H.; Liu, D.; Hao, M.; Li, R.; Yang, Q.; Ming, G.; Liu, H. Identification of Time-Varying Parameters of Distributed Hydrological Model in Wei River Basin on Loess Plateau in the Changing Environment. *Water* **2022**, *14*, 4021. https://doi.org/10.3390/w14244021

Academic Editor: Fi-John Chang

Received: 18 October 2022
Accepted: 5 December 2022
Published: 9 December 2022

1. Introduction

In the context of global climate change, the applicability of traditional runoff simulation and forecasting methods has gradually deteriorated, which brings challenges to hydrometeorological simulation and forecasting. The hydrological simulation of the river basin under the changing environment is mainly affected by the climatic conditions and the underlying surface conditions [1]. Climatic conditions are the driving factors of the water cycle in the basin and the prerequisite for the generation of runoff [2], which directly or indirectly affect the runoff process in the basin through changes in factors such as precipitation, evapotranspiration and temperature [3]. Global warming has become an indisputable fact [4]. The rise in temperature will cause changes in other meteorological elements. Significant changes have taken place in the type, intensity and amount of precipitation in many regions of the world, generally showing that humid regions are becoming more humid, arid regions tend to become more arid and the interannual variability is significantly enhanced [5]. In addition, climatic conditions indirectly affect the hydrological process of the watershed by affecting the growth state of vegetation and its structure [6]. The increase and decrease in watershed runoff is inseparable from the change of the underlying surface conditions. With the promotion of large-scale afforestation and urbanization, surface vegetation coverage conditions and local water and heat flux transfer have undergone dramatic changes [7–10], as well as large-scale water intake [11], water diversion projects and reservoir construction, dispatch operation [12], and so on all leading to sudden changes in natural river runoff. It indicates that the "steady-state" basin assumption in traditional hydrological simulation is facing challenges, so the theory and method of hydrological probability distribution based on the consistency assumption obviously cannot help people accurately reveal the long-term law of water resources and flood evolution in changing environments [13,14]. In the hydrological model, it is generally assumed that the parameters representing the hydrological physical characteristics of the watershed are constant over time, which not only cannot reflect the watershed characteristics correctly, but also seriously weakens the simulation ability of the model. The identification and study of the time-varying characteristics of hydrological model parameters in "unsteady" watersheds, and the establishment of a model parameter estimation method that can reflect the climatic conditions of the watershed and the changing laws of the underlying surface, will improve the simulation and performance of hydrological models in changing environments.

The parameters of the hydrological model are usually closely related to the underlying surface conditions of the watershed, and reflect the hydrological characteristics of the watershed [15]. Since there are significant differences in climatic conditions, geographic locations, vegetation coverage, soil conditions, topography and geological conditions in different watersheds [16,17], the parameter values of the same model will also be very different in different watersheds [18]. In the beginning, the traditional watershed hydrological simulation only considers the spatial variability of the parameters of the hydrological model, but does not consider the dynamic changes of the parameters. Usually, the model parameters are considered to be static and remain unchanged over time in a given period of time. However, under the changing environment, the changes in the characteristics of watershed hydrology are not adequately reflected in the model and such assumptions may no longer be applicable. Many scholars began to question such static assumptions and carried out related researches. Kingumbi et al. [19] simulated the hydrological effect of land-use changes by the MODCOU model in the Merguellil basin in central Tunisia and evaluated the improvement in model representativeness by assigning specific parameters to the production functions in the zones of works of water and soil conservation. Vaze et al. [20] used four different conceptual hydrological models in 61 watersheds in southeastern Australia, and applied the method of segmental calibration to determine the parameter values of the model in each time period, and it was found that climate change during the study period would cause significant dynamic changes of parameters. In 273 watersheds in Austria, Merz et al. [21] identified the dynamic changes of parameters based on the HBV model and found that the changes of model parameters had a strong correlation with environmental meteorological factors such as rainfall, runoff coefficient and potential evapotranspiration. Sun [22] used the THREW model in the upper reach of the Han River, identified that all model parameters had a good correlation with the vegetation index, and proposed a dynamic parameter estimation method based on vegetation coverage, which improved the simulation effect of the model. Under the conditions of increasingly significant changes in the basin environment, the constant model parameters over time will be an important source of simulation errors [23]. Considering the dynamic changes of parameters can significantly improve the simulation effect of the hydrological model for the middle and low water sections of the runoff process [24].

The former studies qualitatively pointed out that the change of river basin environmental factors will cause dynamic changes in model parameters, but did not establish the quantitative relationship between the factors and the parameters. They mostly focused on the influence of climate variability on the parameters, and ignored the changes in the underlying surface condition. In theory, the underlying surface conditions should have a stronger correlation with the model parameters. Therefore, it is necessary to comprehensively analyze the influence of meteorological factors in the watershed and the changes of the underlying surface conditions on the model parameters, establish a time-varying parameter estimation method based on the watershed characteristics, and further enhance the ability of the hydrological model to reflect the changes of watershed characteristics, in order to significantly improve the watershed characteristics. The study will improve the simulation and prediction effects of hydrological models in changing environments. This study takes the upper reach of the Wei River as the study area, analyzes the temporal and spatial evolution of the ecological hydrometeorological elements in the study area, reveals the temporal and spatial variation of these historical sequences and provides a basis for hydrological simulation. The SWAT model is set up in the study area and the time-varying parameter sequence of the hydrological model is obtained by segmental calibration, and the relationship between the model parameters and environmental characterization factors such as precipitation, potential evapotranspiration and the normalized difference vegetation index (NDVI) is analyzed, which provides a preliminary solution for the improvement of the distributed hydrological model. In Section 2, the study area and data are introduced. In Section 3, the methods and hydrological model are described. In Section 4, the simulation results are displayed. Finally, a conclusion is drawn in Section 5.

2. Study Area and Data

2.1. Study Area

The study area is the upper reach of the Beidao Hydrological Station in the Wei River (Figure 1). The study area is located at 103°97′–106°42′ E and 34°17′–36°19′ N, with a catchment area of about 25,000 km^2, accounting for about 19% of the total area of the Wei River Basin. The study area is a Loess hilly and gully area with an altitude of 1079–3934 m. The climate is a continental monsoon climate, with a cold and dry winter and a hot and rainy summer. The average annual precipitation from 1965 to 2017 was 491 mm, and the precipitation was unevenly distributed throughout the year, mostly concentrated in July-September. The annual average temperature is 7.8–13.5 °C, the extreme maximum temperature is 42.8 °C, the extreme minimum temperature is −28.1 °C and the annual potential evapotranspiration is 700–1200 mm.

2.2. Data

In this study, the ground-measured meteorological data include the daily data of 6 meteorological stations and 16 rainfall stations in the study area. The data of the meteorological stations include daily rainfall, temperature, sunshine duration, relative humidity, air pressure and average wind speed. The measured daily runoff data from 2001 to 2017 come from the Beidao Hydrological Station. The location of the stations is shown in Figure 1. The measured flow data and meteorological data come from the *Yellow River Basin Hydrological Yearbook* and the China Meteorological Data Network, respectively. The vegetation data from 2001 to 2017 were derived from the NDVI dataset provided by the MODIS/Terra website. The dataset was the MOD13Q1 product with a spatial resolution of 250 m and a temporal resolution of 16 days.

The DEM data used to build the hydrological model came from the ASTER GDEM data provided by Geospatial Data Cloud (http://www.gscloud.cn, accessed on: 1 October 2022), with a spatial resolution of 30 m. The land use data in 2010 came from the Resource and Environment Data Center of the Chinese Academy of Sciences (http://www.resdc.cn, accessed on: 1 October 2022), and the soil attribute data came from the Harmonized World Soil Database (HWSD), with a spatial resolution of 1000 m. The detailed model construction steps refer to Wu et al. [25].

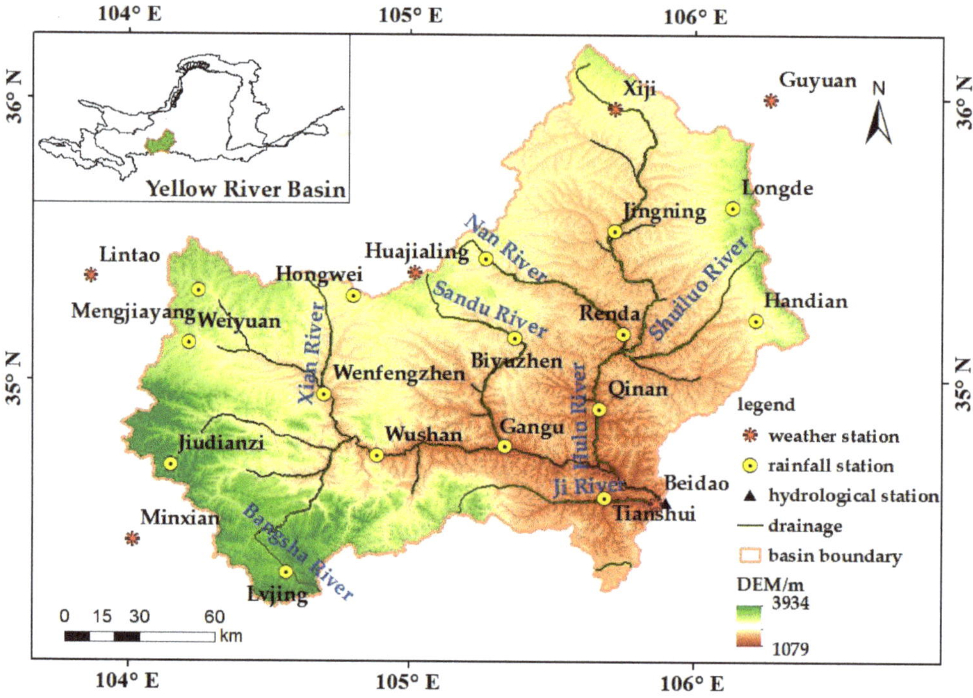

Figure 1. Location of the study area and the stations.

3. Methods

Since there is a lack of recorded solar radiation data at all meteorological stations in the study area, the solar radiation in the watershed was approximated by the number of sunshine hours. The measured meteorological data collected in this study are relatively comprehensive, and the Penman–Monteith method was chosen for the calculation of potential evapotranspiration. For the whole basin average precipitation, potential evapotranspiration, temperature, runoff, vegetation index and other variables in the study area, the linear regression method and the cumulative anomaly method were used to analyze the trend and abrupt change, respectively. The linear regression method is adopted to construct a univariate linear regression equation between the hydrometeorological variable sequence and the time series. The regression coefficient b can intuitively reflect the changing rate of the hydrometeorological sequence, and its positive and negative values indicate the increasing or decreasing trend of the hydrometeorological sequence, respectively. The t-test method ($a = 0.05$) is used to test the significance of the regression coefficient of the equation; the significance of the change trend of the hydrometeorological sequence is evaluated [26]. The formula of linear regression is as follows:

$$x(t) = a + bt \tag{1}$$

where, x is the variable, t is time series, a and b are parameters.

The cumulative anomaly method is a method for evaluating the trend of change according to the increase and decrease in the cumulative anomaly curve. The long-term evolution trend of the hydrometeorological sequence and the approximate time of the sudden change can be evaluated [27]. The formula to calculate the cumulative anomaly value cd at time t is:

$$cd = \sum_{t=1}^{n} (x(t) - x_{mean}) \tag{2}$$

where, $x(t)$ is the variable at t and x_{mean} is the mean of x.

The SWAT model is a physically based distributed hydrological model and is widely used to simulate the hydrological process of the watershed all over the world. It has great advantages in simulating long-term continuous hydrological processes in large-scale complex watersheds. Therefore, since the model was launched, it has been used in Europe, North America, Canada, Asia and other regions with good results [28,29]. The basin SWAT model was constructed, and the DEM data of the basin were used for catchment analysis and river network extraction. The minimum catchment area threshold was set to 400 km^2, and the study area was divided into 39 sub-basins. Sensitivity analysis was performed using the SUFI-2 algorithm, in which six sensitive parameters were closely related to runoff, such as the number of runoff curves (CN2), the effective hydraulic conductivity of the main channel bed (CH_K2), the soil evaporation compensation coefficient (ESCO), the baseflow α factor of riparian regulation and storage (ALPHA_BNK), the Manning coefficient of the main channel (CH_N2) and the saturated hydraulic conductivity of the soil (SOL_K). Since CH_K2 and SOL_K are parameters that characterize the soil characteristics of the basin, they are not suitable to establish a relationship with the meteorological characteristics, and only CN2, ESCO, ALPHA_BNK and CH_N2 are analyzed.

The Nash–Sutcliffe efficiency coefficient (NSE), the coefficient of determination (R^2) and the Kling–Gupta coefficient (KGE) [30–33] are selected to evaluate the simulated daily flow. The formulas are as follows:

$$\text{NSE} = 1 - \frac{\sum_{i=1}^{n}\left(Q_{o,i} - Q_{s,i}\right)^2}{\sum_{i=1}^{n}\left(Q_{o,i} - \overline{Q}_o\right)^2} \tag{3}$$

$$R^2 = \frac{\sum_{i=1}^{n}\left[\left(Q_{s,i} - \overline{Q}_s\right)\left(Q_{o,i} - \overline{Q}_o\right)\right]^2}{\sum_{i=1}^{n}\left(Q_{o,i} - \overline{Q}_o\right)^2 \sum_{i=1}^{n}\left(Q_{s,i} - \overline{Q}_s\right)^2} \tag{4}$$

$$\text{KGE} = 1 - \sqrt{(r-1)^2 + (\beta - 1)^2 + (\gamma - 1)^2} \tag{5}$$

$$\beta = \frac{\mu_s}{\mu_o}, \gamma = \frac{\sigma_s/\mu_s}{\sigma_o/\mu_o} \tag{6}$$

where, n represents the length of the runoff sequence; Q_s and Q_o represent the simulated flow and measured flow; μ and σ are the mean and variance, respectively; r is the linear correlation coefficient between the simulated and measured values. The closer NSE and R^2 are to 1, the better the simulation effect is. The KGE coefficient is a comprehensive index including Cv, correlation coefficient and mean value. The closer the value is to 1, the higher the simulation accuracy.

4. Results

4.1. Spatial Distribution of Meteorological Factors

According to the precipitation, air temperature and potential evapotranspiration data calculated by the Penman–Monteith formula from 1965 to 2017 at the six meteorological stations in the upper reach of the Wei River, the spatial distribution of the annual average precipitation, temperature and potential evapotranspiration was plotted by inverse distance weighting interpolation (IDW), as shown in Figure 2. It shows that the annual average precipitation in the basin has obvious differences in spatial distribution, and generally shows a decreasing trend from the southwest of the basin to the northeast. The spatial distribution of the average temperature in the basin shows an increasing trend from the northwest to the southeast. The temperature in about two-thirds of the area is between 6 and 11 °C. From the perspective of the river flow, the high temperature area is located at the outlet of the downstream watershed, and the temperature in the middle and upper reaches of the watershed is lower. It is closely related to the changes of elevation in the basin. The spatial distribution of the annual average potential evapotranspiration in the basin is extremely uneven, and the overall distribution pattern is that the potential evapotranspiration in the northern part of the basin is slightly larger than that in the southern part of the basin. The spatial distributions of potential evapotranspiration and precipitation are basically opposite; the area with the largest potential evapotranspiration has less precipitation and, on the contrary, the area with the smallest potential evapotranspiration has more precipitation.

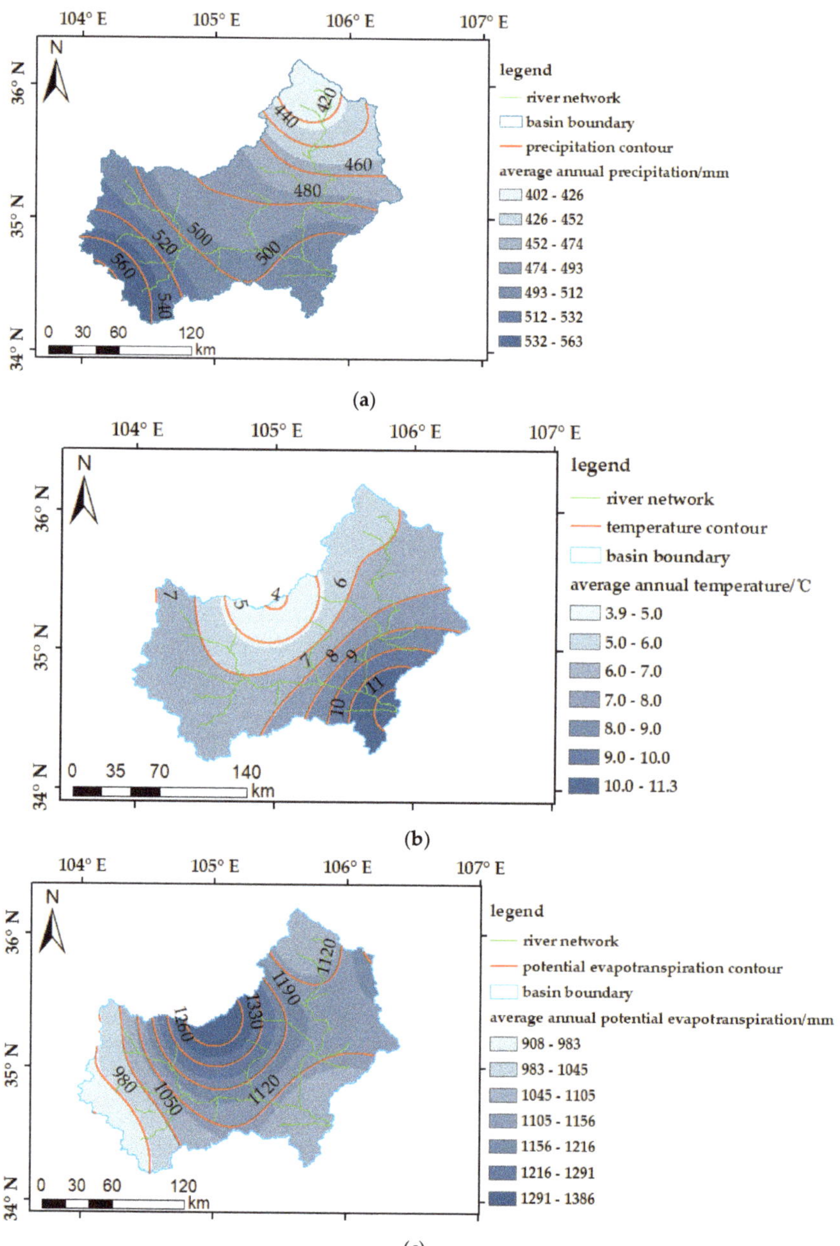

Figure 2. Spatial distribution map of annual average precipitation, temperature and potential evapotranspiration. (**a**) precipitation, (**b**) temperature, (**c**) potential evapotranspiration.

4.2. Interannual Changes of the Environmental Factors

The linear regression method was used to analyze the trend of the annual precipitation, temperature, potential evapotranspiration and runoff depth series at the whole basin scale in the study area from 1965 to 2017, and the significance of the trend results was further tested. The results are shown in Figure 3. It shows that the annual precipitation in the basin shows an insignificant decreasing trend, with a decreasing rate of 0.59 mm/a. The mean annual precipitation in the basin was 491.37 mm, the annual precipitation reached the maximum in 1967, about 704.04 mm, and the minimum in 1982. The annual average temperature in the basin shows a significant growth trend, with an increasing rate of 0.03 °C/a. The mean annual temperature in the basin is 6.62 °C. The annual average temperature reached the maximum in 2016, about 7.98 °C, and was the minimum in 1967. The annual potential evapotranspiration in the basin showed a significant increasing trend, with an increasing rate of 1.87 mm/a. The mean annual potential evapotranspiration in the basin was 1150.72 mm, and the annual potential evapotranspiration reached the maximum in 2016, about 1280.31 mm, and was the minimum in 1967. The annual runoff depth of the Beidao Hydrological Station showed an insignificant trend with an increasing rate of 0.42 mm/a. The mean annual runoff depth of the Beidao Hydrological Station was 26.20 mm, and the annual runoff depth reached the maximum in 2013, about 49.01 mm, and was the minimum in 2002.

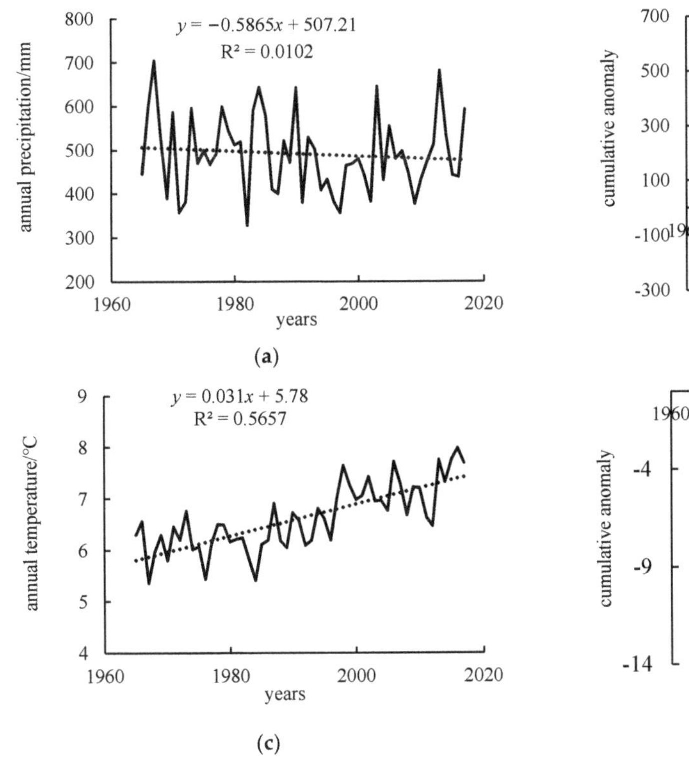

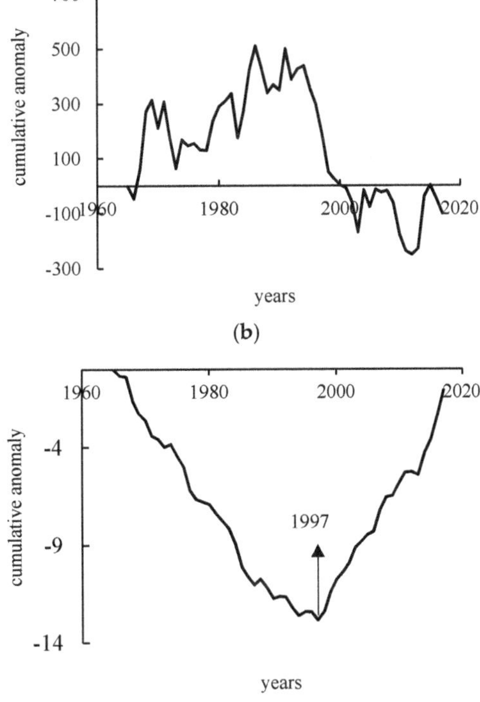

Figure 3. *Cont.*

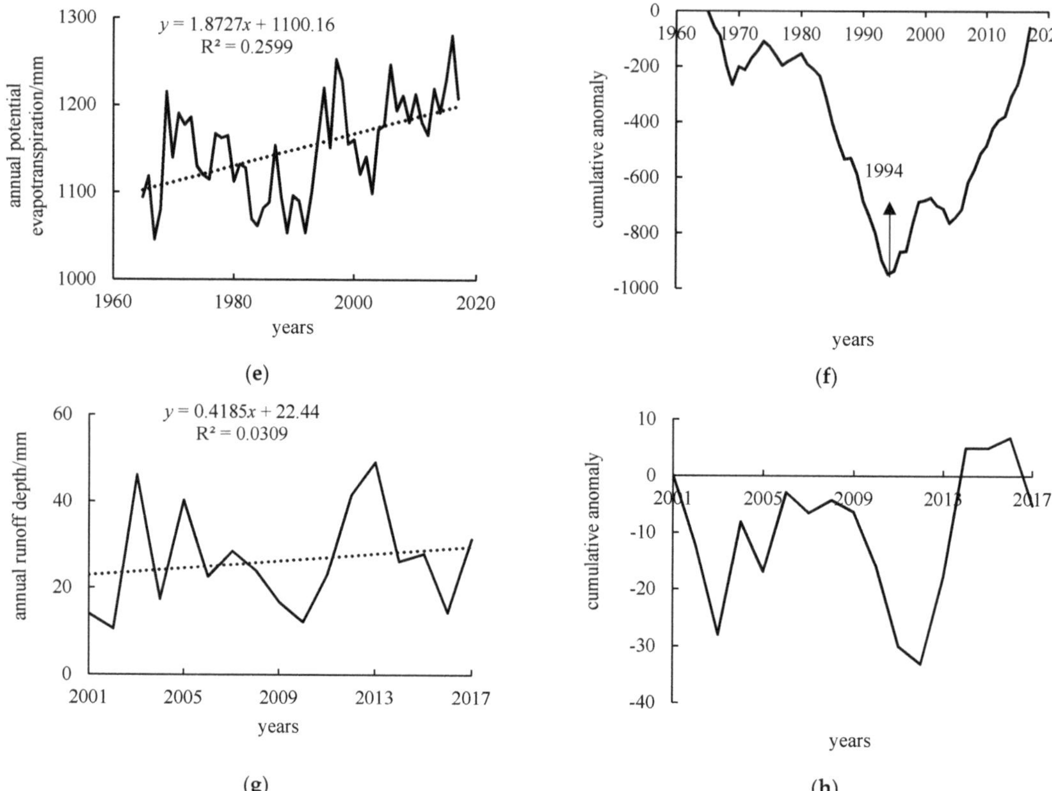

Figure 3. Analysis of trends and abrupt changes of hydrological and meteorological elements in the basin. Note: x = year−1964. (**a**) Annual precipitation trend, (**b**) Analysis of abrupt change in annual precipitation, (**c**) Annual temperature trend, (**d**) Analysis of abrupt change in annual temperature, (**e**) Annual potential evapotranspiration trend, (**f**) Analysis of abrupt change in annual potential evapotranspiration, (**g**) Annual runoff depth trend, (**h**) Analysis of abrupt change in annual runoff depth.

The cumulative anomaly method was used to analyze the mutation of the annual precipitation, temperature, potential evapotranspiration and runoff depth series in the study area from 1965 to 2017. The results are shown in Figure 3. It was identified that the annual precipitation and runoff depth series did not have abrupt changes, the annual mean temperature series had a significant jump around 1997, and the annual potential evapotranspiration series had a significant jump around 1994. The simulation period is 2001–2017 and after the abrupt changes.

Figure 4 shows the changing process of the annual average NDVI series in the study area from 2001 to 2017. It shows that although the average annual NDVI value of the watershed fluctuated during the 17 years, the overall trend is significant increase, which indicates that the vegetation coverage in the study area during the 17 years was increasing continuously. Especially after 2011, the vegetation condition significantly improved. The annual NDVI value of the watershed was between 0.30 and 0.38 from 2001 to 2017. The average annual NDVI value of the watershed increased by 19.7%, and reached the maximum value of about 0.38 in 2013. Figure 5 shows the seasonality of NDVI on the monthly scale in the study area from 2001 to 2017. It shows that the vegetation change in the study area has a significant seasonality across the year. The maximum monthly NDVI in the study area generally occurs from June to August, while the minimum monthly NDVI generally occurs from January to March.

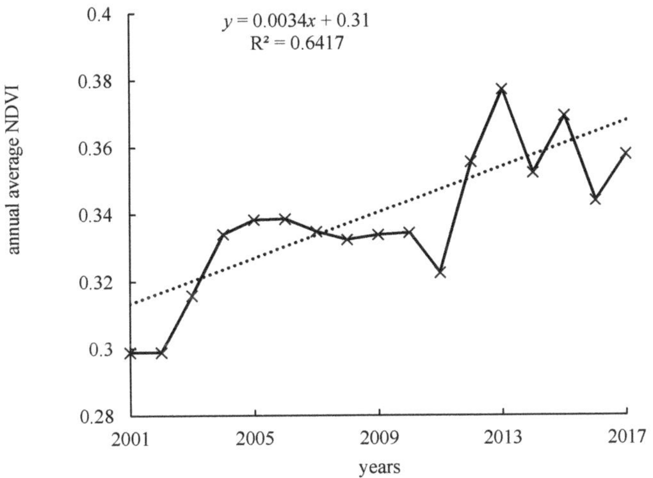

Figure 4. Annual average NDVI in the basin. Note: x = year−2000.

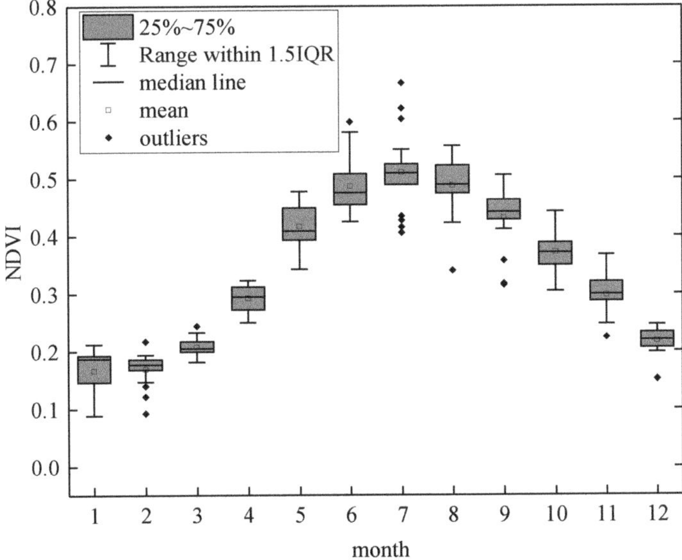

Figure 5. Seasonality of Monthly NDVI.

4.3. Simulation of the Hydrological Process

According to the mutation analysis results of the hydrometeorological data, 2001–2017 was selected as the simulation period of the model, 2001 was set as the warm-up period of the model, 2002–2013 was the calibration period, and 2014–2017 was the verification period. Sensitivity analysis was carried out on the model parameters, among which there are six parameters closely related to runoff, namely the number of runoff curves (CN2), the effective hydraulic conductivity of the main channel bed (CH_K2), the soil evaporation compensation coefficient (ESCO), and the baseflow α factor of riparian regulation and storage (ALPHA_BNK), the Manning coefficient of the main channel (CH_N2) and the saturated hydraulic conductivity of the soil (SOL_K).

The daily runoff at the Beidao Hydrological Station is simulated, and the measured and simulated flow processes in the calibration and verification periods are shown in Figure 6, and the evaluation results of the daily flow simulation are shown in Table 1. It shows that the simulated daily flow process in the calibration period of the Beidao hydrological station is quite good, and the three evaluation indicators are all above 0.54, of which the KGE index reaches 0.70, and the simulation effect is poor in the verification period except for 2014. In general, the SWAT model has been successfully constructed in the study area, and has good applicability in the watershed. Further research is carried out based on this model.

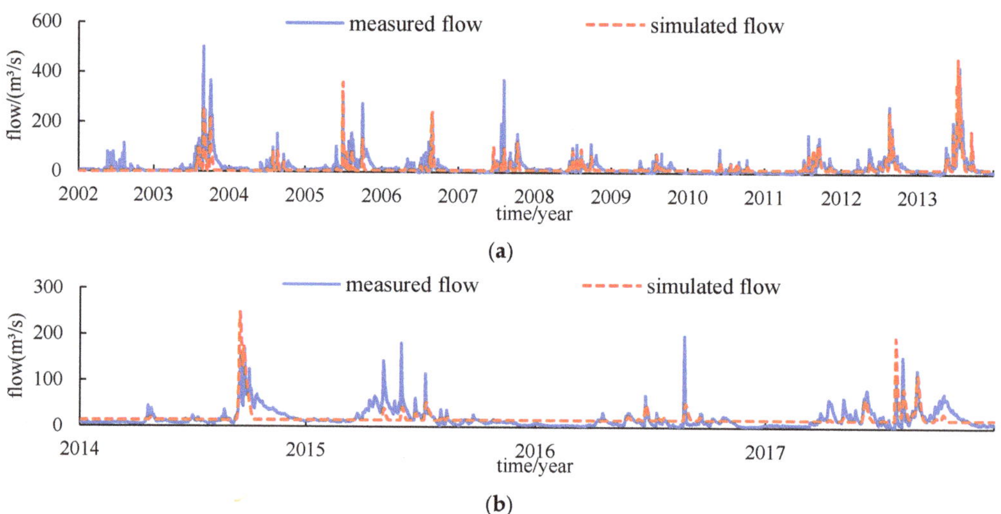

Figure 6. Results of simulated and observed streamflow in calibration period and validation period. (**a**) calibration period. (**b**) verification period.

Table 1. Evaluation result of daily flow simulation value.

Period	R^2	NSE	KGE
Calibration period (2002–2013)	0.58	0.54	0.70
Verification period (2014–2017)	0.31	0.19	0.54

4.4. Time-Varying Parameters of the Hydrological Model

In order to obtain the time-varying parameters of the SWAT model, the SUFI-2 algorithm was used to calibrate the model parameters in each year from 2001 to 2017, and the previous year of each calibration period was used as the model warm-up period. Then, the daily runoff process of the study area in each year was simulated by the SWAT model with one parameter set of this year. The Nash–Sutcliffe efficiency coefficient, R^2 and KGE were used to evaluate the simulation accuracy of the runoff in each period. The results are shown in Table 2. It shows that the daily runoff simulation accuracy in the other 10 years met the requirements, except for the poor simulation in 2002, 2004, 2008, 2009, 2016 and 2017.

Table 2. Evaluation result of simulated daily flow of each year.

Year	R^2	NSE	KGE
2002	0.33	0.26	0.53
2003	0.74	0.73	0.77
2004	0.34	0.34	0.44
2005	0.56	0.53	0.71
2006	0.56	0.55	0.67
2007	0.42	0.42	0.52
2008	0.37	0.33	0.55
2009	0.30	0.15	0.55
2010	0.61	0.59	0.67
2011	0.53	0.52	0.58
2012	0.68	0.66	0.82
2013	0.64	0.62	0.79
2014	0.50	0.42	0.70
2015	0.45	0.44	0.62
2016	0.36	0.35	0.47
2017	0.29	0.12	0.53

The hydrological model was calibrated year by year in 1-year increments. Four time-dependent model parameter series, CN2, ESCO, ALPHA_BNK and CH_N2, were obtained for the study area. The relationship between model parameters and environmental characterization factors, i.e., precipitation, potential evapotranspiration and NDVI in the basin is plotted, as shown in Figures 7–10. It shows that the four parameters are fluctuating in different degrees during the study period. There was no obvious trend in CN2 and CH_N2. The soil evaporation compensation coefficient (ESCO) and ALPHA_BNK showed a downward trend, but the magnitude of change in ALPHA_BNK was small and negligible. The smaller the value, the more evapotranspiration water can be obtained from the lower soil layer, which is consistent with the conclusion that the watershed evaporation is increasing year by year, as shown in Figure 8.

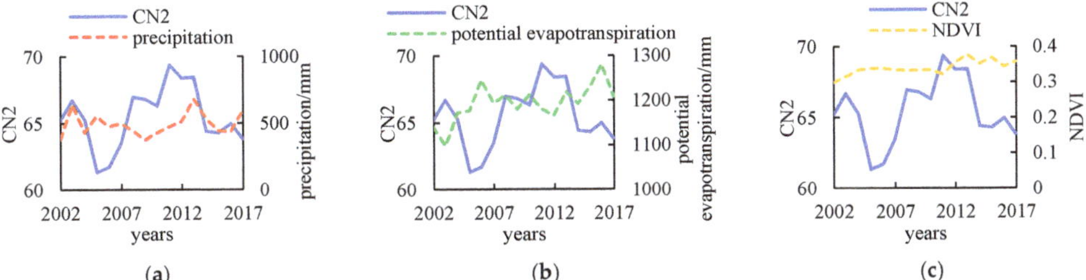

Figure 7. Time-varying process of CN2 to precipitation, potential evapotranspiration and NDVI. (**a**) Precipitation, (**b**) potential evapotranspiration, (**c**) NDVI.

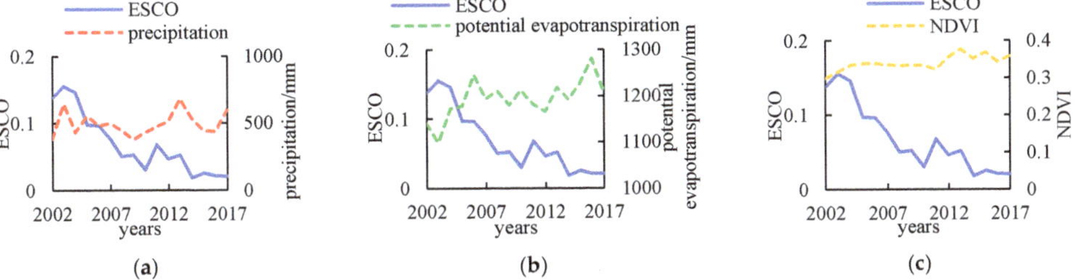

Figure 8. Time-varying process of ESCO to basin precipitation, potential evapotranspiration and NDVI. (**a**) Precipitation, (**b**) potential evapotranspiration, (**c**) NDVI.

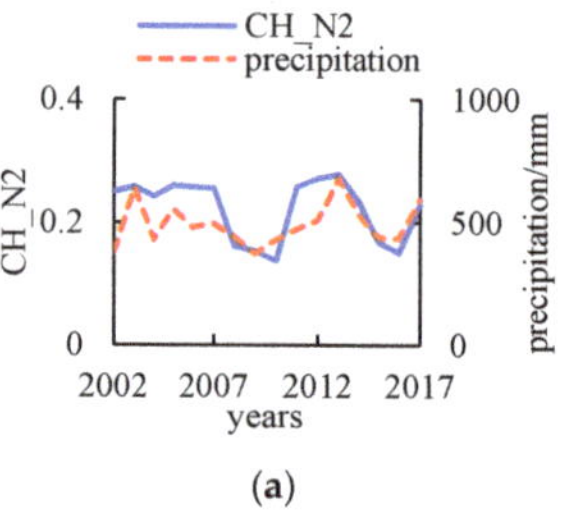

(a)

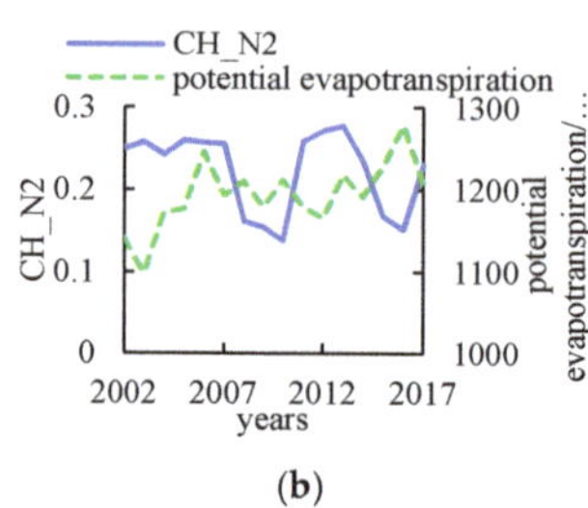

(b)

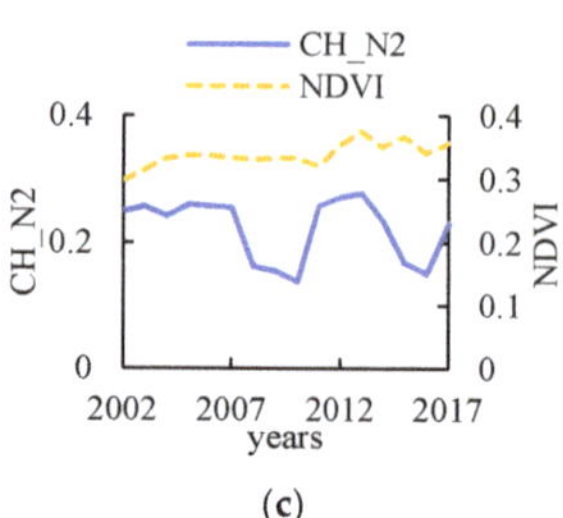

(c)

Figure 9. Time-varying process of CH_N2 to basin precipitation, potential evapotranspiration and NDVI. (**a**) Precipitation, (**b**) potential evapotranspiration, (**c**) NDVI.

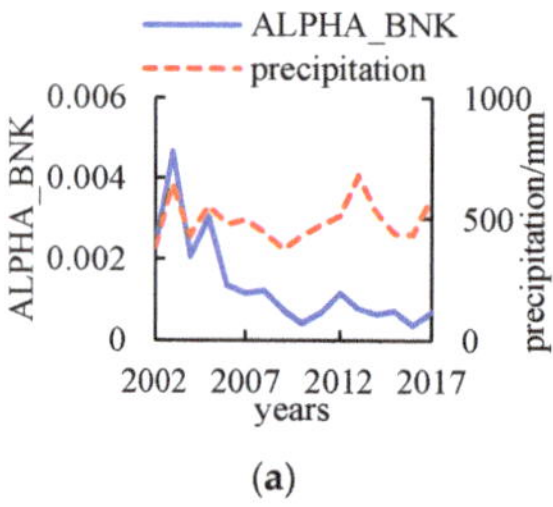

(a)

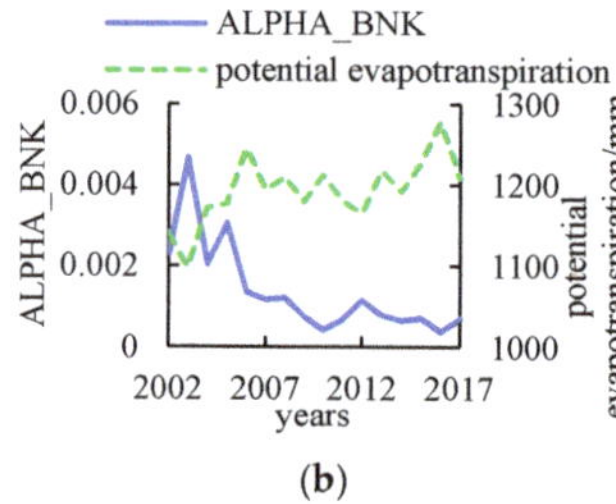

(b)

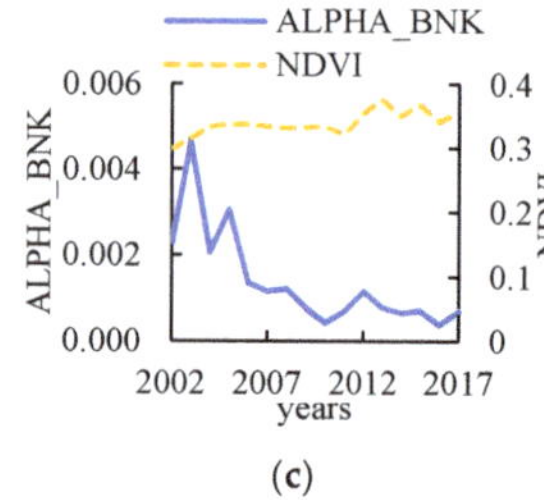

(c)

Figure 10. Time-varying process of ALPHA_BNK to basin precipitation, potential evapotranspiration and NDVI. (**a**) Precipitation, (**b**) potential evapotranspiration, (**c**) NDVI.

4.5. Analysis of the Relationship between Model Parameters and Environmental Factors

The scatter plots of the hydrological model parameters and environmental representation factors of precipitation, potential evapotranspiration and NDVI were shown to identify the relationship between them, and the results are shown in Table 3, Figures 11–14. Figure 11 shows that there is a weak negative correlation between CN2 and potential evapotranspiration. The number of runoff curves (CN2) changes inversely with the change of potential evapotranspiration, and has no obvious correlation with precipitation and NDVI. In Figure 12, ESCO showed a certain negative correlation with potential evapotranspiration and NDVI, and the correlation coefficient passed the 0.05 significance level test. The soil evaporation compensation coefficient (ESCO) decreased with the increase in potential evapotranspiration and NDVI, and had no significant correlation with precipitation. The smaller the value of ESCO, the more evaporative water the model can obtain from the lower soil layer, resulting in the reduction of surface runoff, interflow and subsurface runoff, especially the most obvious reduction in surface runoff. In Figure 13, CH_N2 has a certain correlation with precipitation and potential evapotranspiration. It increases with the increase in precipitation and decreases with the increase in potential evapotranspiration. There is no obvious correlation with NDVI. The correlation coefficient between CH_N2 and precipitation passes the 0.05 significance level test. In Figure 14, ALPHA_BNK showed a certain negative correlation with potential evapotranspiration and NDVI, and the correlation coefficient passed the 0.05 significance level test. The baseflow α factor of riparian regulation and storage (ALPHA_BNK) decreased with the increase in potential evapotranspiration and NDVI, and had a weak positive correlation with precipitation. The baseflow α factor of riparian regulation and storage is a response index reflecting the discharge rate of subsurface runoff to the river discharge at the river bank, and the river discharge is positively correlated with it.

Table 3. The R^2 of each parameter with precipitation, potential evapotranspiration and NDVI.

R^2	Precipitation	Potential Evapotranspiration	NDVI
the number of runoff curves (CN2)	0.0032	0.0634	0.0003
the soil evaporation compensation coefficient (ESCO)	0.0008	0.4463 *	0.4402 *
the Manning coefficient of the main channel (CH_N2)	0.3364 *	0.2062	0.0009
the baseflow α factor of riparian regulation and storage (ALPHA_BNK)	0.1036	0.5453 *	0.2563 *

Note: * Indicates passing the 0.05 significance level test.

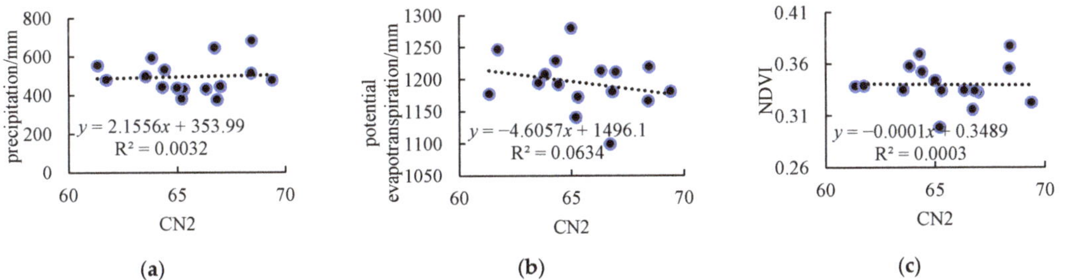

Figure 11. Correlation of CN2 to watershed precipitation, potential evapotranspiration, NDVI. (**a**) Precipitation, (**b**) potential evapotranspiration, (**c**) NDVI.

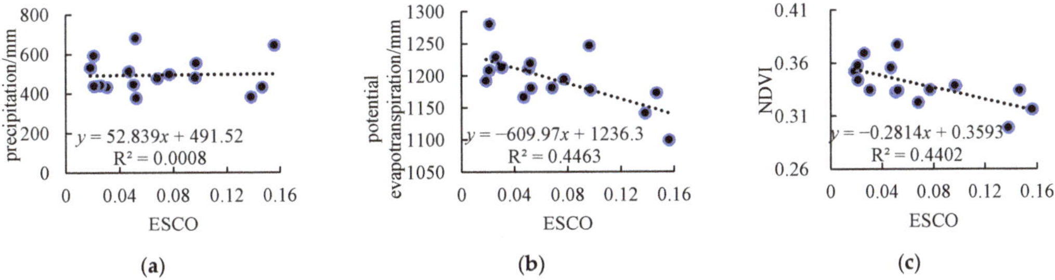

Figure 12. Correlation of ESCO to watershed precipitation, potential evapotranspiration, NDVI. (**a**) Precipitation, (**b**) potential evapotranspiration, (**c**) NDVI.

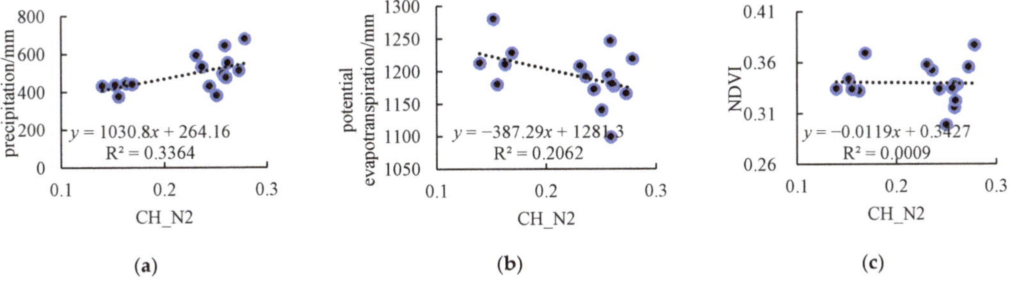

Figure 13. Correlation of CH_N2 to watershed precipitation, potential evapotranspiration, NDVI. (**a**) Precipitation, (**b**) potential evapotranspiration, (**c**) NDVI.

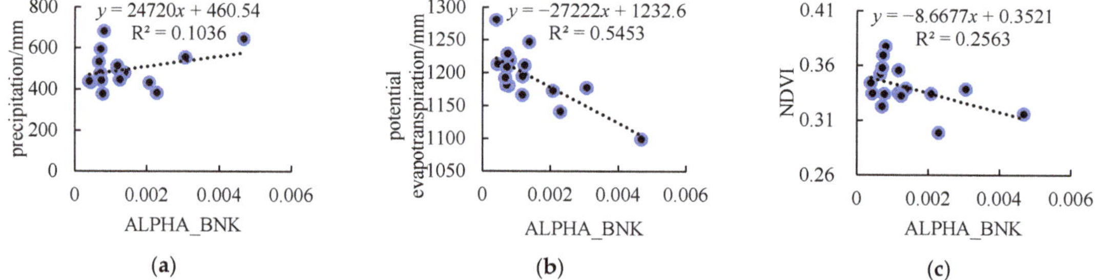

Figure 14. Correlation of ALPHA_BNK to watershed precipitation, potential evapotranspiration, NDVI. (**a**) Precipitation, (**b**) potential evapotranspiration, (**c**) NDVI.

In summary, all four parameters of the SWAT model in the upper reach of the Wei River Basin show a certain correlation with potential evapotranspiration, and some parameters have a correlation with precipitation or NDVI, which indicates that the changes of the hydrological model parameters in the upper reach of the Wei River Basin are closely related to the dynamic changes of potential evapotranspiration. Therefore, potential evapotranspiration can be used as a representative factor of the dynamic change of model parameters.

5. Conclusions

In this study, the upper reach of the Wei River is selected as the study area, and the temporal and spatial evolution of the ecological hydrometeorological elements in the study area is revealed using the linear regression method and the cumulative anomaly method. The time-varying parameter series of the hydrological model is obtained based on the SWAT model in each year. The time-varying characteristics of the model parameters and environmental indicators such as precipitation, potential evapotranspiration and NDVI were analyzed, and the response relationship between them was identified.

In addition to the decreasing trend of annual precipitation in the basin, the hydrometeorological elements such as annual potential evapotranspiration, annual average temperature, annual runoff depth and annual average NDVI all showed an increasing trend. The annual precipitation and annual runoff depth series in the basin did not change abruptly, but the annual mean temperature series and the annual potential evapotranspiration series jumped significantly around 1997 and 1994, respectively.

Except for the six years of 2002, 2004, 2008, 2009, 2016 and 2017 in 2002–2017, the daily runoff simulation accuracy in the other 10 years met the requirements. The four model parameters of CN2, ESCO, CH_N2 and ALPHA_BNK all fluctuated with time during the study period. The number of runoff curves (CN2) and CH_N2 had no obvious trend changes, while ESCO and ALPHA_BNK showed a downward trend. Because the magnitude of change in ALPHA_BNK was small, it can be neglected, and the change of ESCO is consistent with the conclusion that the evaporation in the basin increases year by year.

All model parameters show a certain correlation with potential evapotranspiration in the basin, which indicates that the changes of the hydrological model parameters in the upper reach of the Wei River Basin are closely related to the dynamic changes of potential evapotranspiration. Therefore, potential evapotranspiration can be used as an environmental factor to characterize the dynamic changes of the model parameters.

Author Contributions: Conceptualization, H.W., D.L., M.H. and R.L.; Methodology, H.W., D.L. and M.H.; Software, H.W. and D.L.; Validation, H.W. and D.L.; Formal analysis, H.W., D.L. and M.H.; Investigation, H.W. and D.L.; Resources, H.W. and D.L; Data Curation, H.W. and D.L.; Writing—Original Draft, H.W., D.L., M.H. and R.L.; Writing—Review and Editing, H.W., D.L., M.H., R.L., Q.Y., G.M. and H.L.; Visualization, H.W. and D.L.; Supervision, D.L. and M.H.; project administration, D.L. and M.H. All authors have read and agreed to the published version of the manuscript.

Funding: This research is supported by the National Natural Science Foundation of China (Grant Nos. 42071335, 52109031 and 41874017), the science and technology fund of the Second Monitoring and Application Center, China Earthquake Administration (KJ20220212) and the Natural Science Basic Research Program of Shaanxi-Han to Wei Water Transfer Joint Foundation (2022JC-LHJJ-13).

Institutional Review Board Statement: Not applicable.

Informed Consent Statement: Not applicable.

Data Availability Statement: Not applicable.

Conflicts of Interest: The authors declare no conflict of interest. The funders had no role in the design of the study; in the collection, analyses, or interpretation of data; in the writing of the manuscript, or in the decision to publish the results.

References

1. Tang, Q.H. Global change hydrology: Terrestrial water cycle and global chang. *Sci. China Earth Sci.* **2020**, *63*, 459–462. [CrossRef]
2. Campbell, J.L.; Driscoll, C.T.; Pourmokhtarian, A.; Hayhoe, K. Streamflow responses to past and projected future changes in climate at the Hubbard Brook Experimental Forest, New Hampshire, United States. *Water Resour. Res.* **2011**, *47*, 1198–1204. [CrossRef]
3. Zhang, L.; Liu, D.; Zhang, H.; Huang, Q.; Meng, X. Impact of climate change and human activities on runoff variation in Beiluo River basin. *J. Hydroelectr. Eng.* **2016**, *35*, 55–66.
4. IPCC. Climate Change 2013: The Physical Science Basis. In *Contribution of Working Group I to the Fifth Assessment Report of the Intergovernmental Panel on Climate Change*; Cambridge University Press: Cambridge, NY, USA, 2013.
5. Dore, M.H.I. Climate change and changes in global precipitation patterns: What do we know? *Environ. Int.* **2005**, *31*, 1167–1181. [CrossRef] [PubMed]
6. Jingyi, D.I.N.G.; Wenwu, Z.H.A.O.; Jun, W.A.N.G. Scale effect of the impact on runoff of variations in precipitation/vegetation: Taking northern Shaanxi loess hilly-gully region as an example. *Prog. Geogr.* **2015**, *34*, 1039–1051.
7. Brown, A.E.; Western, A.W.; McMahon, T.A.; Zhang, L. Impact of forest cover changes on annual streamflow and flow duration curves. *J. Hydrol.* **2013**, *483*, 39–50. [CrossRef]
8. Coe, M.T.; Latrubesse, E.M.; Ferreira, M.E.; Amsler, M.L. The effects of deforestation and climate variability on the streamflow of the Araguaia River, Brazil. *Biogeochemistry* **2011**, *105*, 119–131. [CrossRef]
9. Collier, C.G. The impact of urban areas on weather. *Q. J. R. Meteorol. Soc.* **2006**, *132*, 1–25. [CrossRef]
10. Cheng, C.K.M.; Chan, J.C.L. Impacts of land use changes and synoptic forcing on the seasonal climate over the Pearl River Delta of China. *Atmos. Environ.* **2012**, *60*, 25–36. [CrossRef]
11. Weiskel, P.K.; Vogel, R.M.; Steeves, P.A.; Zarriello, P.J.; DeSimone, L.A.; Ries, K., III. Water use regimes: Characterizing direct human interaction with hydrologic systems. *Water Resour. Res.* **2007**, *43*, W04402. [CrossRef]
12. Patterson, L.A.; Lutz, B.; Doyle, M.W. Climate and direct human contributions to changes in mean annual streamflow in the South Atlantic, USA. *Water Resour. Res.* **2013**, *49*, 7278–7291. [CrossRef]
13. Milly, P.C.; Betancourt, J.; Falkenmark, M.; Hirsch, R.M.; Kundzewicz, Z.W.; Lettenmaier, D.P.; Stouffer, R.J. Stationarity is dead: Whither water management? *Science* **2008**, *319*, 573–574. [CrossRef]
14. Lihua, X.I.O.N.G.; Shuonan, L.I.U.; Bin, X.I.O.N.G.; Wentao, X.U. Impacts of vegetation and human activities on temporal variation of the parameters in a monthly water balance model. *Adv. Water Sci.* **2018**, *29*, 625–635.
15. Liu, S.N. *Temporal Variation of the Parameters in a Monthly Water Balance Model*; Wuhan University: Wuhan, China, 2019.
16. Wang, J.; Wang, Z.; Cheng, H.; Kang, J.; Liu, X. Land Cover Changing Pattern in Pre-and Post-Earthquake Affected Area from Remote Sensing Data: A Case of Lushan County, Sichuan Province. *Land* **2022**, *11*, 1205. [CrossRef]
17. Kang, J.; Wang, Z.; Cheng, H.; Wang, J.; Liu, X. Remote Sensing Land Use Evolution in Earthquake-Stricken Regions of Wenchuan County, China. *Sustainability* **2022**, *14*, 9721. [CrossRef]
18. Deng, C. *Study of Temporal Variability of Model Parameters in Monthly Water Balance Modeling*; Wuhan University: Wuhan, China, 2017.
19. Kingumbi, A.; Bargaoui, Z.; Ledoux, E.; Besbes, M.; Hubert, P. Hydrological stochastic modelling of a basin affected by land-use changes: Case of the Merguellil basin in central Tunisia. *Hydrol. Sci. J.* **2007**, *52*, 1232–1252. [CrossRef]
20. Vaze, J.; Post, D.A.; Chiew FH, S.; Perraud, J.M.; Viney, N.R.; Teng, J. Climate non-stationarity-validity of calibrated rainfall–runoff models for use in climate change studies. *J. Hydrol.* **2010**, *394*, 447–457. [CrossRef]

21. Merz, R.; Parajka, J.; Blöschl, G. Time stability of catchment model parameters: Implications for climate impact analyses. *Water Resour. Res.* **2011**, *47*, 2144–2150. [CrossRef]
22. Sun, Y. *Time Instability of Hydrological Model Parameters under Changing Environment*; Tsinghua University: Beijing, China, 2015.
23. Brigode, P.; Oudin, L.; Perrin, C. Hydrological model parameter instability: A source of additional uncertainty in estimating the hydrological impacts of climate change? *J. Hydrol.* **2013**, *476*, 410–425. [CrossRef]
24. Wallner, M.; Haberlandt, U. Non-stationary hydrological model parameters: A framework based on SOM-B. *Hydrol. Process.* **2015**, *29*, 3145–3161. [CrossRef]
25. Wu, H.Z.; Liu, D.F.; Huang, Q.; Zheng, H.W.; Zou, H.; Ye, N.; Lin, M. Study on accuracy evaluation and substitutability of multiple precipitation products on Loess Plateau. *J. Hydroelectr. Eng.* **2021**, *40*, 31–40.
26. Jiang, Y.; Xu, Z.X.; Wang, J. Comparison among five methods of trend detection for annual runoff series. *Shuili Xuebao* **2020**, *51*, 845–857.
27. Zhang, Y.H.; Song, X.F. Techniques of abrupt change detection and trends analysis in hydroclimatic time-series: Advances and evaluation. *Arid. Land Geogr.* **2015**, *38*, 652–665.
28. Hotchkiss, R.H.; Jorgensen, S.F.; Stone, M.C.; Fontaine, T.A. Regulated river modeling for climate change impact assessment: The Missouri river. *J. Am. Water Resour. Assoc.* **2000**, *36*, 375–386. [CrossRef]
29. Yang, J.; Reichert, P.; Abbaspour, K.C.; Xia, J.; Yang, H. Comparing uncertainty analysis techniques for a SWAT application to the Chaohe Basin in China. *J. Hydrol.* **2008**, *358*, 1–23. [CrossRef]
30. Chaefli, B.; Gupta, H.V. Do Nash values have value. *Hydrol. Process. Int. J.* **2007**, *21*, 2075–2080. [CrossRef]
31. Moriasi, D.N.; Arnold, J.G.; Van Liew, M.W.; Bingner, R.L.; Harmel, R.D.; Veith, T.L. Model evaluation guidelines for systematic quantification of accuracy in watershed simulations. *Trans. ASABE* **2007**, *50*, 885–900. [CrossRef]
32. Gupta, H.V.; Kling, H.; Yilmaz, K.K.; Martinez, G.F. Decomposition of the mean squared error and NSE performance criteria: Implications for improving hydrological modelling. *J. Hydrol.* **2009**, *377*, 80–91. [CrossRef]
33. Kling, H.; Fuchs, M.; Paulin, M. Runoff conditions in the upper Danube basin under an ensemble of climate change scenarios. *J. Hydrol.* **2012**, *424*, 264–277. [CrossRef]

MDPI
St. Alban-Anlage 66
4052 Basel
Switzerland
Tel. +41 61 683 77 34
Fax +41 61 302 89 18
www.mdpi.com

Water Editorial Office
E-mail: water@mdpi.com
www.mdpi.com/journal/water

www.ingramcontent.com/pod-product-compliance
Lightning Source LLC
LaVergne TN
LVHW071244170726
843515LV00006BA/1328

* 9 7 8 3 0 3 6 5 7 1 1 5 7 *